Springer Complexity

Springer Complexity is a publication program, cutting across all traditional disciplines of sciences as well as engineering, economics, medicine, psychology and computer sciences, which is aimed at researchers, students and practitioners working in the field of complex systems. Complex Systems are systems that comprise many interacting parts with the ability to generate a new quality of macroscopic collective behavior through self-organization, e.g., the spontaneous formation of temporal, spatial or functional structures. This recognition, that the collective behavior of the whole system cannot be simply inferred from the understanding of the behavior of the individual components, has led to various new concepts and sophisticated tools of complexity. The main concepts and tools – with sometimes overlapping contents and methodologies – are the theories of self-organization, complex systems, synergetics, dynamical systems, turbulence, catastrophes, instabilities, nonlinearity, stochastic processes, chaos, neural networks, cellular automata, adaptive systems, and genetic algorithms.

The topics treated within Springer Complexity are as diverse as lasers or fluids in physics, machine cutting phenomena of workpieces or electric circuits with feedback in engineering, growth of crystals or pattern formation in chemistry, morphogenesis in biology, brain function in neurology, behavior of stock exchange rates in economics, or the formation of public opinion in sociology. All these seemingly quite different kinds of structure formation have a number of important features and underlying structures in common. These deep structural similarities can be exploited to transfer analytical methods and understanding from one field to another. The Springer Complexity program therefore seeks to foster cross-fertilization between the disciplines and a dialogue between theoreticians and experimentalists for a deeper understanding of the general structure and behavior of complex systems.

The program consists of individual books, books series such as "Springer Series in Synergetics", "Institute of Nonlinear Science", "Physics of Neural Networks", and "Understanding Complex Systems", as well as various journals.

Understanding Complex Systems

Series Editor

J.A. Scott Kelso
Florida Atlantic University
Center for Complex Systems
Glades Road 777
Boca Raton, FL 33431-0991, USA

Understanding Complex Systems

Future scientific and technological developments in many fields will necessarily depend upon coming to grips with complex systems. Such systems are complex in both their composition (typically many different kinds of components interacting with each other and their environments on multiple levels) and in the rich diversity of behavior of which they are capable. The Springer Series in Understanding Complex Systems series (UCS) promotes new strategies and paradigms for understanding and realizing applications of complex systems research in a wide variety of fields and endeavors. UCS is explicitly transdisciplinary. It has three main goals: First, to elaborate the concepts, methods and tools of self-organizing dynamical systems at all levels of description and in all scientific fields, especially newly emerging areas within the Life, Social, Behavioral, Economic, Neuro- and Cognitive Sciences (and derivatives thereof); second, to encourage novel applications of these ideas in various fields of Engineering and Computation such as robotics, nano-technology and informatics; third, to provide a single forum within which commonalities and differences in the workings of complex systems may be discerned, hence leading to deeper insight and understanding. UCS will publish monographs and selected edited contributions from specialized conferences and workshops aimed at communicating new findings to a large multidisciplinary audience.

M.A. Aziz-Alaoui C. Bertelle (Eds.)

Emergent Properties in Natural and Artificial Dynamical Systems

With 85 Figures

Springer

Professor M.A. Aziz-Alaoui
Applied Math Laboratory
University of Le Havre
25 Rue Ph. Lebon
BP 540, Le Havre, Cedex, France
E-mail: aziz.alaoui@univ-lehavre.fr

Professor C. Bertelle
Computer Sciences Laboratory
University of Le Havre
25 Rue Ph. Lebon
BP 540, Le Havre, Cedex, France
E-mail: cyrille.bertelle@univ-lehavre.fr

ISSN 1860-0840
ISBN 978-3-642-07113-3 e-ISBN 978-3-540-34824-5

Springer is a part of Springer Science+Business Media
springer.com

Printed in The Netherlands

Cover design: Erich Kirchner, Heidelberg

"I think the next century (21th) will be
the century of complexity"
– Stephen Hawking

In memory of Marc Rouff

Preface

The aim of this book is to study emergent properties arising through dynamical processes in various types of natural and artificial systems. It is related to **multidisciplinary approaches** in order to obtain several representations of complex systems using various methods to extract emergent structures.

Complex systems are a new scientific frontier which has emerged in the past decades with the advance of modern technology and the study of new parametric fields in natural systems. **Emergence** besides refers to the appearance of higher levels of system properties and behaviour that even if obviously originated in the collective dynamics of system's components – are neither found in nor directly deductible from the lower level properties of this system. Emergent properties are properties of the 'whole' not possessed by any of the individual parts making up this whole. Self-organization is one of the major conceptual keys to study emergent properties in complex systems. Many scientific fields are concerned with self-organization for many years. Each scientific field describes and studies these phenomena using its own specific tools and concepts. In physics, self-organized vortex formation (von Karman streets), sand ripples, etc, are observed for a long time. Biology and life sciences are probably the major domains where self-organization regularly appears, like in immune system, neuronal systems, social insects, etc. Mathematics develops studies on the formalisation of self-organized order, for instance in synchronized chaotic systems. Computer science develops innovative models using distributed systems like multiagent systems or interactive networks, where some kind of dynamical organizations are detected during the simulations. Chemistry proposes a formalisation of dissipative structures on open and complex systems. Dissipative structures become a basis of conceptual developments towards some general theories that define the new science of complexity.

This book is organized as follows. In the first part, we introduce some general approaches on complex systems and their trajectories control. We address the important problem of controlling a complex system, because there is now evidence in industrial applications that transforming a complicated man-made system into a complex one is extremely beneficial as far as performance improvement is concerned. A well defined control law can be set so that a complex system described in very general terms can be made to behave in a prescribed way. Handling systems self-organization when passing from complicated to complex, rests upon the new paradigm of passing from classical trajectory space to more abstract task space.

The second part introduces natural system modeling. Bio-inspired methods and social insect algorithms are the tools for studying emergent organizations. Self-organized fluid structures are presented for ecosystem modeling. Automata-based models allow to simulate the dectected structures in terms of stabilization. DNA patterns of self-organization in complex molecular systems are presented.

In the third part we address chaotic dynamical systems and synchronization problem. An innovative geometrical methodology is given for modeling complex phenomena of oscillatory burst discharges that occur in real neuronal cells. Emergence of complex (chaotic) behaviour in synchronized (non chaotic) dynamical systems is also described.

The fourth part focuses on decision support systems. Automata-based systems for adaptive strategies are developed for game theory. This model can be generalized for self-organization modeling by genetic computation. Adaptive decentralized methods for medical system diagnosis are presented as emergent properties from a multiagent system. Outlines of a decision support system for agricultural water management are described as emergent results from complex interactions between a multiagent system and a constraint programming system.

Finally, part 5 and 6 deal with technological complex systems leading to invariable properties. This invariance which is an emergent result charaterizes self-organization. Spline functions are shown to be efficient and adapted formal tools to study and control these invariance properties in technological complex systems. Methodologies are given to control complex systems, by considering their whole behaviour, using feed-back processes based on final output analysis. Applicative studies on flexible robotic systems are shown.

Le Havre, France,
April 2006

M.A. Aziz-Alaoui
Cyrille Bertelle

Contents

Part I

General Introduction

From Trajectory Control to Task Space Control – Emergence of Self Organization in Complex Systems

Michel Cotsaftis

ECE
53 rue de Grenelle
75007 Paris, France
mcot@ece.fr

Summary. A consequence of very fast technology development is the appearance of new phenomena in man made systems related to their large number of heterogeneous interacting components. Then because of resulting larger complexity over passing human operator capability, the system can no longer be only guided and controlled at trajectory level. A larger and more global delegation should be given the system at decision making, and it is proposed here to manage it at task level usually corresponding to well identified sequences in system operation. To succeed in this transfer attention has to be paid to the fact that there are in general many trajectories for one prescribed task. So a new and completely transparent link should be established between trajectory and task controls, both acting at their own levels in the system. The corresponding double loop control is developed here, and consists mainly in an asymptotically stable functional control acting at trajectory level as a whole, and explicit in terms of main system power bounds guaranteeing robustness inside a ball the size of which is the manifold generated by all system trajectories for task accomplishment. At higher level a new decision control based on trajectory utility for succeeding in the task is proposed, the role of which is to maintain system dynamics inside the selected trajectory manifold corresponding to task. With this two step control, human operator role is eased and can be more oriented toward higher coordination and maintenance management.

Key words: complex systems, functional trajectory control, task utility, fixed point theorem.

1 Introduction

A new situation has gradually emerged from the extraordinary advance in modern technology which took place in the last decades soon after World War II. Aside unsurpassed performance routinely reached by each elementary component, new more and more extended systems are nowadays conceived and realized. They gather very heterogeneous elements each in charge of a part of the final global action of the system, and for economic efficiency they are more and more integrated in the design, in the sense that their mutual interactions are extremely strong and are considerably conditioning the final system output. Such a situation is quite new and stresses the prominent role these interactions are playing in system operation. As a direct consequence, a new emerging requirement for the system is, under actual working conditions, a larger delegation as concerns the definition of its trajectory. As explained later, the reason is that the trajectory escapes from operator capability, contrary to the classical case where the input control is set to make the system to follow a prescribed trajectory. The delegation could take different aspects and levels [46] and, in highest stage, should completely replace human intelligence. This later stage is by far not realistic today and a much more modest step is considered here. Observation of living creatures clearly shows the importance of well identified sequences, separable in as many tasks during the development of their current life. In the same way, man made machine activity can also be separated in units corresponding to the accomplishment of a specific task, a combination of which is often needed for reaching the goal assigned from the beginning to the machine. In general there are many trajectories associated to a task, so an important problem is in the possibility for the system to define its own trajectory once the elementary task has been assigned. Then as a consequence the system will be only operated in task space rather than in "classical" trajectory space. When switching gradually to this intricate and still not well known situation, new problems are arising, and it is only very recently that consciousness of a method reassessment for approaching such problems became clear. It is very important to realize that the present step is the last one to date of a very long series of human actions aiming at transferring always more capability to tools. From the very early beginning of his origin, Human kind understood the needs for developing tools to extend the action of his hand. Since then, their continuous development associated with the invention of always better quality actuators and sensors has been allowing in the last two centuries the ascent of industrial civilization. A further considerable advance took place forty years ago when the mechanical devices in charge of the necessary information flux inside the system have been replaced by electronic more powerful ones, much easier to operate safely in different environments and carrying much more information, giving rise to Mechatronics [61]. This is summarized on Fig. 1, where block 1 is corresponding to the millennium long development of base system, block 2 describes the structure of controlled system with recent mechatronic components, and block 3 is representing the present next step toward system independence.

In parallel to this transformation of man made systems, research has been reaching new domains where the behavior of studied objects is itself strongly depending on interactions between elementary components, whether in new plasma state of matter [1], in new chemical or mechanical aggregates [2], in

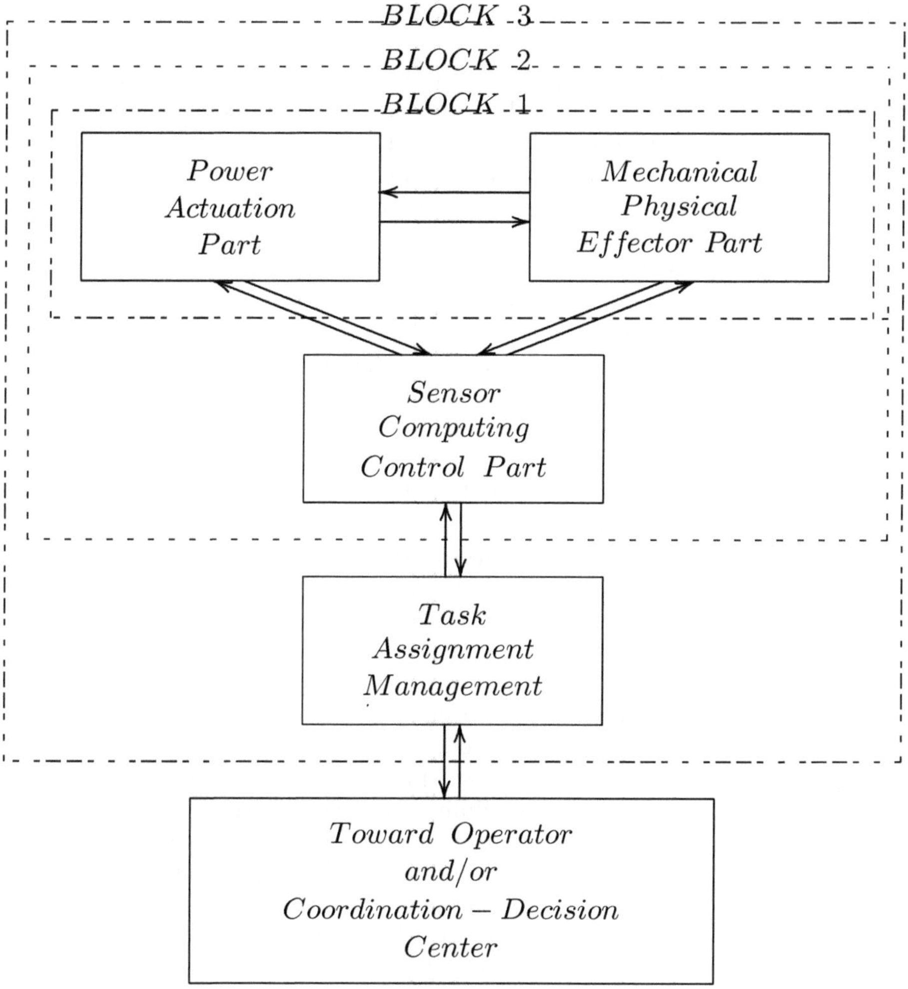

Fig. 1. System structure evolution with main added component parts

biology [3,5], or in more classical physical states of matter in unusual parameter domains [6]. In all cases, the strength of interactions completely masks the elementary interaction between basic components, and the research of final system behavior is, due to importance of nonlinearities, generally outside the range of application of classical methods. This is understandable inasmuch as the system often reaches its stage after exhibiting a series of branching along which it was bifurcating toward a new global state the features of which are not usually amenable to a simple local study, being remembered that the branching phenomenon is resting upon a full nonlinear and global behavior.

2 Discussion

Various attempts have been proposed so far to deal with these phenomena and these states, both in Applied Mathematics and in Control methods. In first class mainly oriented toward analysis and description of "chaotic" state, a large body of results [10,26,65,69,70,71] has shown that the later represents the general case of non linear and non integrable systems [27], and that it is reached for high enough value of (nonlinear) coupling parameter. In second class are belonging extensive new control methods often (improperly) called "intelligent" [14,15, 22,29,30,31,53,57,84], supposed to give the concerned systems the ability to behave in a much flexible and appropriate way. The great defect of these very interesting analysis is that they are resting on a questioned premise which does not fit the nature of the researched solution. They all rest upon the postulate that system trajectory can be followed as in the classical "mechanical" case of a rocket launched to reach the moon, and moreover that it is in some sense possible to act on this trajectory by appropriate means. In present case on the contrary, the very strong interaction between components in natural systems induces as observed in experiments a wandering of the trajectory which from simple observation point of view becomes indiscernable from neighboring ones [42], and only manifolds can be identified and separated [41,48]. So even if it could be tracked, there is in reality no way to modify one specific system trajectory by acting on it at this level because there is no information related to it from system point of view, a well known phenomenon since Maxwell´s devil in Thermodynamics [82]. A very similar situation occurs in modern technology applications where the new challenge is to give now the system the possibility to decide its trajectory for a fixed task assignment, being easily understood that there are in general many allowed trajectories for a given task. In both cases there is a shift to something different, whether imposed by the mathematical structure which generates a manifold instead of a trajectory, or needed for fulfilling technical requirements in task execution under imposed (and often tight) economic constraints. This already corresponds to a very important qualitative jump in the approach of natural and technical highly nonlinear systems, which requires proper tools for being correctly handled. It should be mentioned that another approach has also been recently proposed [16,21,23,25,38], where complex systems are considered as a set of strongly interacting "agents" possessing a certain number of properties, and one looks for "emergence" * of global behavior without precluding the existence of any general law. This "weak" approach, where the global behavior is "thermodynamically" reconstructed from elementary interactions, is markedly different from the present analysis in its very essence and will not be discussed here.

The striking point is that Nature has been confronted with this issue a few billion years ago when cells have emerged from the primitive environment. They exhibit the main features engineers try today to imbed in their own constructions, mainly a very high degree of robustness resulting from massive parallelism and high redundancy. Needless to say, the high degree of accomplishment manifested by cells makes the analysis for understanding their mechanism in order to take advantage of it a formidable challenge in itself, as it should be understood that cells are today at the last step of a very long evolving process. In summary, the outcome of the previous short discussion is the emergence of a new type of demand which can be mathematically expressed in the following simple way : How to pass from space time local trajectory control to more global manifold control? . It is intended in the following to give some clues to the problem which is the formulation of a new paradigm, and to show that it can be properly embedded into the formalism of recent advances in modern functional analysis methods.

The first step is to clarify the consequences of the enhanced interaction regime between components in a system as concerns its control. Instead of usual physics "contemplative" observation, the approach from a control point of view is also applicable to natural systems provided some sensitive enough parameters are considered as control inputs driving system dynamics, ie are elements of a control space, and generate the system under study when they are fixed at their actual value in the system. In the same way, the control inputs in man made systems will be considered as elements of a more general control space, which, in both cases, can be defined on a reasonable and useful physical base. Then the classical control problem [67,68], with typical control loops guaranteeing convergence of system output toward a prescribed trajectory fixed by an independent entity, shifts to another one where the system, from the only task prescription, has to generate its own trajectory in the manifold of the realizable ones. A specific type of internal organization has to be set for this purpose which not only gives the system the necessary knowledge of the outside world, but also integrates its new features at system level [11,15]. This in particular means that the new controller should be able to handle the fact that the trajectory is not prescribed as before but belongs to a manifold which has to be related in a meaningful way to the task.

3 Method

To proceed, it is first necessary to understand the consequences for the control of the wandering nature of system trajectory. The main point is that as neighboring trajectories are becoming indiscernable, the information content in their observation vanishes, so there is no information loss at this level when abandoning their detailed observation. But as the number of degrees of freedom (d.o.f) has not changed (or is even becoming larger at branching), the conclusion is that the system is reaching a new stage where some of the d.o.f are now taken care of by internal system reorganization. They are moreover in sufficient number so that the only necessary inputs to drive the system are the ones defining the manifold on which the dynamics are now taking place. This system self-organization which reduces the possible action on it from the outside is absolutely fundamental [4,8,9,17,18,19, 35,60,73-78]. It expresses the simple fact that when reaching a higher interactive level between its components, the system can no longer stand with all its d.o.f controlled from outside, but takes itself part of this control in a way which allows system existence to be continued. It should be stressed that this is completely different from usual textbook case of control with a smaller dimension input control than system dimension. Here again there is no restriction of control inputs number. But because some inputs are just changing from outside to inside the system, they cannot be maintained unless they are in conflict with the new internal organization which fully determine them already. This situation is extremely important because it corresponds to the general case of natural systems, and is accompanying the change into a so-called dissipative structure [34].

The systems in this reorganized (and partly self-controlled) situation will be called complex systems [13,20,28,33,36,37,58,59,62, 63,78,79,80] (from Latin root – cum plexus : tied up with), by opposition to simple ones which can be completely controlled from outside. The remarkable fact is that past some level of complication (note the meaning of the word is different from complexity) by extension of their components number, natural systems necessarily become more complex once interactions get large enough. So when understanding this natural exchange, it becomes now possible to take advantage of it by accepting the compromise to reduce the outer control action to the only d.o.f left free after system reorganization, knowing that there is no loss of information about the system in the process which leads to hierarchize system structure [24]. On the other hand, technical systems with large number of d.o.f have to be constructed for accomplishment of complex enough tasks, and it becomes more and more evident that the resulting high degree of complication makes the control of such systems very fragile, if not strictly impossible when interactions are becoming large enough for some actual operating parameters under the pressure of "economic" constraint.

Mathematically speaking, the very common mixed system representation with a random background noise does no longer apply [82]. It corresponds to the often obtained simple stability result strongly depending on the maximum gain acceptable by system actuators and here over passed, calling for another more dynamically detailed one to be developed [13]. It is suggested here that recognizing the situation described above represents the only possibility for man made systems to exist in a useful and manageable way. This in turn implies a different approach to system control mainly based on a compromise in order to avoid too severe constraints on system dynamics, as this is the only workable way for the system to maintain its existence. The new control scheme becomes more general in the sense that design parameters, which were fixed independently in previous scheme, should now be determined by the final pre-assigned behavior when including the internal loops corresponding to d.o.f linkages making the system complex. In other words, there is no more clear distinction between design and control parameters. This global approach, often called optimal design, should be developed in adequate functional frame as only manifolds, and not trajectories, are now accessible. On the other hand it guarantees locally asymptotic stability of trajectories within a robustness ball the size of which is fixed by the parameters of an equivalence class including the system at hand as an element.

The next step is to find the link at system level between task definition and the manifold of (self-organized) trajectories the system can only follow. For living organisms, this property is hardwired into the brain representation of the environment and of the living being itself from experience through their sensory and nervous systems. This self-consciousness takes all possible levels of importance depending of the richness of living individual development. It can be very modest as for planar worms, small aquatic animals existing from 400 million years with only seven neurons, but already able to perform a very large range of motions and to produce many different reactions. On the other hand, an important observation is that living creatures have broad range sensors and that they filter for each task the relevant information (which will be called "useful" here) needed to guide their trajectory. These two properties will be reproduced at system level by first defining a functional of system trajectories expressed in terms of the only invariants characterizing the manifold on which they are lying, and by constructing a functional control law which only acts at manifold (or larger) level. This is possible by considering now a complete trajectory as a "point" in a functional space, $x = x(.)$, rather than the usual succession of positions $x(t_j)$ for each time t_j. In this view, stability is obtained as the belonging of $x(.)$ to a pre-selected function space expressed through a fixed point condition in this space.

Then derivatives of the functional with respect to task parameters give system sensitivity to the task and provide the filter matrix selecting the relevant manifold for task accomplishment by the system. A control law can be found in explicit analytical form by extension of Yakubovic-Kalman-Popov (YKP) criterion, written for linear upper bound of system non-linearity [50,51,52,55], to general non-decreasing more accurate bound [12]. It gives explicit asymptotic stability limit and defines the robustness equivalence class within which the property holds. Other more general expressions are also found by application of fixed point theorem which can be shown to contain as applications most published results since Lyapounov and Poincaré pioneering work [39,40]. The simplest trajectory functional is the potential (the Lagrangian in conservative case) from which trajectories are defined as the zeros of its functional derivative. It may reduce to system invariant in the case there exists only one, as shown by elementary example of temperature in Thermodynamics. From these elements, the control structure can be constructed with its various parts.

4 Results

Following the scheme developed above, the task controller structure corresponding to block 3 in Fig. 1 to be developed now comprises as a main

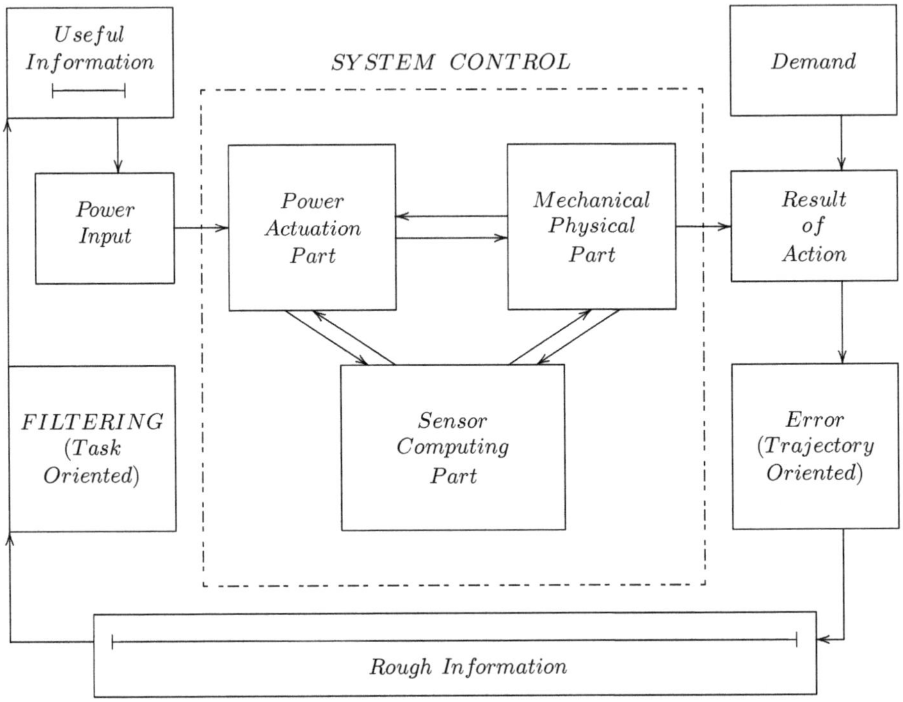

Fig. 2. System structure with main component parts and information filtering for task orientation

element the information filter selecting the task relevant information from system sensors, see Fig. 2. The basic postulate is here that the knowledge of both information and power fluxes at each instant completely determines a system. But in the laws of Physics, the information flux is rigidly linked to power flux, so with this last knowledge the system will just follow the "natural" evolution corresponding to Physics laws, to which there corresponds a set of "natural" trajectories in geometrical space characterizing the "inanimate" world. In present problem however, the system should have the possibility to escape from previous situation in order to modify its behavior for adapting to adverse effects, whether of internal or external origin, opposing to task accomplishment. To solve this problem the way followed by Nature at early beginning of Earth existence has been to allow the splitting of information and power fluxes by creating molecules able to store an information content independently, out of which existence of living creatures able to perform their own tasks has been made possible. Similarly, even if usual system control at level of block 2 proves with asymptotic stability to be usually efficient for restoring prescribed trajectory [32,54], further increase of system performance requires a new degree of autonomy for building its own strategy. Of course a seemingly simple possibility would be to give human operator the required information for efficient drive of the system [47]. Due to ergonomic constraints and to various limits on technical response times, this action has a limited potential as it does not properly delegate to system level enough freedom for full exploitation of system technical capabilities. Consequently the problem is not only to give the system proper information collected from sensors now utilized in various displays by human operator, but more importantly to put them in a format compatible with system own dynamics. Comparison between human eye, extremely useful to human operator due to the very remarkable brain organization, and machines with camera and recognition software stresses the importance to verify that the delivered information matches with system possibility of action in the same way as human operator knows how to interpret vision signals for determining the proper action. Though not possible at general goal level, it can be shown that the only necessary step is to manage system organization so that delivered information meets with each specific task sub-level. In present case this rests upon the concept of "useful" information [56] which has to be internally filtered by the system for retaining the only elements relevant for actual action, see Fig. 2.

Observation shows that existence of events rests upon removal of a double uncertainty : the usual quantitative one related to occurrence probability and the qualitative one related to event utility for goal accomplishment. So events may have same probability but very different utility, which may explain why extra information on top of existing ones have sometimes no impact on reaching the goal. Calling u_j and p_j the utility and the probability of event E_j, its associated useful information $\mathcal{I}(u_j; p_j)$ will be defined as

$$\mathcal{I}(u_j, p_j) = k\, u_j \log p_j \tag{1}$$

from both additivity of information for simultaneous events and proportionality of information to utility with k the Boltzmann constant. Total useful information associated with a process corresponding to trajectory $\mathcal{T}$ is the sum of all useful information from each contributing element $\mathcal{I}(\mathcal{T}) = \sum_j \mathcal{I}(u_j, p_j)$. Usual entropy calculation is recovered when all events have same utility for goal accomplishment, certainly true in Thermodynamics where all molecules are totally interchangeable and thus indistinguishable. The only invariant corresponding to this equivalence class is the energy (or the temperature). So thermodynamic systems can only be separated and thus controlled with respect to their temperature. Similarly in present case, some internal system dynamical effects may be events which cannot be distinguished between one another in general case. Such is the case of flexion and torsion link deformations in a robotic system, which are layered [66] on invariant surfaces determined by the value of bending moment M at link's origin. Using their observation to improve system dynamical control is not possible, in the same way as observing individual molecule motion in a fluid would not improve its global control. So increasing the amount of information from sensors as commonly developed is not the solution. Only relevant information has to be collected, and this justifies why raw sensor information has to be filtered so that only useful information for desired task accomplishment is selected. This is precisely the remarkable capability of living systems to have evolved their internal structure so that this property is harmoniously embedded at each level of organization corresponding to each level of their development. To evaluate the utility u, consider Lyapounov function $\mathcal{L}(\mathcal{T})$ selecting those trajectories $\mathcal{T}$ of the manifold $\mathcal{M}(\mathcal{T})$ inside allowable workspace domain $\mathcal{D}$ and meeting task requirement, and finally expressed in terms of problem parameters and system invariants of these trajectories. Then the quantity

$$u = \nabla\mathcal{L}(\mathcal{T})/|\mathcal{L}(\mathcal{T})| \qquad (2)$$

automatically selects as most useful the elements against which the system exhibits larger sensitivity. As indicated later, sensitivity threshold has to be fixed so that resulting uncertainty ball is inside attractor ball corresponding to asymptotic stability of system trajectories. Note that elimination of elements to which the system is insensitive, just after sensors data collection and prior to any calculation in control unit, enormously reduces the computation load. As for the probability term p, considering trajectories all equiprobable in $\mathcal{D}$ and indistinguishable in local neighboring cell corresponding to non directly controllable internal phenomenon such as flexion and torsion modes for deformable bodies, one would simply get $p = \Delta/\mathcal{D}(\mathcal{T})$. With this supervisory information filtering and appropriate control, the system will asymptotically follow any trajectory in D(T) depending on initial conditions. Other more selective probability based on trajectory smoothness and power consumption optimization can be likewise be defined. In all cases one should note that useful information $\mathcal{I}$ can be made completely explicit once the system is fixed. When during the course of its motion the system is constrained by adverse effects, their interaction with the system are changing its dynamical equations and the domain $\mathcal{D}(\mathcal{T})$ is no longer fully available. Then the remaining domain $\mathcal{D}_c(\mathcal{T})$ is determined by a new Lyapounov function $\mathcal{L}_c$ with the controller guaranteeing asymptotic stability of new system trajectories. From it a new utility function u_c from eqn(2) can be formed and the new constrained useful information $\mathcal{I}_c$ can similarly be made explicit. So as it stands, useful information is a class property characterizing from their invariants a family of possible trajectories in workspace which have their representation determined from dynamic and /or general geometric properties.

Next because of the two new conditions of 1)-operation at task level and 2)-trajectory non distinguishability in general complex system, the controller should work at a level avoiding local trajectory detail. Rather than continuing with usual state space, an interesting step is to consider each trajectory as a point in functional space, and to define a functional control solely guaranteeing the trajectory to belong to a selected functional space corresponding to researched properties. For finite dimensional nonlinear and time dependent systems

$$\frac{dx_s}{dt} = F_s(x_s(t), u(t), d(t), t) \qquad (3)$$

where $F_s : \mathbf{R^n} \times \mathbf{R^m} \times \mathbf{R^p} \times \mathbf{R}^1_+ \rightarrow \mathbf{R^n}$ is a $\mathbf{C}^1$ function of its first two arguments, $x_s(t)$ the system state, $u(t) \in \mathbf{U}$ the control input, and $d(t) \in \mathbf{D}$ the disturbance acting upon the system, explicit expressions for the controller $\mathcal{C}_F$ can be obtained in terms of system global characteristics, see Appendix, and exhibit robust asymptotic stability inside the desired function space under mild conditions by application of fixed point theorem [43]. Fixing specific exponential convergence decay [44] and extension to unknown systems [45] are more generally possible. The main feature is that the controller $\mathcal{C}_F$ takes the form

$$u(t) = -Ke + \Delta u \tag{4}$$

ie is the sum of a linear PD type part and an ad-hoc nonlinear part depending on the distance of actual trajectory to allowed manifold trajectories $\mathcal{M}(\mathcal{T})$, and where e is the trajectory error with respect to nominal one belonging to $\mathcal{M}(\mathcal{T})$. The role of additional u is to counteract the effect of nonlinear and disturbed remaining parts in system equations, and to define an asymptotic stability ball within which all system trajectories are guaranteed to stay inside selected manifold $\mathcal{M}(\mathcal{T})$. This is very clear when writing the upper bound on derivative of adapted Lyapunov function along system trajectories

$$\frac{dL}{dt} \leq -k\mathcal{L} + \epsilon M_s(t) g_s(\mathcal{L}, t) f(\mathcal{L}, t) \tag{5}$$

see eqn(A9) of Appendix, which is bounded by the sum of an attractive spring resulting from controller action and a bounding generally repulsive force representing system effect. The controller is furthermore functionally robust as it only implies a global bounding function of nonlinear terms, which means that all systems with same bounding function will be asymptotically controlled by $\mathcal{C}_F$. With this controller $\mathcal{C}_F$ working at trajectory level $\mathcal{T}$ it is possible to design the block diagram of task oriented control displayed on Fig. 3. Independent of lower level controllers inside local subsystem such as actuator and effector boxes to be tuned aside, it mainly implies, on top of the functional controller loop (1) guaranteeing required trajectory following, a second higher level loop (2) of decisional nature based on information Ic which verifies that the system is actually following a trajectory belonging to the manifold $\mathcal{M}(\mathcal{T})$ corresponding to the assigned task, and opens a search toward this class when for instance conditions are changing. The search can be conducted by interpolation within a preloaded neural network covering the different possible situations over the workspace. The present scheme is thus organized to meet the condition that actual trajectory followed by the system is not necessarily unique, as there is in general only a class of trajectories associated to a task. No more specific conditions are required on allowed trajectories provided controller parameters are such that these trajectories are all inside the asymptotic robustness ball for this (explicit) trajectory controller. So controlled system dynamics are defining a trajectory which is followed until it would escape without controller from acceptable manifold corresponding to

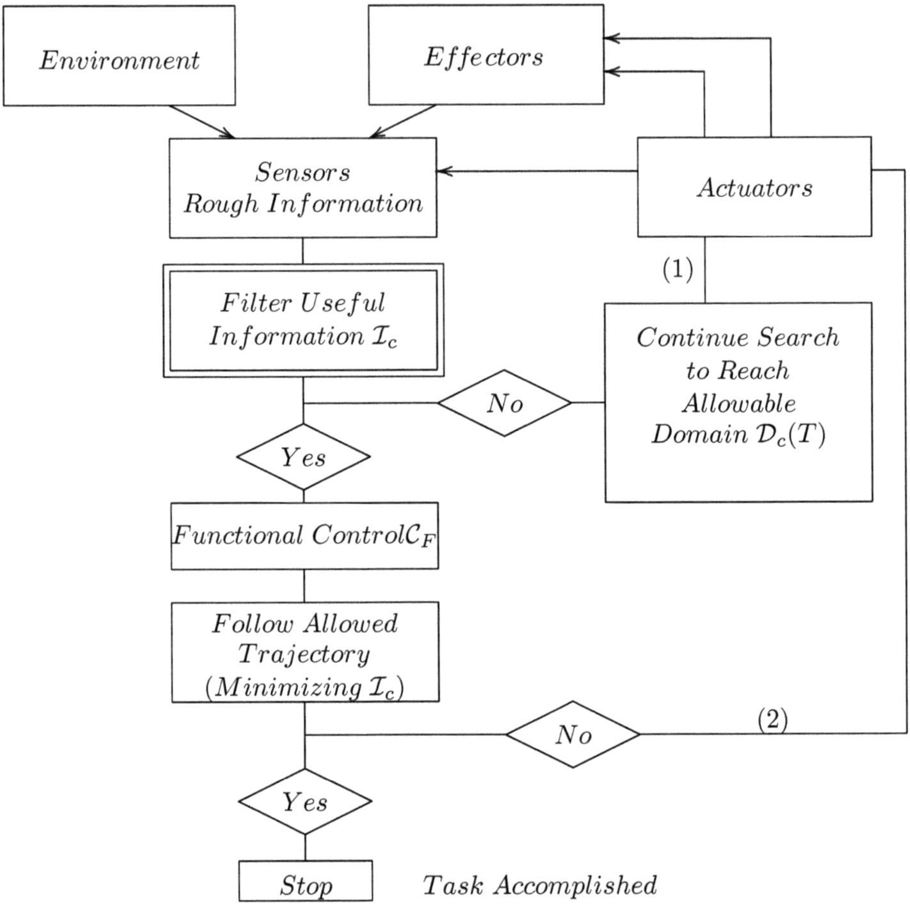

Fig. 3. Structure of filtering block of Fig. 2 with task oriented control

the defined task and selected by useful information $\mathcal{I}_c$. A constraint may be further added to limit or to minimize useful information $\mathcal{I}_c$ if required, but the correlative restriction of acceptable trajectory manifold occurs at the expense of system adaptiveness, so a smooth constraint is often more appropriate if any.

The important point is that the various elements in the chart of Fig. 3 have been constructed so that 1)- they are expressible in terms of system parameters (or can be reconstituted if unknown) and 2)- they are linked together in completely coherent way through consideration of trajectory as a whole appearing as the good "unit", both qualities needed to create adequate link between the two loops for transferring decisional power, ie "intelligence", to the system. In this sense the system is given its own task consciousness as it does not obey here a strict step by step outer command.

5 Conclusion

As largely documented, the structure of natural systems mostly belongs to the broad class of complex structures when interaction between system components becomes strong enough. There results an internal self-organization the role of which is to minimize the dependence of the system with respect to the outside world. This is seen from system trajectory which becomes more erratic in state space, and so cannot be distinguished from neighboring ones. Only manifolds corresponding to system invariants can be separated in general, indicating that the system has been reducing the number of outer control inputs, and is taking the other control inputs under their own dynamics. Such a compromise is a natural trend expressed by the general principle of autonomy of complex systems, which states that they naturally evolve as a dissipative structure toward the state giving them the least dependence on outside world compatible with boundary conditions. This is culminating with living systems which, because the previous compromise forces them to maintain a metabolism driving them to a finite life, have extended the principle to complete species by reproduction.

It is suggested here to take advantage of this phenomenon for man made systems which are becoming complex enough, and, following the natural bottom-up approach described here, to upgrade their control from trajectory to task control more adapted to their new structure. The construction of the new controller is made possible in two steps by developing an explicit base trajectory control of functional nature, which is asymptotically stable and robust enough to cover the manifold of possible trajectories. Secondly, introducing the concept of "useful" information, a task functional is expressed in terms of system parameters, which defines compatible trajectory manifold. From them a double loop has been constructed giving the system the possibility to accomplish the task for any allowed trajectory. The main point here is that the system determines itself its own path from its own elements, and not, as often the case, from preloaded and/or tele-ported procedure. In this sense it has gained more independence in its behavior and, similar to very advanced living systems, is able to operate autonomously at its own level.

References

1. B.B Kadomtsev (1992) Self Organization and Transport in Tokamak Plasma, *Plasma Phys. and Nucl. Fus.*, 34(13):1931-1938
2. P. Bautay, I.M. Janosi (1992) Self Organization and Anomalous Diffusion, *Physica A*, 185(1-4):11-18
3. E. Ben Jacob, H. Schmueli, O. Shochet, A. Tannenbaum (1992) Adaptive Self Organization during Growth of Bacterial Colonies, *Physica A*, 187(3-4):378-424
4. W. Ebeling (1993) Entropy and Information in Processes of Self Organization: Uncertainty and Predictability, *Physica A*, 194(1-4):563-575

5. S.Camazine, J.L. Deneubourg, N.R. Franks, J. Sneyd, G. Theraulaz, E. Bonabeau (2002) Self organization in Biological Systems, Princeton Univ. Press, Princeton, New-Jersey
6. J. Holland(1995) Hidden Order : How Adaptation Builds Complexity, Helix Books, Addison Wesley, New-York
7. M. Waldrop (1993) Complexity : the Emerging Science at the Edge of Order and Chaos, Touchstone, NY
8. S. Kauffman (1993) At Home in the Universe : the Search for Laws of Self organization and Complexity, Oxford Univ. Press, New-York
9. M.E.J. Newman (1997) Evidence of Self Criticality in Evolution, *Physica D*, 107(2-4):293-296
10. Yu.L. Klimontovitch (2002) Criteria for Self Organization, *Chaos, Solitons and Fractals*, 5(10):1985-1995
11. S. Nolfi, D. Floreano (2000) Evolutionary Robotics : the Biology, Intelligence and Technology of Self-organizing Machines, The MIT Press, Cambridge, Mass.
12. M. Cotsaftis (2003) Popov Criterion Revisited for Other Nonlinear Systems, *Proc. ISIC 03 (International Symposium on Intelligent Control)*, Oct 5-8, Houston,USA
13. M. Cotsaftis(1997) Comportement et Contrôle des Systèmes Complexes, Diderot, Paris
14. P.J. Antsaklis, K.M. Passino (1993) An Introduction to Intelligent and Autonomous Control, Kluwer Acad. Publ., Norwell, MA
15. A. Kusiak (1990) Intelligent Manufacturing Systems, Prentice Hall, Englewood Cliffs, New Jersey
16. E. Bonabeau, M. Dorigo, G. Theraulaz (1999) Swarm Intelligence : from Natural to Artificial Systems, Oxford Univ. Press, New-York
17. P. Bak (1996) How Nature Works : the Science of Self Organized Criticality, Springer, Berlin
18. G. Nicolis, I. Prigogine (1997) Self Organization in Non Equilibrium Systems, Wiley, New-York
19. H. von Foerster, G. Zopf, (eds.) (1992) : Principles of Self-Organization, Pergamon, New York
20. S.Y. Anyang (1998) Foundations of Complex System Theories in Economics, Evolutionnary Biology and Statistical Physics, Cambridge Univ. Press, Cambridge, Mass.; R. Albert, A-L Barabasi(2002) Statistical Mechanics of Complex Networks, *Rev. of Modern Phys.*, 74:47
21. H. Nwana (1996) Software Agents : an Overview. *The Knowledge Engineering Review*, 11(3):205-244
22. S. Russell, P. Norvig (1995) Artificial Inte-ligence : a Modern Approach, Prentice-Hall
23. A.D. Linkevich (2001) Self-organization in Intelligent Multi-agent Systems and Neural Networks, *Nonlinear Phenomena in Complex Systems*, Part I : 4(1):18-46; Part II : 4(3):212-249
24. G. Nicolis (1986) Dynamics of Hierarchical Systems, Springer, Berlin
25. M.J. Wooldridge, N.R. Jennings (1995) Intelligent Agents : Theory and Practice, *the Knowledge Engineering Review*, 10(2):115-152
26. M.D. Mesarovic, Y. Takahara (1975) General System Theory : Mathematical Foundations, Acad Press, New-York
27. A. Goriely (2001) Integrability and Nonintegrability of Dynamical Systems, World Scientific Publ.

28. L. O. Chua (1998) CNN : a Paradigm for Complexity, World Scientific Publ.
29. D.P. Bertsekas, J.N. Tsitsiklis(1996) Neuro Dynamic Programming, Athena Scientific Press, Belmont, MA
30. Y.Y. Chen (1989) The Global Analysis of Fuzzy Dynamical Systems, PhD Thesis, Univ. of California, Berkeley
31. L.X. Wang (1994) Adaptive Fuzzy Systems and Control : Design and Stability Analysis, Prentice-Hall, Englewood Cliffs, NJ
32. K. Zhou, J.C. Doyle, K. Glover (1996) Robust and Optimal Control, Prentice-Hall, Englewood Cliffs, NJ
33. G. Nicolis, I. Prigogine (1989) Exploring Complexity : an Introduction, W.H. Freeman and Co, NY
34. D. Kondepudi, I. Prigogine (1997) Modern Thermodynamics : from Heat Engines to Dissipative Structures, J. Wiley and Sons, NY
35. S.A. Kauffman (1993) The Origins of Order : Self-Organization and Selection in the Universe, Oxford Univ. Press
36. B. Parker (1996) Chaos in the Cosmos : the Stunning Complexity of the Universe, Plenum Press
37. R. Serra, M. Andretta, M. Compiani, G. Zanarini (1986) Introduction to the Physics of Complex Systems (the Mesoscopic Approach to Fluctuations, Nonlinearity and Self-Organization), Pergamon Press
38. G. Weiss (1999) Multi-agent Systems. A Modern Approach to Distributed Artificial Intelligence, the MIT Press, Cambridge, Mass.
39. H. Poincaré (1892-1899) Les Méthodes Nouvelles de la Mécanique Céleste, 3 , Gauthier-Villars, Paris
40. A.M. Lyapounov (1907) Le Problème Général de la Stabilité du Mouvement, Ann. Fac. Sciences Toulouse
41. M. Cotsaftis (1998) Vision Limitation for Robot Deformation Control, *Proc. 5th Intern. Conf. on Mechatronics and Machine Vision in Practice (M2VIP)*, Nanjing, China:393
42. M. Cotsaftis(1996) Recent Advances in Control of Complex Systems, Survey Lecture, *Proceedings ESDA'96*, Montpellier, France, ASME, I:1
43. E. Zeidler(1986) Nonlinear Functional Analysis and its Applications, I, Springer-Verlag, New-York
44. M. Cotsaftis (1999) Exponentially Stable and Robust Control for Dynamical Systems, *Proceedings 1999 IEEE Hong-Kong Symposium on Robotics and Control*, II:623
45. M. Cotsaftis (1998) Robust Asymptotically Stable Control for Unknown Robotic Systems, *Proceedings 1998 Symposium on ISSPR*, Hong Kong, I:267
46. T.B. Sheridan (1992) Telerobotics, Automation and Human Supervisory Control, The MIT Press, Cambridge, Mass.
47. T. Sakaki, Y. Inoue, S. Tachi (1994) Tele-existence Virtual Dynamic Display Using Impedance Scaling with Physical Similarity, *Proc. 1993 JSME Intern. Conf. on Advanced Mechatronics*, Tokyo, Japan:127
48. M. Cotsaftis (1998) Application of Energy Conservation to Control of Deformable Systems, *Proceedings 3rd Workshop on Systems Science and its Application*, Beidaihe, China:42
49. J. Appell, P.P. Zabrijko (1990) Nonlinear Superposition Operators, Cambridge University Press, Mass.

50. G.A. Leonov, I.V. Ponomarenko, V.B. Smirnova (1996) Frequency Domain Methods for Nonlinear Analysis : Theory and Applications, World Scientific Publ., Singapore
51. S. Lefschetz (1965) Stability of Nonlinear Control Systems, Academic Press, NY
52. M.A. Aizerman, F.R. Gantmacher (1964) Absolute Stability of Regulator Systems, Holden-Day, San Franscisco
53. M. Brown, C. Harris : Neuro Fuzzy Adaptive Modelling and Control, Prentice Hall, Englewood Cliffs, NJ, 1994
54. G.E. Dullerud, F. Paganini (2000) A Course in Robust Control Theory : a Convex Approach, Springer-Verlag, New-York
55. J.C.S. Hsu, A.U. Meyer (1968) Modern Control Principles and Applications, McGraw-Hill, New-York
56. M. Cotsaftis (2002) On the Definition of Task Oriented Intelligent Control, *Proceedings ISIC'02 Conf.*, Vancouver, Oct. 27-30
57. H.O. Wang, K. Tanaka, F. Griffin (1996) An Approach to Fuzzy Control of Nonlinear Systems : Stability and Design Issues, *IEEE Trans. On Fuzzy Systems*, 4(1):14-23
58. R. Feistel, W. Ebeling (1989) Evolution of Complex Systems, Kluver, Dordrecht
59. H. Haken (1988) Information and Self-Organization, Springer, Berlin
60. Yu.L. Klimontovich (1999) Entropy, Information, and Criteria of Order in Open Systems, *Nonlinear Phenomena in Complex Systems*, 2(4):1-25
61. W. Bolton (1999) Mechatronics, 2nd ed., Addison-Wesley, UK
62. M. Gell-Mann (1995) What is Complexity, *Complexity J.*, 1(1):16;H. Morowitz (1995) The Emergence of Complexity,1(1):4; M. Gell-Mann (1994) The Quark and the Jaguar – Adventures in the simple and the complex, Little, Brown and Company, New-York
63. G. Nicolis, I. Prigogine (1992) A la Rencontre du Complexe, PUF, Paris
64. V.G. Majda (1985) Sobolev Spaces, Springer-Verlag, New-York
65. M. Hirsch, C. Pugh, M. Shub(1977) Invariant Manifolds, Lecture Notes in Math. 583, Springer-Verlag, Berlin
66. M. Cotsaftis (1997) "Global Control of Flexural and Torsional Deformations of One-Link Mechanical System", *Kybernetika*, 33(1):75
67. S. Arimoto (1996) : Control Theory of Nonlinear Mechanical Systems : a Passivity Based and Circuit Theoretic Approach, Oxford Univ. Press, Oxford, UK
68. J.H. Burl (1999) Linear Optimal Control, H_2 and H_∞ Methods, Addison-Wesley Longman, Menlo Park, CA
69. R.C. Hilborn (1994) Chaos and Nonlinear Dynamics, Oxford Univ. Press, Oxford, UK
70. T. Kapitaniak (1998) Chaos for Engineers : Theory, Applications and Control, Springer, Berlin
71. I. Prigogine (1993) Les Lois du Chaos, Nouvelle Bibliothèque Scientifique, Flammarion, Paris
72. D.R. Smart (1980) Fixed Point Theorems, Cambridge Univ. Press, Mass.
73. R. Badii and A. Politi (1997) Complexity: Hierarchical Structures and Scaling in Physics, Cambridge University Press, Cambridge, Mass.
74. B. Goodwin (1994) How the Leopard Changed Its Spots: The Evolution of Complexity, Weidenfield and Nicholson, London
75. S. Johnson(2001) Emergence, Penguin, New-York

76. S. Wolfram(1994) Cellular Automata and Complexity, Collected Papers, Addison-Wesley
77. F.E. Yates, ed. (1987) Self-Organizing Systems: The Emergence of Order , Plenum Press
78. R.K. Standish (2004) On Complexity and Emergence, *Complexity International*, 9:1-6
79. G. Parisi : Complex Systems : a Physicist's Viewpoint, Internet arxiv:cond-mat/0205297
80. C.R. Shalizi : Methods and Techniques of Complex Systems Science : an Overview, Internet arxiv:nlin.AO/0307015
81. R. Axelrod, M.D. Cohen (1999) Harnessing Complexity : Organizational Implications of a Scientific Frontier, Free Press, New-York
82. H.C. von Baeyer (1998) Maxwell's Demon, Random House
83. S. Chandrasekhar (1943) Stochastic Problems in Physics and Astronomy, *Rev. Mod. Phys.*, 15(1):1-89
84. B. Kosko (1991) Neural Networks and Fuzzy Systems: A Dynamical Systems Approach to Machine Intelligence, Prentice-Hall, Englewood Cliffs, NJH.

6 Appendix

To get the analytical form of the controller $\mathcal{C}_F$, eqn(3) with eqn(4) gives for error e the form

$$\frac{de}{dt} = A(t)e + B_0(t)\Delta u + G_s(e, x_s, \Delta u, d, t) \tag{6}$$

when splitting the linear part. It is supposed that the nonlinear part satisfies the bounding inequality

$$\|G_s(e, x_s, \Delta u, d, t)\| \leq M_s(t) g_s(\|e(t)\|, t) \tag{7}$$

and that the linear gain K in eqn(4) is such that the linear part of the system is asymptotically stable (for time independent matrices A and B_0 this means that $A = A_0 - B_0 K$ is Hurwitz). Supposing there exists positive definite matrices P and Q such that $PA + ATP + dP/dt = Q$, one can define the (positive definite) Lyapunov function

$$\mathcal{L}(x_s) =< ePe > \tag{8}$$

using the bra ket formalism. Its derivative along system trajectories is given by

$$\frac{d\mathcal{L}(e)}{dt} = - < eQe > + < ePB\Delta u > + < ePG_s > \tag{9}$$

Chosing from eqn(A2) the controller form

$$\Delta u = -\beta M_s g_s \frac{\|B^T Pe\|}{\|Pe\| + \epsilon f} \tag{10}$$

where β and ϵ are two parameters and $f = f(.)$ is the driving function to be defined later, eqn(A4) transforms after substitution into the inequality

$$\frac{d\mathcal{L}(e)}{dt} \leq - < eQe > +\epsilon M_s g_s f + M_s g_s \frac{\|Pe\|^2 - \beta \|B^T Pe\|^2}{\|Pe\| + \epsilon f} \tag{11}$$

Taking

$$\beta \geq \frac{|\lambda_{max}(P)|^2}{|\lambda_{min}(B^T P|^2} \tag{12}$$

the last term on the RHS can be made negative by taking noting $\lambda(X)$ the eigenvalue of matrix X, in which case eqn(A6) simplifies to the first two terms in the RHS of eqn(A6) ie

$$\frac{d\mathcal{L}(e)}{dt} \leq - < eQe > +\epsilon M_s g_s(\|e(t)\|, t) f(e, t) \tag{13}$$

showing the opposite action of the two terms as discussed in the text. If $f(.)g_s(.)$ is upper bounded over the whole interval, the RHS of eqn(A8) is negative above some threshold value. On the other hand, because $g_s(.)$ is for regular $f(.)$ a higher order term at the origin by construction, the RHS of eqn(A8) is negative close to the origin. So there may exist an interval on the real line where $d\mathcal{L}(e)/dt > 0$ separating two stable domains. If $f(.)g_s(.)$ is growing at infinity, there exists a threshold value above which the system is unstable. So there is an asymptotic stable ball around the origin, and the decay is fixed by the specific functional dependence of the functions $f(.)$ and $g_s(.)$. The next step is to fix the driving term $f(.)$ for determining the upper bound on the time decay of Lyapunov function and finally on the norm of error vector $e(t)$. As $< eQe >$ and $< ePe >$ are equivalent norms, there exists $k > 0$ so that $< eQe > \geq k\mathcal{L}(e)$ and eqn(A8) can be replaced by

$$\frac{d\mathcal{L}}{dt} \leq -k\mathcal{L} + \epsilon M_s(t) g_s(\mathcal{L}, t) f(\mathcal{L}, t) \tag{14}$$

with new dependent variable $\mathcal{L}$, bounded by the solution $Y(t)$ of eqn(A9) with equal sign. Here $f(., t)$ is immediately determined for given functional form of $g_s(., t)$ with prescribed $Y_d(t)$, but a more useful approach is to consider the embedding problem where, for given $M_s(t)g_s(\mathcal{L}, t)$, a correspondence is researched between function spaces $\mathcal{F}$ and $\mathcal{Y}$ to which $f(t)$ and $Y(t)$ respectively belong with $\mathcal{Y}$ fixed by global properties such as continuity and decay for large t.

The most general way is to use substitution theorems relating different function spaces [49] and fixed point theorem [72]. Supposing that $Y(t) \in \mathcal{W}_1^p$ implies $M_s(t)g_s(\mathcal{L}, t) \in \mathcal{W}_1^q$ with Sobolev space $\mathcal{W}_m^n$ [64], application of Holder inequality shows that there should be $\mathcal{F} \subseteq \mathcal{W}_1^n$ with $n^{-1} = p^{-1} - q^{-1}$, $1 \leq p, q, n < \infty$ which relates the decays of $f(t)$ and of $Y(t)$ for large t. One can then define the driving function $f(t) = \mu Y^{\phi}$ now implicit in time through the error bound directly measurable. More specific results can be found for more explicit constraints. For instance when $g_s(Y,t) \leq b(t)Y^s$ with $b(.) \in \mathcal{L}^{\beta}$ and $s^{-1} = q^{-1} - \beta^{-1}$ the bounding equation for $Y(t)$ from eqn(A9) becomes a Bernouilli equation with solution

$$y(t) = \exp(-kt).[1 - G]^{-1/(\theta-1)} \tag{15}$$

where

$$G = (\theta - 1)\frac{\epsilon\mu}{k}Y_0^{\theta-1}\int_0^{kt} \exp(-(\theta-1)u).M_s(u)b(u)du \tag{16}$$

with $\theta = s + \phi$, $y(t) = Y(t)/Y(0)$ and with normalized time $u = kt$, which exhibits a non exponential asymptotic decay for $\theta < 1$ and for any $Y(0)$, and a conditional decay when $\theta > 1$ directly depending on the balance between recalling linear spring and repulsive nonlinear force. The present analysis extends in similar form to systems with unknown dynamics by using nonlinear network representation the parameters of which are adaptively constructed to converge toward system representation.

When the coefficients A, B_0 and the bounding function g_s of eqns (A1,A2) are time independent, the more general (Lurie type) Lyapunov function [51]

$$\mathcal{L}(e) =< ePe > +\lambda \int_0^{De} \Psi(\xi)d\xi \tag{17}$$

can also be used with D an adjustable vector and $\Psi(.)$ a sector function with bound τ , ie such that $\Psi(\xi)[\Psi(\xi) - \tau\xi] \leq 0$. Using extension of Popov criterion [12] the explicit form of driving function $f(\xi) = \xi^{1/2}$ is finally found in eqn(A9) leading to the same bound as in eqn(A10) with $\phi = 1/2$ whereas the additional controller takes the simple form $\Delta u = \Psi(D.e)$ instead of eqn(A5). Analysis of $U = (1 - G)^{-1/(\theta-1)} = U(kt, Y_0, \theta)$ in eqn(A10) shows that when $(\epsilon\mu/k)Y_0^{\theta-1} < 1$, $y(t)$ has no singularity over the complete real line, and two cases can occur. If $\epsilon\mu/k > 1$, Y_0 is always < 1 and larger limit value of Y_0 corresponds to larger exponent. So extension of Popov criterion gives a smaller robustness ball than fixed point result. If $\epsilon\mu/k < 1$, Y_0 can be > 1 in which case conversely, larger limit value of Y_0 corresponds to smaller exponent. So extension of Popov criterion is equivalent to fixed point result when taking the smallest allowable exponent value 1/2. When $(\epsilon\mu/k)Y_0^{\theta-1} > 1$, there exists a critical time kt_c for which $y(t)$ is singular, showing finite time Lagrange instability. Only finite time bounded-ness is possible in this case, and given a ball $\mathcal{B}$ it is interesting to determine the largest initial ball $\mathcal{B}(0, Y_0)$ so that its transform $\mathcal{B}(T, Y(T))$ by eqn(A10) satisfies $\mathcal{B}(T, Y(T)) \subseteq \mathcal{B}$ with largest T corresponding to smallest input-output amplification factor. In the same way, smallest exponent corresponds to largest T and largest Y_0 when $\epsilon\mu/k < 1$ whereas a compromise has to be found when $\epsilon\mu/k > 1$. So in all cases an adapted controller exists in explicit form, and is very robust as it only depends on global bounding functions for system equations. These may change with the task, so here the system has a consistency link between the task assignment and the nature of its response through the controller guiding system trajectory whatever it starts from toward any trajectory belonging to the task manifold.

Part II

Natural Systems Modeling

Competing Ants for Organization Detection

Application to Dynamic Distribution

Alain Cardon[1], Antoine Dutot[2], Frédéric Guinand[2], and Damien Olivier[2]

[1] LIPVI – University of Paris VI
8, rue du Capitaine Scott
75015 Paris, France
Alain.Cardon@lip6.fr

[2] LITIS – University of Le Havre
25 rue Philippe Lebon
76600 Le Havre, France
Antoine.Dutot@univ-lehavre.fr

Summary. A simulation application may be modeled as a set of interacting entities within an environment. Such applications can be represented as a graph with a one-to-one mapping between vertices and entities and between edges and communications. As for classical applications, performances depend directly on a good load balancing of the entities between available computing devices and on the minimization of the impact of the communications between them. However, both objectives are contradictory and good performances may be achieved if and only if a good trade off is found. Our method for finding such a trade off leans on a bio-inspired method. We use competitive colonies of numerical ants, each one depositing colored pheromones, to find organizations of highly communicating entities.

Key words: complex systems, self-organization, artificial ants, dynamic graph, community structures, dynamic load balancing

1 Introduction

We present a new heuristic for the dynamic load balancing problem, when both the application and the computing environment change during the execution. From the computing environment point of view, the generalization of wireless technologies together with the emergence of grid-like and peer-to-peer environments allows us to imagine next HPC (High Performance Computing) environments as collections of heterogeneous computing devices appearing and disappearing during the computation. Within such a context the availability of dynamic load balancing strategies able to manage both application dynamics and environment changes might be of primary importance .

There exist many application classes whose structural as well as numerical characteristics may change during their execution. Simulation is one of these classes. Simulation covers a wide range of domains: from life and earth sciences to social sciences. Many applications rest on a mathematical analysis and formulation of the global process to be simulated with systems of differential equations. An alternative or a complementary way of modeling problems for simulation purposes consists in modeling the environment, each entity moving and evolving within this environment, and the mechanisms needed for computing entities evolution, their activities and their interactions. Entities may be living beings, particles, or may represent non-material elements (vortices, winds, streams...). Individual-based models fall into this category. Within such models, interactions between the environment and the individuals are explicit as well as interactions between individuals themselves. During the simulation, some individuals may appear and/or disappear without prior notice, and interactions may change, leading to unpredictable sets of highly communicating entities. Due to the dynamics of such applications, the distribution of the entities has to be continuously examined and some migration choices may be proposed in order to improve both performances and emergent organization detections.

In this paper, a bio-inspired algorithm (based on ants) named $AntCO^2$ is described that advises on a possible better location of some entities according to the trade off between load balancing and minimization of communication overhead. The paper is organised as follows: the next section described in detail the considered models of applications and of computing environments. In section 3 the adaptation of the ant approach to our problem is presented, and some experiments are reported in section 4, and we conclude in the last section.

2 Problem Formulation

We are interested in the problem of the allocation and migration of the tasks/-entities/agents/elements of applications on parallel and distributed computing environments. Our method was primarily designed for managing dynamic characteristics of both the application and the execution environment.

2.1 Application Model

Ecological simulations are especially interesting from our point of view since very few is known about the development and the evolution of elements and about their interactions before the execution. Every piece of the simulation needing special processing or interactions is called an entity. Thus, we are in front of complex systems, in which there are numerous interacting entities. As the system evolves, communications between entities change. Entities and their interactions appear and disappear creating stable or unstable organizations. Many individual-based models (IBMs) are often considered, either as an alternative or as a complement to state variable models [22]. Each entity, in IBMs, is simulated at the level of the individual organism rather than the population level, and the model allows individual variation to be simulated [3]. Thus, during the simulation process some entities may appear or disappear, interactions may meet, leading to an important increase of the number of communications between them (for instance a shoal of sardines passing a group of tuna). While highly dynamic, the whole system may be modeled using a classical non oriented graph whose structural and numerical characteristics may change during the simulation. Within such a graph, the vertex u is associated to an element of the simulation that may be a biological entity or any environmental element, and an edge $e = (u, v)$ materializes a communication or an interaction between two elements of the simulation.

$AntCO^2$ aims to provide some kind of load-balancing for dynamic communication-based graphs and the simulation is coupled with it, in such way that both applications remain independent up to a certain degree. In one side we have the simulation and in the other side $AntCO^2$ whereas it goal is to provide a dynamic suggestion on how and where to migrate entities. In other words, $AntCO^2$ offers a service to the simulation.

2.2 Execution Environment Model

In addition to supercomputers and network of workstations, the last decade have seen the emergence of a new generation of high-performance computing environments made of many computing devices characterized by a high degree of heterogeneity and by frequent temporary slowdowns. These environments – computational grids, peer to peer environments, and networks of mobile computing devices – work almost always in a semi-degraded mode. The sudden failure of a computing device may entail problems about the recovery of elements executed on that device. However, research of solutions for that problem is out of the scope of our work. We assume that fault tolerance during the execution is managed either by the platform used for the implementation or by the application itself. If such a failure occur, we only focus on the reallocation of elements that have lost the computing device they were allocated to.

We also suppose that some events, occurring during the execution, may be automatically recorded and applied to the graph. This can be achieved using functionalities proposed by some programming environments [9]. Relevant events are: arrival and disappearance of elements, communications (source and destination elements, time and/or volume and/or frequency of exchanged data), hardware failures (links, computing devices). An interface is defined between simulation and $AntCO^2$ (see [15]).

2.3 Previous Works on Dynamic Load Balancing

The problem may be formulated as the on-line allocation, reallocation and migration of tasks to computing resources composing a distributed environment that may be subject to failure. Another way of formulating this problem needs the description of the different communications that may occur between elements. On the one hand, communications between elements located on the same computing resource are supposed negligible. This hypothesis refers to data locality in parallel processing. On the other hand, communications between elements located on distinct computing resources constitute the source of the communication overhead. The latter are called *actual communications* in the sequel. Then, our objective may be expressed as the following trade-off: to reduce the impact of actual communications while preserving a good load-balancing between computing resources.

There exist many variants of this problem which has been extensively studied, and two main classical approaches may be distinguished. The static one consists in computing at compilation time an allocation of the tasks using the knowledge – assumed to be known prior to the execution – about the application and about the computing environment. This problem is known as the mapping problem [6]. A global optimization function, representing both load-balancing and communication overhead minimization, should be minimized. Strategies for achieving this objective often leans on graph partitioning techniques using optimization heuristics like simulated annealing, tabu search or evolutionary computing methods. However this approach is not suitable as-is in our context since we do not have information at compilation time.

The second approach is dynamic load-balancing. Many works have been dedicated to the special case of independent tasks [11,16,31,34]. When communications are considered, the problem is often constrained by some information about the precedence relation between tasks. In [24], for instance, the proposed heuristic, named K1, for scheduling a set of tasks with some dynamic characteristics takes into consideration precedence tasks graphs with inter-tasks communication delays, but, the execution of the task set is supposed to be repeated in successive cycles and the structure of the precedence task graph remains fixed, and only numerical values (execution times and communication delays) may change during the execution. When no information is available about the precedence relation between tasks, Heiss and Schmitz have proposed a decentralized approach based on a physical analogy by considering tasks as particles subject to forces [23]. Their application model is similar to ours, but they do not consider changes that may happen to the execution environment. Moreover, their approach is based on local negotiations between neighbors resources and may spread out, depending of the load.

The key point for achieving a good trade-off lies in the capacity of combining local decisions (communicating tasks forming clusters) with global ones (load balancing). Partitioning of the application graph seems to be a very suitable method for that goal, since clustering may be driven by the objective of minimizing the communications, and the load balancing may be achieved by fixing the number of clusters. In general, finding an exact solution to k-partitioning problem of this kind is believed to be NP-hard [19]. A wide variety of heuristic has been used to generally provide acceptably good solutions, for example the Kernighan-Lin algorithm [26]. In the last few years, alternatives using natural principles such as simulated annealing, tabu search, genetic algorithms, ant algorithms ([2,25,27,29]) have shown great potential. Partitioning a dynamic application graph seems to be out of reach of traditional clustering approaches. Moreover, when local changes occur in the application graph, it seems more profitable to modify/improve the current solution than to perform a brand new clustering. This implies to consider the global solution as emergent from the set of local solutions rather than as the result of a driven process. All these considerations have led us to consider an artificial collective intelligence approach for our problem.

2.4 Problem Definition

The considered simulations are constituted by a number n at time t of intercommunicating entities, which we wish to distribute over a number of p processing resources. Entities do not necessarily communicate with all the others and their numbers and the communications vary during time. The pattern of communications can be represented by a dynamic graph (network) in which entities are mapped one-to-one with vertices and communications with edges.

Definition 1 (Dynamic graph). *A dynamic graph is a weighted undirected graph* $G(t) = (\mathcal{V}(t), \mathcal{E}(t))$ *such that:*

- $\mathcal{V}(t)$ *is the set of vertices at time* t.
- $\mathcal{E}(t)$ *is the set of edges at time* t. *Each edge* e *is characterized by:*
  - *a weight* $w^{(t)}(e) \in \mathbb{N}^+$.

The problem is to distribute at anytime in such a way to balance the load on each processing resources and at the same time minimizing the actual communications (see 2.3). To solve this problem, we search a partition of the graph $G(t)$.

Definition 2 (Graph partition). *Let* $G(t) = (\mathcal{V}(t), \mathcal{E}(t))$ *a dynamic graph, a partition* $\mathcal{D}(t)$ *of* $G(t)$ *is composed by* k *disjointed subsets* $\mathcal{D}_i(t)$ *of* $\mathcal{V}(t)$ *called domains with :*

$$k > 0, \bigcup_{i=1..k} \mathcal{D}_i(t) = \mathcal{V}(t)$$

The set of edges connecting the domains of a partition (i.e. edges cut by the partition) $\mathcal{D}(t)$ *is called an edge-cut denoted by* $\mathcal{B}(t)$.

Our objective is to find a k-partition, at anytime, which evenly balances the vertex weight among the processing resources whilst minimizing the total weight of $\mathcal{B}(t)$. The number of domains k must be greater or equal to p the number of processing resources (for example the partition found with three colors ($p = 3$ and four domains ($k = 4$) on the graph of Figure 1).

Our method uses organizations inside the communication graph to minimize communications. To each organization corresponds a domain $\mathcal{D}_i(t)$. Detected organizations in the communication graph may, but do not necessarily correspond to entity organizations in the application. These detected organizations appear, evolve and disappear along time.

3 Colored Ant System

Algorithms using numerical ants are a class of meta-heuristics based on a population of agents exhibiting a cooperative behaviour [30]. Ants are social insects which manifest a collective problem-solving ability [12]. They continuously forage their territories to find food [20] visiting paths, creating bridges, constructing nests, etc. This form of self-organization appears from interactions that can be either direct (e.g. mandibular, visual) or indirect. Indirect communications arise from individuals changing the environment and other responding to these changes: this is called *stigmergy*[3]. There are two forms of stigmergy, sematectonic stigmergy produces changes in the physical environment – building the nest for example – and stigmergy based on signal which uses environment as support. Thus, for example, ants perform such indirect communications using chemical signals called *pheromones.* The larger the quantity of pheromones on a path, the larger the number of ants visiting this path. As pheromone evaporates, long paths tend to have less pheromone than short ones, and therefore are less used than others (binary bridge experiment [21]). Such an approach is robust and well supports parameter changes in the problem. Besides, it is intrinsically distributed and scalable. It uses only local information (required for a continuously changing environment), and find nearly optimal solutions. Ant algorithms have been applied successfully to various combinatorial optimization problems like the Travelling Salesman Problem [13], routing in networks [8, 33], clustering [18], coloring [10], graph partitioning [28], and more recently DNA sequencing [4], Dynamic load balancing falls into the category of optimization problems. As reported in [7], Ant Colony Optimisation (ACO) approaches may be applied to a given problem if it is possible to define: the problem representation as a graph, the heuristic desirability of edges, a constraint satisfaction, a pheromone updating rule and, a probabilistic transition rule. Our approach adds to the load balancing

[3] PP. Grassé, "La théorie de la stigmergie: Essai d'interprétation du comportement des termites constructeurs", in Insectes Sociaux, 6, (1959), p. 41-80, introduced this notion to describe termite building activity.

problem, organization detection and placement of these to reduce network communication. Even if we can find similarities with ACO, we will see in the following that our algorithm, having no satisfaction constraint, depart from this.

Let us first motivate our choice for coloring ants and pheromones.

3.1 Colored Ants

The method proposed by Kuntz et al. [28] for graph partitioning is able to detect clusters within a graph and it also has the ability of gathering vertices such that:

1. if they belong to the same cluster they are gathered in the same place,
2. the number of inter-cluster edges is minimized and,
3. distinct clusters are located in different places in the space.

Points 1. and 2. are relevant for our application, however, additional issues have to be considered:

1. the number of clusters may be different of the number of processing resources,
2. the sum of the sizes of the clusters allocated to each processing resource has to be similar,
3. and their number as well as the graph structure may change.

For the first issue, in [7], the authors mentioned the possibility of setting parameters of the KLS algorithm in order to choose the number of clusters to build. Unfortunately, for our application, we do not know in advance what should be the best number of clusters for achieving the best load balancing, as shown in Figure 1, where elements of the two opposite and not directly linked clusters should be allocated to the same processing resource.

As the number of processing resources available at a given moment is known, this problem may be identified as a competition between processing resources for computing as many elements as possible. This may be directly included in the ant approach by considering competing ant colonies. In our work, a distinct color is associated to each computing resource, and an ant colony is attached to each resource. Each ant is also colored, and drops pheromones of that color.

So, in addition to the classical collaborative nature of ant colonies we have added a competition aspect to the method. In fact, this reflects exactly the trade-off previously discussed about dynamic load-balancing. On the one hand, the collaborative aspect of ants allows the minimization of communications by gathering into clusters elements that communicate a lot, while, on the other hand, the competition aspect of colored ants allows the balancing of the load between computing resources.

The technical issues about the management of ants, colors and pheromones are described in detail in the next sections.

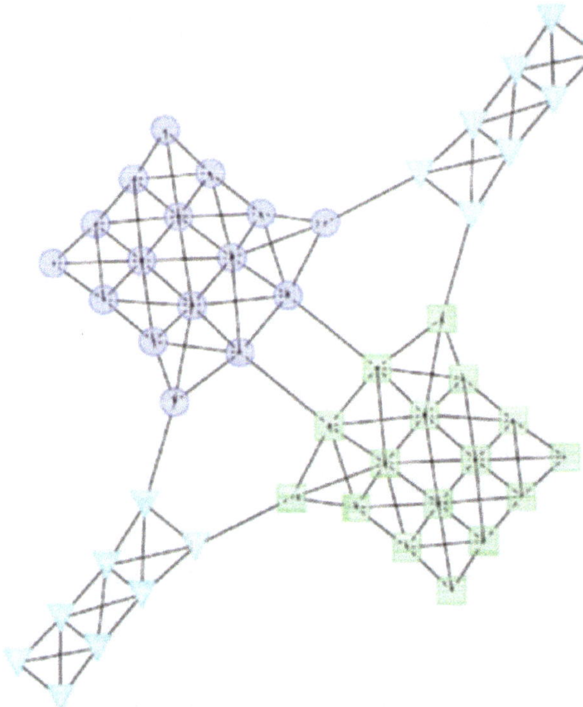

Fig. 1. *Example for which the best number of clusters is different of the number of processing resources (left to right).*

3.2 Graph Description and Notations

As previously mentioned, we consider an undirected graph whose structural as well as numerical characteristics may be subject to changes during the execution. Colored ants walk within this graph, crossing edges and dropping colored pheromones on them.

Definition 3 (Dynamic Communication Colored Graph). *A dynamic colored communication graph is a dynamic graph* $G(t) = (\mathcal{V}(t), \mathcal{E}(t), \mathcal{C}(t))$ *such that:*

- $\mathcal{V}(t)$ *is the set of vertices at time* t*. Each vertex* v *is characterized by:*
  - *a color* $c \in \mathcal{C}(t)$*,*
- $\mathcal{E}(t)$ *is the set of edges at time* t*. Each edge* e *is characterized by:*
  - *a weight* $w^{(t)}(e) \in \mathbb{N}^+$ *that corresponds to the volume and/or the frequency and/or the delay of communications between the elements at each end of edge* e*.*
  - *a quantity of pheromones of each color.*

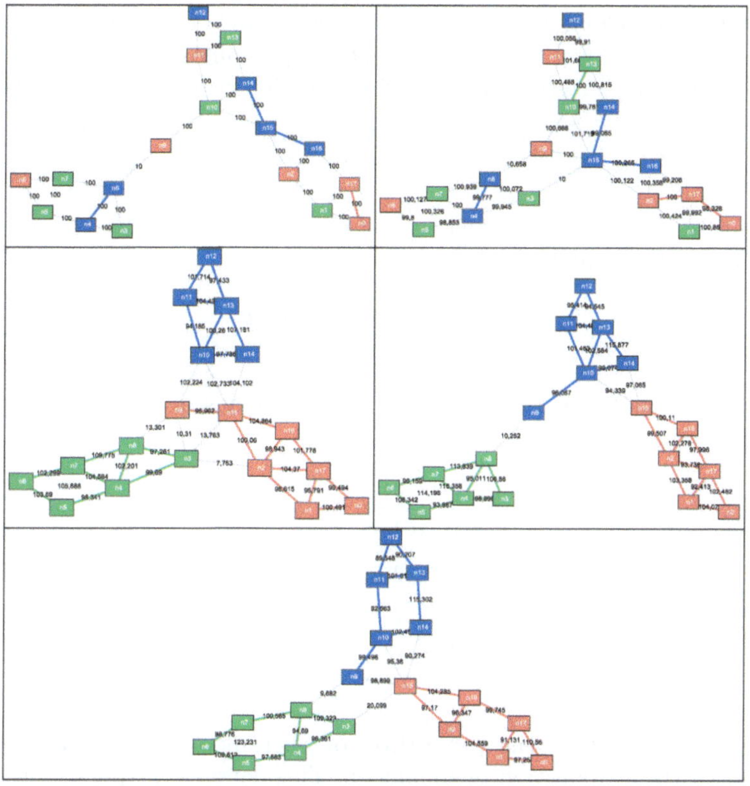

Fig. 2. *Example of a dynamic colored communication graph at five stages of its evolution.*

- *$\mathcal{C}(t)$ is a set of p colors where p is the number of available processing resources of the distributed system at time t.*

The Figure 2 shows an example of a dynamic colored communication graph at several steps of its evolution, where the proposed method described in the following, changes the color of vertices if this improve communications or processing resource load. The algorithm tries to color vertices of highly communicating clusters with the same colors. Therefore a vertex may change color several times, depending on the variations of data exchange between entities.

3.3 Pheromones Management

We denote by $\mathcal{F}(t)$ the population of ants at time t, and $\mathcal{F}_c(t)$ the set of ants of color c at time t. Pheromones are dropped on edges by ants crossing them. Pheromones are colored. An ant x of color c crossing an edge e between steps

$t-1$ and t will drop a given quantity of pheromone of color c. This quantity is denoted by $\Delta_x^{(t)}(e,c)$, and the quantity of pheromones of color c dropped by ants when they crossed edge e during time interval $]t-1,t]$ is equal to:

$$\Delta^{(t)}(e,c) = \sum_{x \in \mathcal{F}_c(t)} \Delta_x^{(t)}(e,c) \tag{1}$$

The total quantity of pheromones of all colors dropped by ants on edge e during $]t-1,t]$ is equal to

$$\Delta^{(t)}(e) = \sum_{c \in \mathcal{C}(t)} \Delta^{(t)}(e,c) \tag{2}$$

If $\Delta^{(t)}(e) \neq 0$, the rate of dropped pheromones of color c on e during $]t-1,t]$ is equal to:

$$K_c^{(t)}(e) = \frac{\Delta^{(t)}(e,c)}{\Delta^{(t)}(e)} \text{ with } K_c^{(t)}(e) \in [0,1] \tag{3}$$

The quantity of pheromone of color c on the edge (e) at time t is denoted by $\tau^{(t)}(e,c)$. At the beginning $\tau^{(0)}(e) = 0$ and this value changes according to the following recurrent equation:

$$\tau^{(t)}(e,c) = \rho\tau^{(t-1)}(e,c) + \Delta^{(t)}(e,c) \tag{4}$$

Where $\rho \in]0,1]$ is the proportion of pheromones which has not been removed by the evaporation phenomenon. This denotes the persistence of the pheromones on the edges .

$\tau^{(t)}(e,c)$ may be considered as a reinforcement factor for clustering vertices based on colored paths. However, due to the presence of several colors, this reinforcement factor is corrected according to $K_c^{(t)}(e)$ that represents the relative importance of the considered color with respect to all colors. This corrected reinforcement factor is noted:

$$\Omega^{(t)}(e,c) = K_c^{(t)}(e)\tau^{(t)}(e,c)$$

Then, if we denote by $\mathcal{E}_u(t)$ the set of edges connected to vertex u at time t, the color $\xi^{(t)}(u)$ of vertex u is obtained from the main color of its incident edges:

$$\xi^{(t)}(u) = \arg\max_{c \in \mathcal{C}(t)} \sum_{e \in \mathcal{E}_u(t)} \tau^{(t)}(e,c) \tag{5}$$

3.4 Ants Moving and Population Management

Ants move according to local information present in the graph. Each processing resource is assigned to a color. Each vertex gets its initial color from the processing resource it was allocated to. Initially the number of ants of a given color is proportional to the processing resource power that they represent.

Population Management

The process is iterative, between two steps, each ant crosses one edge and reaches a vertex. When there are too few ants, evaporation makes pheromones disappear and the method behaves as a greedy algorithm. If there are too many ants, pheromones play a predominant role and the system efficiency may decreases. Furthermore, the population is regulated with respect to the number of processing resources and to the number of entities.

Initially, our algorithm creates a fixed number of ants per vertex, which depends on processing resources power. Vertices are, in the same way, proportionally associated to processing resources. Then, during the execution, our method tries to keep the population density constant in the graph. When some new vertices are created, new ants are created in order to maintain the population density, and some ants are removed from the population when some vertices disappear. When one new processing resource becomes available, the number of colors increases by one. However, the population has to be modified in order to take into account this new color. This is done by changing the color of an equal number of ants of each current color into the new color. Symmetrically, when a processing resource, associated to color c, disappears from the environment, either because of a failure or because this resource is out of reach in case of wireless environments, all ants of color c change their color into the remaining ones. The creation and the removing of edges have no effect on the population.

Ants Moves

The moving decision of one ant located on a vertex u is made according to its color and to the concentration of the corresponding colored pheromones on adjacent edges of u. Let us define $p^{(t)}(e, c)$ the probability for one arbitrary ant of color c, located on the vertex u, to cross edge $e = (u, v)$ during the next time interval $]t, t+1]$. If we denote $w^{(t)}(e)$ the weight associated to this edge at time t, then:

$$\begin{cases} p^{(t)}(e,c) = \dfrac{w^{(0)}(e)}{\sum\limits_{e_i \in \mathcal{E}_u(t)} w^{(0)}(e_i)} & \text{if } t = 0 \\ p^{(t)}(e,c) = \dfrac{(\Omega^{(t)}(e,c))^{\alpha}(w^{(t)}(e))^{\beta}}{\sum\limits_{e_i \in \mathcal{E}_u(t)} (\Omega^{(t)}(e_i,c))^{\alpha}(w^{(t)}(e_i))^{\beta}} & \text{if } t \neq 0 \end{cases} \tag{6}$$

The parameters α and β (both > 0) allow the weighting of the relative importance of pheromones and respectively weights. However, if ants choices were only driven by this probability formula, there would be no way for avoiding oscillatory moves. So, we introduce a handicap factor in equation (6) $\eta \in]0, 1]$ aiming at preventing ants from using previously crossed edges. The idea is very similar to the tabu list. We introduce this constraint directly into the probability formula. Each ant has the ability to remember the k last edges it has crossed. These edges are stored into a list: $\mathcal{W}_x$ with $\text{card}(\mathcal{W}_x) < M$ (M constant threshold). Then, the value of η for an ant x considering an edge (u, v) is equal to:

$$\eta_x(v) = \begin{cases} 1 \text{ if } v \notin \mathcal{W}_x \\ \eta \text{ if } v \in \mathcal{W}_x \end{cases} \tag{7}$$

For the ant x, the probability of choosing edge $e = (u, v_i)$ during time interval $]t, t+1]$ is equal to:

$$p_x^{(t)}(e,c) = \frac{(\Omega^{(t)}(e,c))^{\alpha}(w^{(t)}(e))^{\beta}\eta_x(u)}{\sum\limits_{e_i \in \mathcal{E}_u(t)} (\Omega^{(t)}(e_i,c))^{\alpha}(w^{(t)}(e_i))^{\beta}\eta_x(v_i)} \tag{8}$$

To complete this, we introduce a demographic pressure to avoid vertices that already contain too many ants, so that they are better spread in the graph. It is modeled by another handicap factor $\gamma(v)$ on vertices that have a population greater than a given threshold. This factor is introduced in the formula (8). Given $N(v)$ the ant count on the vertex v and N^* the threshold.

$$\gamma(v) = \begin{cases} 1 & \text{if } N(v) \leq N^* \\ \gamma \in]0,1] & \text{else} \end{cases} \tag{9}$$

The formula (8) becomes :

$$p_x^{(t)}(e,c) = \frac{(\Omega^{(t)}(e,c))^{\alpha}(w^{(t)}(e))^{\beta}\eta_x(u)\gamma(u)}{\sum\limits_{e_i \in \mathcal{E}_u(t)} (\Omega^{(t)}(e_i,c))^{\alpha}(w^{(t)}(e_i))^{\beta}\eta_x(v_i)\gamma(v_i)} \tag{10}$$

Ants Way of Life

The algorithm is based on an iterative process. During the time interval $]t, t+1]$, each ant may either, hatch, move, or die.

An ant of color c, located on a vertex u dies if the proportion of color c on adjacent edges is under a threshold. If $\phi \in [0, 1]$ is the threshold, then, the ant of color c located on u dies if:

$$\frac{\sum\limits_{e \in \mathcal{E}_u(t)} \tau^{(t)}(e,c)}{\sum\limits_{c_i \in \mathcal{C}(t)} \left(\sum\limits_{e \in \mathcal{E}_u(t)} \tau^{(t)}(e,c_i) \right)} < \phi \tag{11}$$

A new ant is then created in another location.

This "jumping" mechanism improves the global algorithm behavior. For instance, when the graph becomes unconnected, some ants may be prisoners of isolated clusters, and the die-and-hatch sequence allows them to escape, as illustrated on Figures 3(a) and 3(b). The mechanism, while keeping population constant, also avoids locked situations (grabs, overpopulation, starvation) (see Fig. 4) occurring when the system meets local minima. Moreover, this improves the reactivity of our algorithm that runs continuously, not to find the best solution for a static graph but, for providing anytime solutions to a continuously changing environment.

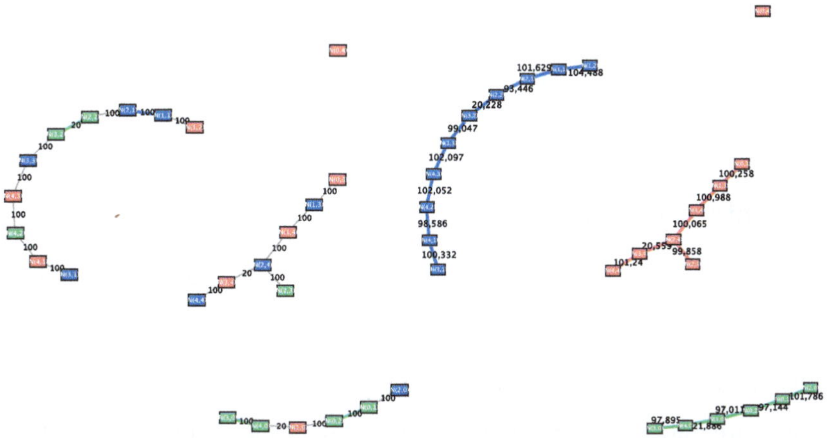

(a) *At start.* (b) *Later on, after some ants jumped.*

Fig. 3. *Example of a disconnected dynamic graph.*

4 Experiments and Results

Dynamic load balancing falls into the category of distributed time-varying problems. It seems difficult to perform a comparison to optimal solutions on dynamic graphs. This is why the first part of the performance analysis is dedicated to the comparison of allocations computed by our method for some classes of static graphs. The second part of the analysis focuses on the reactivity and on the adaptability of our algorithm for some relevant dynamic graphs.

Before entering into details, some performance measures are defined in the next section.

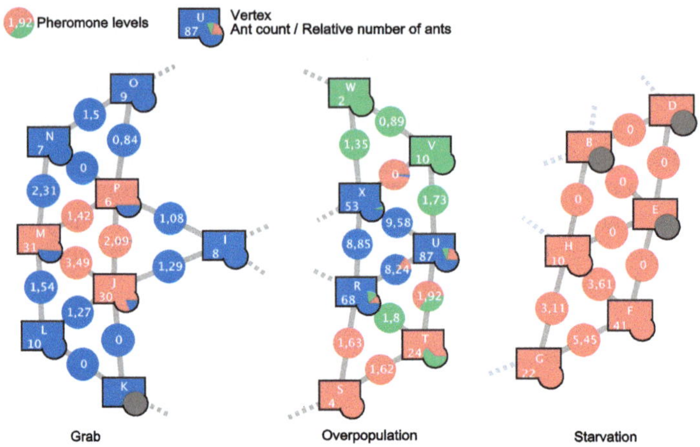

Fig. 4. *Problems solved by jumping mechanism.*

4.1 Quality Analysis

Two measures are relevant for defining the quality of a dynamic load balancing:

- The global costs of communications;
- The load-balancing of the application.

They are antagonist. So, in order to evaluate our solution we first define two quality criteria r_1 and r_2.

The first criterion $r_1 \geq 0$ corresponds to the ratio of actual communications (see section 2.3) over the total number of communications occurring in the graph. Solutions are better when r_1 is close to zero.

$$r_1 = \frac{\text{actual communications}}{\text{all the communications}}$$

The second criterion r_2 measures how the load is balanced among the available processing resources, independently of the communications. For each color c, we have $\mathcal{V}_c(t)$ the set of vertices having color c at time t. Then we have:

$$r_2 = \frac{\min(card(\mathcal{V}_c(t)))}{\max(card(\mathcal{V}_c(t)))}$$

The load-balancing is better when r_2 is close to 1.

In case of static graphs, these criteria, enable us to store the best solutions obtained so far.

Since we seek to find organizations, r_1 is considered to be a more important criterion. Indeed, it is explicitly defined in ant behavior, unlike r_2 that is implicitly optimized by competition mechanisms.

4.2 Static and Dynamic Analysis

The main objective of our approach is not to compute load-balancing for static graphs, however, several tests have been made on different kinds of static graphs to check the validity of our approach. We assume that the computation time required by every element, during one time step, is the same, and we do not take into account migration costs. Each vertex is randomly assigned to an initial processing resource at start. No assumption is made about this initial mapping. For each presented class, figures show both the colored graph solutions, and the evolution of r_1 and r_2 criteria. The abscissa is the number of iterations. For most graphs, parameter used where $\alpha = 1$, $\beta = 3$, $\rho = 0.86$, $N^* = 10$, $\phi = 0.3$, $\eta = 0.0001$, $M = 4$. Changes in these parameters are indicated when needed.

Random Graphs

We tested two kinds of random graphs [17], one is purely random in its topology with uniform weights, and another with the same topology that in addition defines "communications structures". The first graph was created by connecting a given number of vertices randomly with the degree distribution following a Poisson law, the second was made by creating a spanning tree on the first graph, cutting some edges in the tree to create a forest, and giving to each remaining edges of this tree larger communication weights.

Figure 5 shows coloration and criteria evolution for the first graph. Though sometimes nodes with large degree perturb the algorithm, a good distribution is found (r_2 almost always greater than 0.8). The second criterion r_1 is not close to 0 since there are no organizations in the graph.

The Figure 6 uses the same graph topology, but contains organizations of all sizes under the form of communications (Figure 6(a) shows the colored spanning tree with high weights communications only for the random graph of Figure 6(b)). The r_1 criterion is better than for Figure 5 due to the presence of organizations. The slight diminution of r_2 compared to the previous graph comes from the fact organizations introduced by natural clusters do not always define groups of equal size.

Some slight coloration problems appear. They are due to the high number of low communication edges that perturb the algorithm.

Complete Graphs and Graphs with Small Degree Vertices

We also consider complete graphs in order to measure how our algorithm balances the load between resources.

The first graph whose r_1 and r_2 evolution is shown on Figure 7, is a complete one with 100 nodes. This graph is unweighted and it therefore does not present organizations. We see that we have a good r_2 criterion, though the algorithm is quite perturbed by the degree of each node. Naturally, r_1 cannot

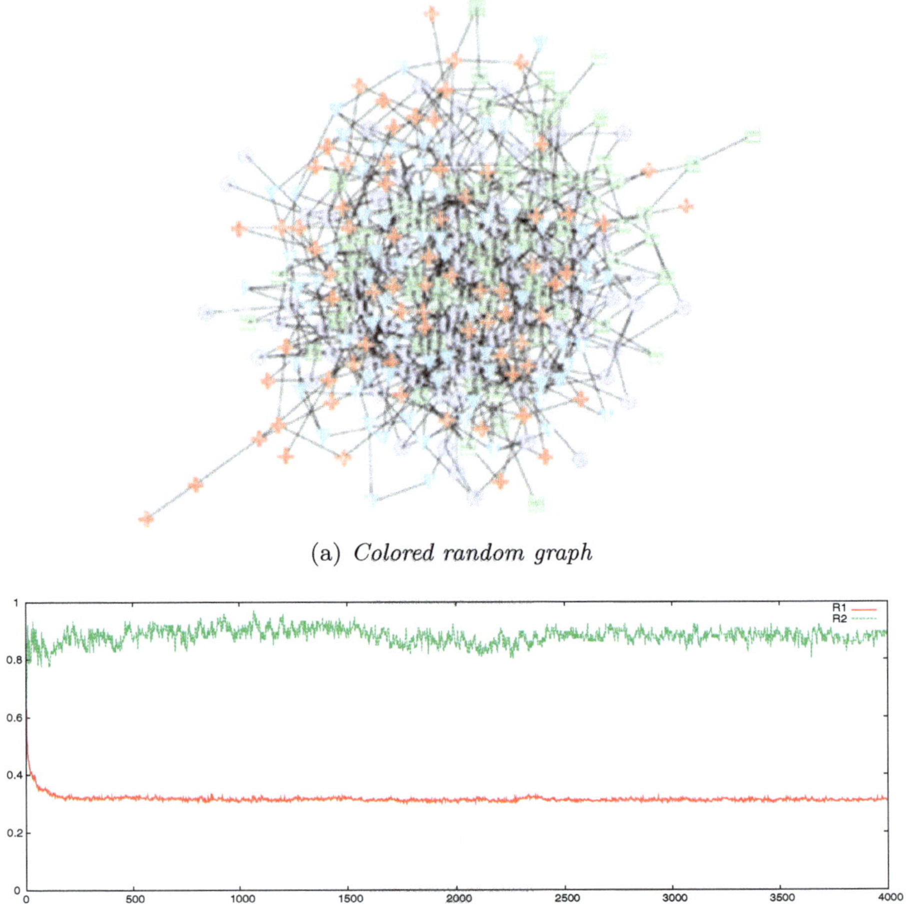

(a) *Colored random graph*

(b) *Evolution of r_1 and r_2*

Fig. 5. *A random graph with uniform weights and 4 colors (300 vertices, 602 edges).*

be made small. Indeed the value of r_1 depends on the distribution, and only on it, as all nodes play the same role (same degree and same weight). Therefore it is not really meaningful. As an information on the diagram 7, a dashed blue line indicates the theoretical best r_1 value if all clusters had had the same size (this optimal is crossed by our r_1 value on the graph since all clusters do not have the same size, the plotted "best" r_1 is only informative). This value is:

$$\frac{n(p-1)}{p(n-1)}$$

Else, if clusters are not all the same size, r_1 is the optimal and is equal to:

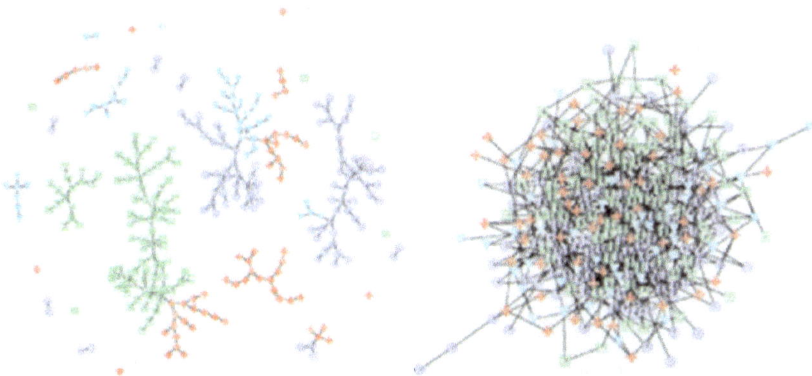

(a) *Spanning tree* (b) *Colored weighted random graph*

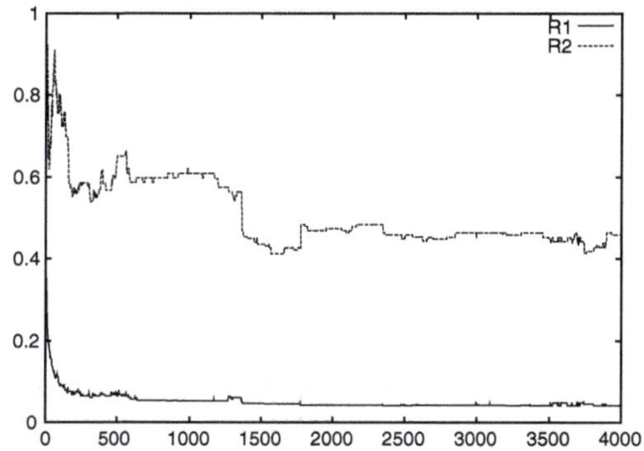

(c) *Evolution of r_1 and r_2 on the random graph*

Fig. 6. *A random graph with "communication structures" following a spanning tree, and 4 colors (300 vertices, 602 edges).*

$$\frac{\sum_{j=1}^{p}\left(card(\mathcal{V}_j(t))\left(\sum_{i=1,i\neq j}^{p} card(\mathcal{V}_i(t))\right)\right)}{n(n-1)}$$

The second graph whose r_1 and r_2 evolution is shown on Figure 8 is almost the same, but we introduced, as for random graphs, a communication structure under the form of weights following a spanning tree in the graph. The graph therefore contains high weights and low weights with a large interval between the two. The r_1 criterion is better than for the first graph. As an indication, the theoretical best r_1 value if all clusters had had the same size is indicated as a dashed blue line. This number is computed as follows. Let $\bar{h}$ and $\bar{l}$ be the average high and low weights respectively. Let $\mathcal{E}_h(t)$ and $\mathcal{E}_l(t)$ be the set

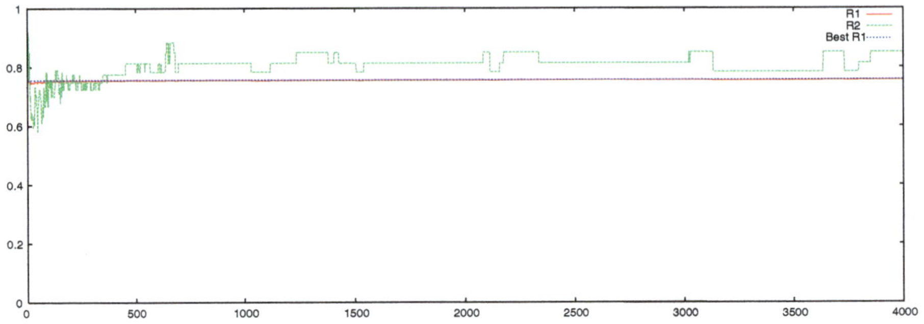

Fig. 7. *A complete graph with uniform weights and 4 colors (100 vertices, 4950 edges).*

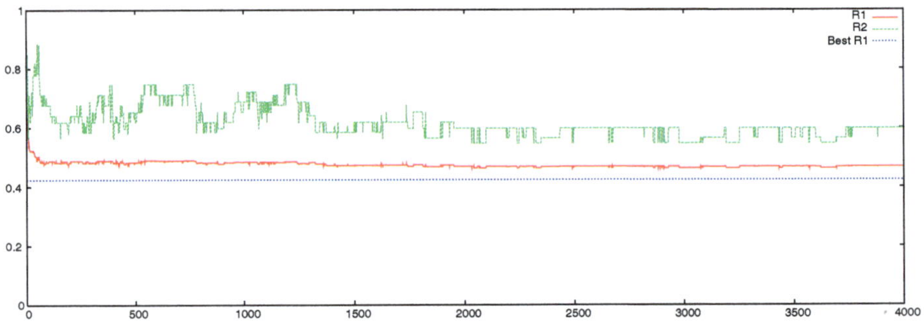

Fig. 8. *A complete graph with "communication structures" following a spanning tree, and 4 colors (100 vertices, 4950 edges).*

of edges at time t having high and low weights respectively. Let $\mathcal{A}_h(t)$ and $\mathcal{A}_l(t)$ be the set of actual communication edges at time t having high and low weights respectively. Then the optimal assumes that $\mathcal{A}_l(t)$ has the smallest possible number of elements (for the graph we use, it is zero):

$$\frac{card(\mathcal{A}_h(t))\bar{h} + card(\mathcal{A}_l(t))\bar{l}}{card(\mathcal{E}_h(t))\bar{h} + card(\mathcal{E}_l(t))\bar{l}}.$$

Notice that no particular structure has been given to the spanning tree, notably, communication cluster are not all the same size. This impacts on r_2.

Scale-Free Graphs

These graphs are characterized by the fact that a restrictive number of vertices have many connections with weakly connected vertices. The degree of each vertex follows a power law [1]. These networks are interesting because of their omnipresence in nature. Human interactions in society as well as computer networks or protein interaction networks are some examples of such scale-free graphs [32].

Figure 9 shows the algorithm ability to find organizations of communicating entities. In this example the r_2 value is due to the specific conformation of the graph. The graph forms *natural* clusters which influence how ants travel through the graph: they have far less chances to cross edges between natural clusters except if pushed-of them by competition (as ants tend to favor the r_1 criterion explicitly in their behavior whereas r_2 is only implicitly optimized by the fact several colonies are in competition, as explained above).

Dynamic Graphs

Our algorithm is very well-suited for dynamic graphs processing. The relevance of our method lies in the distributed nature of solutions computations. A change in the input (a change in the dynamic graph, the number of computing resources) occurring at any time during the computation, will entail only local changes, changes taken into account by the method that still continue its task of organizations detection. Indeed, at the base component level, the dynamic graph is constantly being reconfigured. On the contrary, taken as a whole, long lasting organizations appear. They are the image of organizations appearing in the distributed application the graph is a representation of. Such organizations often have a longer lifetime than the average duration of edges or vertices of the graph. Inside these organizations, communication is higher both in terms of volume and connectivity. These two last points are criteria used by ants to form clusters.

Following are several experiments we made with dynamic graphs. For some tests, we used program that simulate the application by creating a graph and then applying events to it. Events are the appearance or disappearance of an edge, a vertex or a processing resource, but also functions that change weights on edges. For others theses we used an ecosystem application where entities have a boid-like behaviour.

In Figures 10 and 11, the graph representation is as follows: vertices are rectangles, and edges are shown with a pie chart in the middle that indicates relative levels of pheromones with the maximum pheromone level numbered. Vertices are labeled by their name at the top with underneath at the left the total number of ants they host and at the right a pie chart indicating the relative number of ants of each color present on this vertex.

The first experiment, already shown in Figure 2 and detailed in Figure 10 is a small graph (18 vertices), where three main communication clusters appear. These clusters are linked at the center by low communication edges that appear and disappear. Inside the clusters some edges also appear and disappear. For this experiment we used parameters $\alpha = 1.0$, $\beta = 4.0$, $\rho = 0.8$ $\phi = 0.3$, $\eta = 0.0001$, $N^* = 10$ and $M = 4$ vertices. These parameters will be the same for all other experiments.

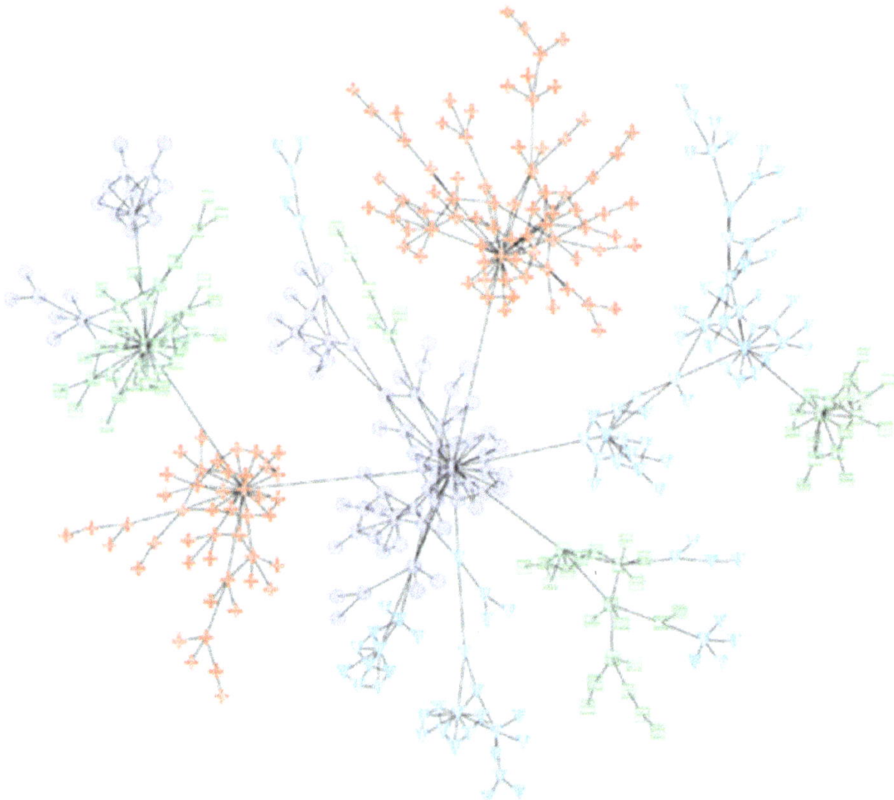

(a) *Colored scale-free graph*

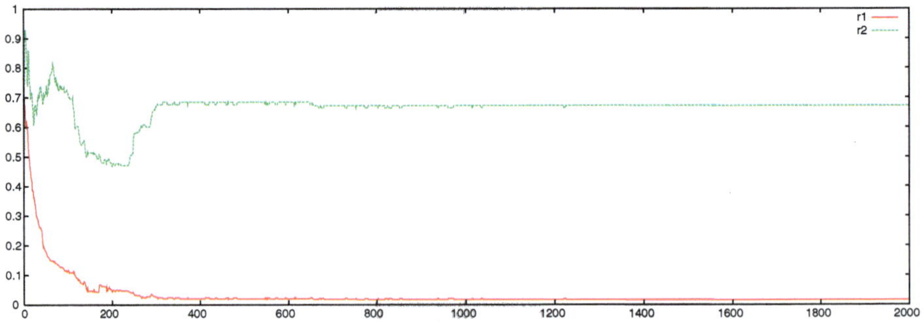

(b) *Evolution of* r_1 *and* r_2

Fig. 9. *A weighted scale-free graph with 4 colors (391 vertices, 591 edges).*

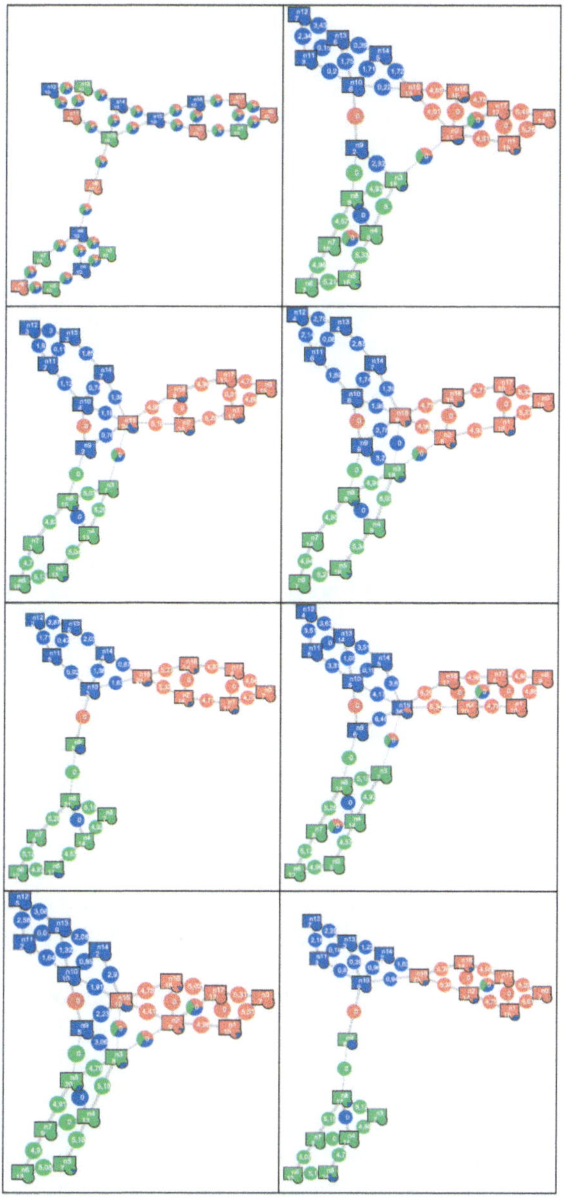

Fig. 10. *Experiment 1.*

The second experiment used a bigger graph (32 vertices) that continuously switch between three configurations. Six snapshots of the graph are presented and show that clusters remains stable across reconfigurations (Figure 11).

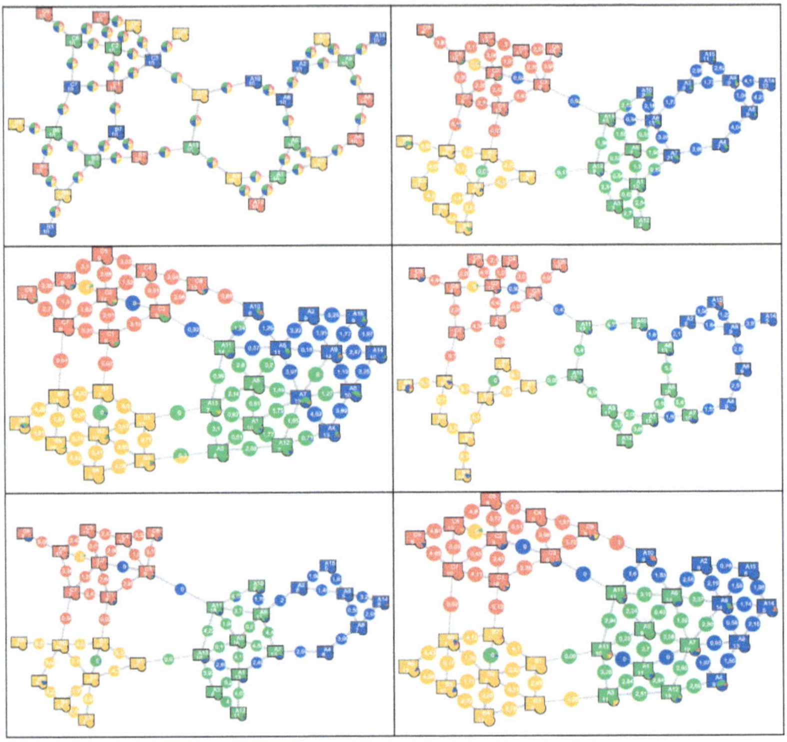

Fig. 11. *Experiment 2.*

The third experiment uses a grid like graph. A small graph travels through the first one. It represents a stable organisation which interacts with the grid. It is continuously connected to the grid but these connection change, so that the small graph moves along the grid (see Figure 12).

The small graph keeps the same color along the experiment while it crosses different domains of the grid having a distinct color because communications in the small graph are stronger. Therefore, the organization formed by the small graph is not perturbed by its interactions with the grid. This graph could represent an aquatic simulation application where a fish school passes in an environment (the grid being the environment and the small graph being the fish school). Interactions in the fish school are based on the fish vision area and are more important, and durable, inside the school than with the environment.

The fourth experiment is made using an ecosystem simulation where entities have a boid-like behavior, made of three rules:

- avoidance: they try to stay at a small distance of perceived boids,
- cohesion: they try to fly toward the average position of all perceived boids,
- alignment: they try to match velocity with perceived boids.

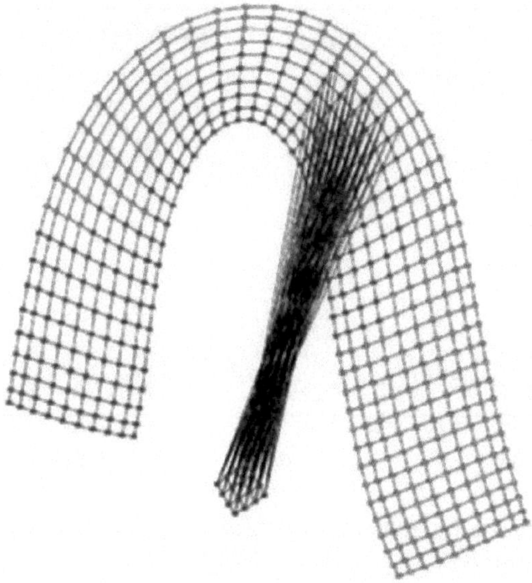

Fig. 12. *Experiment 3.*

These rules create one or several groups of boids. Furthermore, boids try to avoid predators introduced in the simulation. Predators cut boids groups in sub-groups.

Each boid is modeled by a vertex in the dynamic graph. When a boid enters in the field of view of another this creates an interaction (boids reacting to others in their field of view) and therefore an edge in the graph. Figures 15 and 16 show both the boids simulation and the corresponding colored dynamic graph. On Figure 16 boids group formed, and the graph shows the corresponding clusters, detected by $AntCO^2$ as shown by colors.

We compared $AntCO^2$ with two other distribution strategies: random and grid. The first assigns a color to a boid randomly following an uniform law. As the number of boids stays the same during the experiment, the load balancing is perfect and stays the same. A contrario interactions between boids on distinct computing ressources are much more probable.

In the other strategy, we divide the environment in cells as a grid. Grid cells are mapped to processing ressources. When a boid is in a cell it is running on the processing ressource of this cell. In the same way, interactions between boids which are on two different sub-grids generate actual communications.

Figures 13 and 14 show the r_1 and r_2 criteria evolution respectively for the three strategies on a boid simulation using 200 boids during 5000 steps.

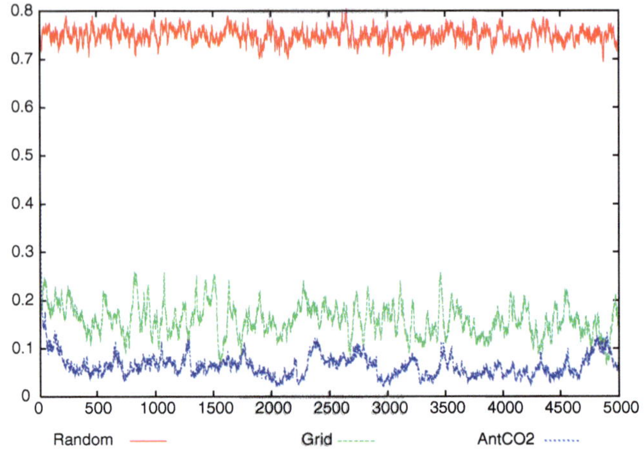

Fig. 13. *R_1 evolution on a 200 boids application using three distribution strategies: random, grid and AntCO^2.*

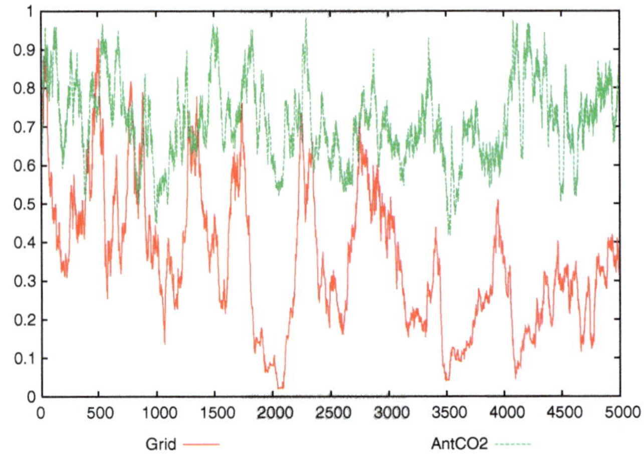

Fig. 14. *R_2 evolution on a 200 boids application using two distribution strategies: grid and AntCO^2 (random always give $r_2 = 1$).*

5 Conclusion

In this paper we presented a colored ant algorithm allowing to detect and distribute dynamic organizations. The algorithm offers advices for entity migration in a distributed system taking care of the load and communication balancing. We described a base colored algorithm, observed its behaviour with static and dynamic graphs and provided methods to manage them.

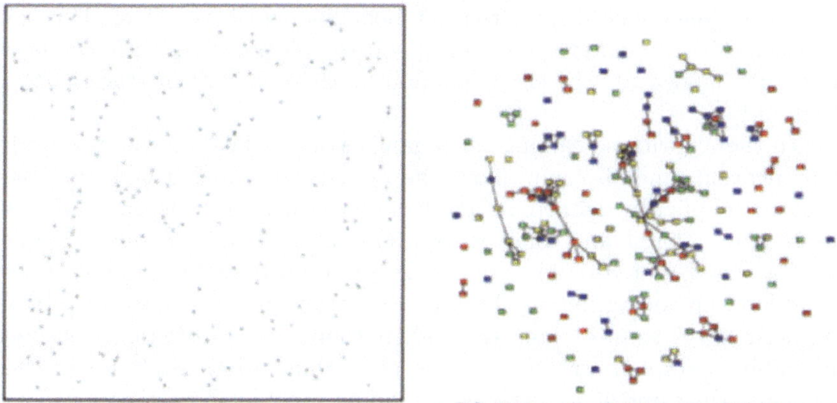

(a) Simulation

(b) Colored relation graph between entities of the simulation

Fig. 15. *At start of the simulation.*

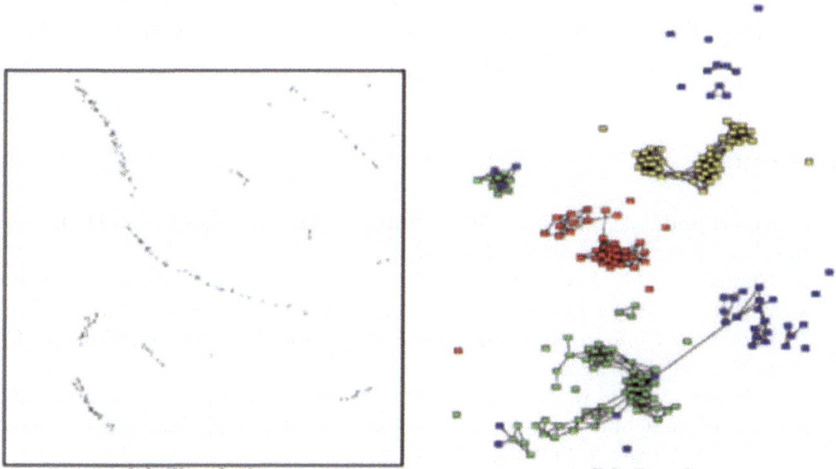

(a) Simulation

(b) Graph

Fig. 16. *Later on in the simulation.*

Our algorithm handles dynamic graphs. Two properties of the algorithm allow this: *positive feedback* maintain paths in the graph between highly correlated vertices, *negative feedback* isolate these communities. The first is controlled by ants (pheromone drops), while the other is completely driven by the environment (evaporation, edge deletion, weights). Negative feedback is what makes our algorithm truly adaptive to dynamic graphs, allowing to forget bad communities introduced by the dynamics.

Organizations emerge from the ant behavior which is not explicitly implemented. These organizations correspond to solutions, which is the reason why we don't need an objective function at the contrary of traditional ant systems [14].

Since the algorithm searches for organizations, it favours the r_1 criterion, communication minimisation, above the r_2 criterion, load balancing. The r_1 criterion is explicitly defined in ant behavior whereas r_2 is only implicitly expressed by colored ant competition. This can be seen in our example 9 (scale-free graph), organizations in the graph are sometimes antagonist to an optimal load balancing, merely because organization size is not predictable.

Experimental results show that when there are no organizations (random graphs, complete graphs...), the algorithm indeed finds no artifacts, and favours the implicit r_2 criterion providing good load distributions.

We started development on a heuristic layer allowing to handle some constraints tied to the application, like entities that cannot migrate (e.g. bound to a database), but also information peculiar to the application.

This work takes place within the context of aquatic ecosystem models [5], where we are faced to a very large number of heterogeneous auto-organizing entities, from fluids representatives to living creatures with their specific behaviour.

References

1. R. Albert and A.L. Barabaśi (2002) Statistical mechanics of complex networks. *Reviews of modern physics*, 74:47–97
2. R. Banos, C. Gil, J. Ortega, and F.G. Montoya (2003) Multilevel heuristic algorithm for graph partitioning. In G.R. Raidl et al, editor, *Applications of Evolutionary Computing*, volume 2611, pages 143–153. Lecture Notes in Computer Science, Springer
3. D. J. Barnes and T. R. Hopkins (2003) The impact of programming paradigms on the efficiency of an individual-based simulation model. *Simulation Modelling Practice and Theory*, 11(7-8):557–569
4. C. Bertelle, A. Dutot, F. Guinand, and D. Olivier (2002) Dimants: a distributed multi-castes ant system for DNA sequencing by hybridization. In *NETTAB 2002*, pages 1–7, AAMAS 2002 Conf, Bologna (Italy)
5. C. Bertelle, V. Jay, and D. Olivier (2000) Distributed multi-agents systems used for dynamic aquatic simulations. In D.P.F. Müller, editor, *ESS'2000 Congress*, pages 504–508, Hambourg
6. S. H. Bokhari (1981) On the Mapping Problem. *IEEE Transactions on Computers*, 30:207–214
7. E. Bonabeau, M. Dorigo, and G. Theraulaz (1999) *Swarm Intelligence – From natural to Artificial Systems.* Oxford University Press
8. G. Di Caro and M. Dorigo (1997) Antnet: A mobile agents approach to adaptive routing. Technical report, IRIDIA, Université libre de Bruxelles, Belgium
9. D. Caromel, W. Klauser, and J. Vayssiere (1998) Towards seamless computing and metacomputing in java. In Geoffrey C. Fox, editor, *Concurrency Practice*

and Experience, volume 10, pages 1043–1061. Wiley & Sons, Ltd., http://www-sop.inria.fr/oasis/proactive/.
10. D. Costa and A. Hertz (1997) Ant can colour graphs. *Journal of Operation Research Society*, (48):105–128
11. S. K. Das, D. J. Harvey, and R. Biswas (2002) Adaptative load-balancing algorihtms using symmetric broadcast networks. *Journal of Parallel and Distributed Computing (JPDC)*, 62:1042–1068
12. J.-L. Deneubourg and S. Goss (1989) Collective patterns and decision making. *Ethology Ecology and Evolution*, 1(4):295–311
13. M. Dorigo and L.M. Gambardella (1997) Ant colony system: A cooperative learning approach to the traveling salesman problem. *IEEE Transactions on Evolutionary Computation*, 1(1):53–66
14. M. Dorigo, V. Maniezzo, and A. Colorni (1996) The ant system: optimization by a colony of cooperating agents. *IEEE Trans. Systems Man Cybernet.*, 26:29–41
15. A. Dutot, R. Fisch, D. Olivier, and Y. Pigné (2004) Dynamic distribution of an entity-based simulation. In *JICCSE 04 – Jordan International Conference on Computer Science and Engineering*, Alt-Salt, Jordan
16. D. L. Eager, E. D. Lazowska, and J. Zahorjan (1986) A comparison of receiver-initiated and sender-initiated adaptive load sharing. *Performance evaluation*, 6:53–68
17. P. Erdös and A. Rényi (1959) On random graphs. *Pubiones Mathematicaelicat*, 6:290–297
18. B. Faieta and E. Lumer (1994) Diversity and adaptation in populations of clustering ants. In *Conference on Simulation of Adaptive Behaviour*, Brighton
19. M.R. Garey and D.S. Johnson (1979) *Computers and Intractability, a Guide to the Theory of NP-Completeness*. W.H. Freeman and Compagny
20. D.M. Gordon (1995) The expandable network of ant exploration. *Animal Behaviour*, 50:995–1007
21. S. Goss, S. Aron, J.-L. Deneubourg, and J. M. Pasteels (1989) Self-organized shortcuts in the argentine ant. *Naturwissenchaften*, 76:579–581
22. V. Grimm (1999) Ten years of individual-based modelling in ecology: what have we learned and what could we lear in the future ? *Ecological Modelling*, 115(2-3):129–148
23. H.-U. Heiss and M. Schmitz (1995) Decentralized dynamic load balancing: The particles approach. *Information Sciences*, 84:115–128
24. Z. Jovanovic and S. Maric (2001) Heuristic algorithm for dynamic task scheduling in highly parallel computing systems. *Future Generation Computer Systems*, 17:721–732
25. P. Kadluzczka and K. Wala (1995) Tabu search and genetic algorithms for the generalized graph partitioning problem. *Control and Cybernetics*, 24(4):459–476
26. B.W. Kernighan and S. Lin (1970) An efficient heuristic procedure for partitioning graph. *The Bell System Technical Journal*, 49(2):192–307
27. P. Korošec, J. Šilc, and B. Robič (2004) Solving the mesh-partitioning problem with an ant-colony algorithm. *Parallel Computing, Elsevier*, 30:785–801
28. P. Kuntz, P. Layzell, and D. Snyers (1997) A colony of ant-like agents for partitioning in vlsi technology. In *Fourth European Conference on Artificial Life*, pages 417–424, Cambridge, MA:MIT Press

29. A. E. Langham and P.W. Grant (1999) Using competing ant colonies to solve k-way partitioning problems with foraging and raiding strategies. In D. Floreano et al., editor, *Advances in Artificial Life*, volume 1674, pages 621–625. Lecture Notes in Computer Sciences, Springer
30. C.G. Langton, editor (1987) *Artificial Life.* Addison Wesley
31. F. C. H. Lin and R. M. Keller (1987) The gradient model load balancing method. *IEEE TOSE*, 13:32–38
32. M. E. J. Newman (2003) The structure and function of complex networks. *SIAM Review 45*, pages 167–256
33. T. White (1997) Routing with swarm intelligence. Technical Report SCE-97-15
34. M. H. Willebeek-LeMair and 1. P. Reeves (1993) Strategies for dynamic load balancing on highly parallel computers. *IEEE Transactions on parallel and distributed systems*, 4(9):979–993

Problem Solving and Complex Systems

Frédéric Guinand and Yoann Pigné*

LITIS – University of Le Havre,
25 rue Philippe Lebon
76620 Le Havre cedex, France
frederic.guinand@univ-lehavre.fr, yoann.pigne@univ-lehavre.fr

Summary. The observation and modeling of natural Complex Systems (CSs) like the human nervous system, the evolution or the weather, allows the definition of special abilities and models reusable to solve other problems. For instance, Genetic Algorithms or Ant Colony Optimizations are inspired from natural CSs to solve optimization problems. This paper proposes the use of ant-based systems to solve various problems with a non assessing approach. This means that solutions to some problem are not evaluated. They appear as resultant structures from the activity of the system. Problems are modeled with graphs and such structures are observed directly on these graphs. Problems of Multiple Sequences Alignment and Natural Language Processing are addressed with this approach.

Key words: problem solving, ant colony optimization, multiple sequences alignment, natural language processing, graph

1 Introduction

The central topic of the work presented in that paper is to propose a method for implicit building of solutions for problems modeled by graphs.

Whatever the considered domain, physics, biology, mathematics, social sciences, chemistry, computer science... there exist numerous examples of systems exhibiting global properties that emerge from the interactions between the entities that compose the system itself. Shoal of fishes, flocks of birds [14], bacteria colonies [9], sand piles [5], cellular automata [15, 30], protein interaction networks [4], city formation, human languages [29] are some such examples. These systems are called *complex systems* [32]. They are opened, crossed by various flows and their compounds are in interaction with themselves and/or with an environment that do not belong to the system itself.

* This work is partially supported by the French Ministry of Higher Education and Research.

They exhibit a property of self-organization that can be defined as an holistic and dynamic process allowing such systems to adapt themselves to the static characteristics as well as dynamic changes of the environment in which their compounds move and act.

The work presented here aims at exploiting that self-organization property for computing solutions for problems modeled by graphs. The central idea consists in the conception of an artificial complex system whose entities move in and act on a graph (the environment) in which we are looking for structures of special interest: solutions of our original problem. In order to answer our expectations, the entities must leave some marks in the environment and these marks should define expected structures. In other words, we want to observe a projection on the graph of the organization emerging at the level of the complex system. For that purpose, the considered complex system has to be composed of entities able to move in the graph, able to interact with each-other and with their environment. The effects of this last kind of interactions materialize the projection of the organization of entities onto the environment.

In ethology, structures and organizations produced by animal societies are studied according to two complementary points of view [25]. The first tries to answer to the question *why* while the second focuses on the question *how*. For the latter, the goal is to discover the link that exists between individual behavior and the structures and collective decisions that emerge from the group. In this context, self-organization is a key concept. It is indeed considered that the global behavior is an emergent process that results from the numerous interactions between individuals and the environment. More precisely, when animal societies are considered, three families of collective phenomena may be distinguished: (a) spatio-temporal organization of individual activities, (b) individual differentiation and (c) phenomena leading to the collective structuration of the environment. This latter phenomenon mainly occur with wasp, termite and ant societies. This explains probably why for about one decade, ants have inspired so much work on optimization [10].

The approach described in this paper is closely related with the question *how*. The more promising collective phenomenon for reaching our aim is the third one: collective structuration of the environment. Knowing previous works on ant colonies and more precisely on Ant System, artificial ants appear obviously as excellent candidates for representing the basic entities of our artificial complex system.

Ants are mobile entities, they can interact directly using antennation and/or trophallaxy, or non directly, using the so-called stigmergy mechanism [12]. Stigmergy may be sign-based or sematectonic. Sign-based stigmergy is illustrated by ants dropping pheromones along a way from the nest to a food source. Sematectonic stigmergy is based on a structural modification of the environment, has illustrated by termites building their nest [13]. In its simplest version, an ant algorithm follows three simple rules. In order to simplify, let us consider that the algorithm operates on a graph: (1) each ant drops a small quantity of pheromone along the path it uses; (2) at a crossroads, the choice of the ant is partially determined by the quantity of pheromones on each outgoing edge: the choice is probabilistic and the larger the quantity of pheromones on one edge, the larger the probability to choose this edge; (3) pheromones evaporate with the time. During the process, some edges are more frequently visited than other ones, making appear at the graph level some paths, groups of edges/vertices, or other structures resulting from the local and indirect interactions of ants.

Artificial ant colonies have been widely studied and applied to optimization problems. The way they have been used makes ant colonies belonging to the general class of metaheuristics. This term refers to high level strategies that drive and modify other heuristics in order to produce solutions better than what could be expected by classical approaches. A metaheuristic may be defined as a set of algorithmic concepts that can be used for the definition of heuristic methods applicable to a large spectrum of problems [22]. A heuristic rests on a scheme that may be a constructive method or a local search method. In the former case, one solution is built using a greedy strategy, while a local search strategy iteratively explores the neighborhood of the current solution, trying to improve it by local changes. These changes are defined by a neighborhood structure that depends on the addressed problem. In [22], the authors note that artificial ants may be considered as simple agents and that the good solutions correspond to an emergent property resulting from the interactions between these cooperative agents. In addition, they describe an artificial ant in ACO (Ant Colony Optimization)

as a stochastic building process that builds step by step a solution by adding opportunely elements to the partial solution. The considered problems are expressed in a classical way, with the definition of a space of solutions, a cost function that should be minimized and a set of constraints. Then, each ant possesses an evaluation function and this evaluation drives the process.

Particle Swarm Optimization (PSO) is another kind of optimization method based on the collective behavior of simple agents. It is close to ant colonies, has been proposed and described in [17]. But once again, in PSO, each particle owns an evaluation function and knows what is the space of solutions. Each particle is characterized, at time t by its position, its speed, a neighborhood, and its best position according to the evaluation function since the beginning of the process. At each step, each particle moves and its speed is updated according to its best position, the position of the particle belonging to the neighborhood showing the better result to the evaluation function.

Our approach, while belonging to the class of population-based methods, is different than both ACO and PSO. In our mind, depending on the model of the problem and on the characteristics of the ants, the role of each individual ant is not to compute a solution (neither complete, nor partial), but the solution has to be observable into the environment as the result of the artificial ants actions. That may be considered as a kind of implicit optimization. We argue that it is possible to compute non-trivial solutions to problems without using any global evaluation function. In what we have done, neither the ants, nor the elements of the environment in which they move evaluate the result of their interactions during the process. The motivations for such an approach are multiple. Building a global evaluation function is sometimes not possible or would required an effort as important as solving the problem itself. Sometimes, the evaluation of the function is not conceivable during the resolution process because of the dynamics of the problem. Some other problems cannot be easily expressed as optimization problems, as in the example developed in Sect. 4.

That's the reason why the behavior of our artificial ants remains simple and only depends on local information. Depending on the application, we can have one or several collaborating or competing colonies. During the process, each ant can drop pheromones, interact with other ants, and modify its environment. The choice between several edges for proceeding with the path depends on the quantity of pheromones dropped by ants belonging to the same colony (attraction) and on the quantity of pheromones dropped by ants of another colony (repulsion). The modification of the environment may be the creation of a vertex or the deletion of an edge on the graph.

In the sequel, the general approach is more precisely stated in Sect. 2. It is illustrated by two case studies: the bioinformatic problem of multiple sequence alignment (Sect. 3) and the search for interpretation trails in texts written in natural languages (Sect. 4). Sect. 5 exposes some difficulties facing this approach, particularly the identification of relevant parameters and characteristics and the sensitivity of the method to critical parameters. To finish, a conclusion draws some short and mid-term perspectives.

2 General Approach

As a starting point, we consider problems that may be modeled by a graph. Evolutionary Computation excepted, in classical metaheuristics (ACO, PSO, GAs, tabu search based methods, GRASP), solutions are explicitly computed and one unavoidable step is the evaluation of these solutions in order to drive further investigations. In EC the individuals are not necessarily solutions of the considered problem, but it is always possible to build a solution from the set of individuals. This feature is not required for us. Indeed, our approach relies on the report that in many cases solutions correspond to structures in the graph. Such structures can be paths, set of vertices, set of edges, partitions of the graph... or any other group of graph elements. We consider this graph as the environment, into which many moving and active entities (our artificial ants) are born, operate, and (sometimes) die. They have the knowledge of neither the environment, nor the final goal of the system, nor the objective function. Their actions are not driven by the problem, only their characteristics may depend on it. Each ant can perform only two actions: to move and to modify the environment. Moving depends on local characteristics of the environment and sometimes on an additional basic goal (turning back nest for instance). Modifications of the environment is used for both purpose: indirect communications between ants and structures forming in the environment. Modifications may be of two types, corresponding to the two stigmergy types: sign-based (the pheromones) and sematectonic (changes affect graph structure). In addition, the modifications of the environment have a feedback effect on the behavior of artificial ants as illustrated on Fig. 1.

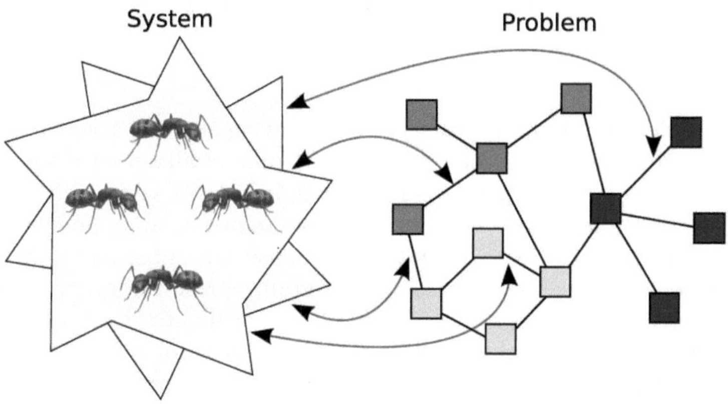

Fig. 1. The general model: The system evolves on the graph of the problem, modifies it and is influenced by its own modifications. Finally solutions are directly observed on the graph.

As mentioned in the introduction, artificial ants societies are complex systems, indeed, they are able to exhibit a self-organizing property that may be translated into, for instance, a set of edges with a large volume of pheromones, a set of vertices of the same color... as many structures that could be solutions of some problems in relation with the original graph.

The general process for problem-solving using a complex system as described in the sequel is made of three steps: (1) identification of the structures in the graph that can be associated to solutions of the original problem, (2) choice of the features required by the system to achieve the goal (that is building of structures), (3) critical parameters tuning. Let us now enter into details.

2.1 Structures Identification

The first step, before the construction of the system itself is to identify, for the graph modeling the considered problem, the structures that are solutions.

Let us illustrate this with some examples described in Table 1.

Table 1.

Problem description	Relevant structure
Mapping	sets of vertices
Routing	set of paths
Partitioning	sets of vertices
TSP	circuits
Shortest path	paths
Multiple sequence alignment	sets of edges
DNA-Sequencing	paths

The problem of mapping consists in allocating sets of tasks to sets of resources. At the end of the process, each resource is allocated a set of tasks. With respect to the graph this corresponds to sets of vertices gathered according to their allocated resources. Multiple sequence alignment will be extensively described in Sect. 3, and Word Sense Disambiguation in Sect. 4. For DNA-Sequencing, there exists several graph formulations. In all cases, a solution corresponds to a path in the graph. According to the model, the path may be either a Hamiltonian path, an Eulerian path or a constrained path. In the case of multi-objective problems, the solutions can be presented as a set of paths or subgraphs that in the best case all belong to the Pareto front.

However, a given problem may be modeled using different graph formulation, thus, it is unlikely to find only one association between a problem and a graph structure.

Figure 2 illustrates how such structures may be observable directly by a user. The first picture (left-hand side) is a path built by ants for the shortest path problem. The original problem associated to the second picture is dynamic load balancing. This time, the structures are sets of colored vertices, to each color corresponds a resource. The method used for obtaining the right-hand side graph is described in the same volume [8]

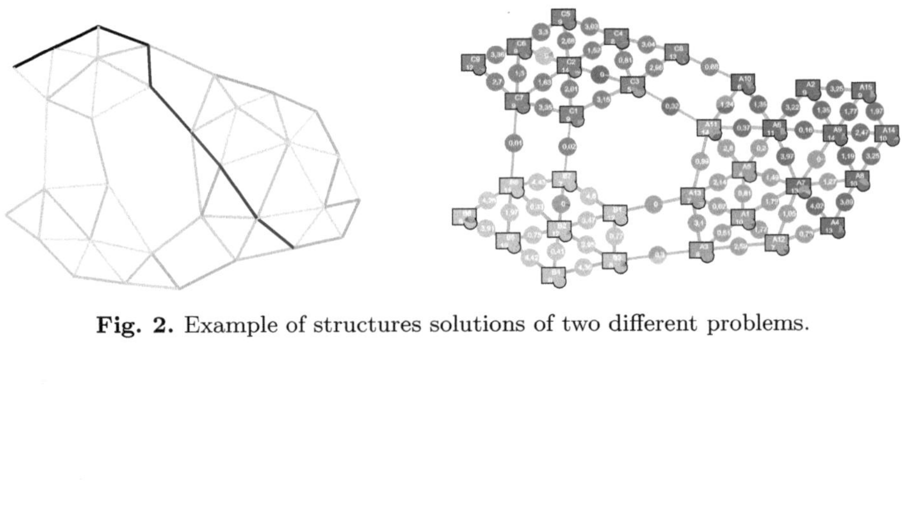

Fig. 2. Example of structures solutions of two different problems.

2.2 The Model

Our choice for a complex system based on artificial ants was their ability to modify their environment. Moreover, ant-based systems have an additional property since the death of any ant do not impede the system to work. Finally, a lot of ants may be used together since they don't directly communicate, but they use the environment instead, thus, the overhead entailed remains acceptable, and the whole system is easily distributed on a computer system (a grid for example) as soon as the environment (the graph) is itself distributed.

As it is classically defined complex systems (CSs) are led by feedback loops. Positive feedback amplify the evolution while negative feedback reduce it. Ant-based systems are bound to the same mechanisms. Positive feedback takes place with stigmergy. When performing walks in the graph, ants modify their environment by structural changes in the graph (sematectonic stigmergy) or by laying pheromones down the graph (sign-based stigmergy). This modification are information for other ants that will later visit the same parts of the graph. These ants will be attracted by the environment's information and will deposit their own information, attracting more and more ants into a positive feedback loop. If nothing stops the mechanism, the system freezes and cannot evolve anymore. To prevent the system from stagnating, negative feedback loops are used. In our approach, it is obtained with evaporation or erosion processes. In the nature, real pheromone do evaporate. This is a brake to previous stygmergic mechanisms.

The ant-based complex system has three kinds of elements: the nests, the colonies and the ants. The conception of the system takes place at two levels: the colony level and the ant level, the nests are only positions from which the ants belonging to the corresponding colony begin their walks.

Nests and Colonies

The system may be constituted by one or several colonies. The choice of the number of colonies is mainly problem-driven. Indeed, if we need only one structure, in most cases one colony is enough. But, in case of solutions made of several structures determined as the result of a competition (partitioning or mapping for instance), then, several colonies should be necessary. However, sometimes, it is necessary that ants are born everywhere in the graph, in such particular cases, there is a one-to-one mapping between nests and vertices.

One or more colonies of ants may operate into the environment. The nest of the colony can take place in the graph on one particular vertex, if the original problem defines a starting point. This is the case for instance when the structure is a path defined by a fixed starting point. However, there are also scenarios for which all vertices are nests, each corresponding to a distinct colony. In this case ants are competing for accessing some resources. It is also possible to design systems in which each vertex is a nest belonging to a supercolony (in the sense of unicoloniality [11] (Argentine ants in southern Europe)).

In all cases, an ant stemmed from one given colony inherit from this colony a set of characteristics (eventually empty), a color, a sensitivity to pheromones... More generally, colonies differentiation allow the use of different heuristics on the same problem. A wide range of global system behavior rise from this. One of the most interesting properties is the use of ant differentiation so as to raise competition or collaborative mechanisms. These mechanisms are useful when dealing with multiple objective problems or when the construction of clusters or subgraphs are desired.

To obtain competition, it is common to assign a colony a color. Ants from one particular color lay down the environment "colored" pheromones of the same color. Ants are attracted by the pheromones with their color and are repulsed by other colored pheromones. this general behavior lead to the coloration of disconnected parts of the graph like clusters or subgraphs.

To obtain collaboration, the same mechanism is used. Ants deposit pheromones of their color and are attracted by it but there is no repulsion with other colors. This mechanism is ideal when dealing with multiple objective problems. One objective represent one color. Each colony enhances its objective while being attracted by all the pheromones.

Ants

First, general characteristics of ants are studied. Then details about the way pheromone trails are constructed and how ants move are considered.

General Characteristics

Each ant is autonomous, and independent from the others, it can thus be executed concurrently and asynchronously. Each ant is reactive to the environment and its local neighborhood is defined as the set of adjacent vertices and edges of the vertex the ant is located on. Local information available from this set of vertices and edges constitute the basic material for the ant to prepare its next move. These information are of two kinds: information belonging to the problem itself if the problem needs a valuation of vertices and edges. The topology of the graph can also indicates constraints from the problem (degree of vertices, directed versus non-directed graph). Second, the information raised from the activity of the ant colony: pheromones and other data deposit on the edges and on the vertices.

Depending on the considered problem, ants are assigned characteristics. These characteristics are more or less problem specific. Some of them can be used with multiple problems and deserve a presentation:

- The assignment of a color to the ants is useful when dealing with partitioning problems or multiple objective problems.

- A tabu list prevent ants from visiting twice an already visited vertex. The size of this list is relevant for many problems.
- A sensitivity to pheromones parameter is usually used to find a balance between the reinforcement of pheromone trails and the diversification of the search. When ants are too much attracted by pheromones they follow already defined trails and don't try to visit unexplored parts of the graph. On the contrary, if they don't pay enough attention to the pheromones, no stigmergy is available.

Moves and Pheromones

Every single ant's general aim is to move in the graph and to lay down some pheromones during their moving. These walks are led by local information and constraints found in the environment.

Ants lay down pheromone trails on their way. The quantity of pheromone deposited is a constant value or calculated with a local rule but is never proportional to any evaluation of a constructed solution. Pheromone is persistent: it continues to exist after the ant's visit. This pheromone will influence other ants in the future, indeed, among the set of adjacent edges of one particular vertex v, the larger the quantity of pheromone on the adjacent edge the more attractive this edge will be for ants located on v.

The choice of the next vertex to move to is done locally for each vertex as follows: Let consider a vertex i who's neighbors vertices are given by the function $neighbors(i)$ as a set of vertices. The probability $P_{i,j}$for one ant located on vertex i to move to neighbor vertex j is:

$$P_{i,j} = \frac{\tau_i}{\sum_{k \in neighbors(i)} \tau_k} \tag{1}$$

Different methods can be used to perform the pheromone lay down. The nature of the lay down and the nature of the pheromones themselves depends on the problem.

Depending on the kind of solution that is expected, these moves will raise different structures.

- If the expected structure is a path or a set of paths in the graph defined by a source vertex and a destination vertex, then ants moves define paths. The Pheromone deposit can be done when going to the destination and when returning to the nest or when returning only.
- If the structures looks like clusters or subgraphs, then the moves don't define paths. Ants have no destination but their behavior is not necessarily different from the previous one. But the existence of concurrent colonies may restrict the movements of ants and may entail a kind of gathering of ants of the same color in the same region. The pheromone deposit will be performed at each step. If ants are assigned a color, then the pheromones they lay down will be colored with the same color.

General Consideration about the Model

The motivation of developing such a system comes from its three main properties: self-organization, robustness and flexibility. The system do not explicitly build structures, instead of that, structures appear as the results of the interactions between ants and between ants and the environment. This is a consequence of the self-organization property.

The system do not compute explicitly solutions but rather relies on the notion of structures, thus, when changes occur within the environment, the whole process remains the same since no objective function is evaluated by ants, in other words, the behavior of ants changes in no way. Thus, intrinsically, this approach owns a robustness property.

Finally, the system is flexible. Indeed, ants are autonomous and independent entities. If one or several ants disappear from the system, this one keeps on working. This property is very interesting because it allows a way of working based on flows of entities stemmed from colonies that may be subject to mutations (practically changes in some characteristics).

3 Multiple Sequence Alignment

Multiple Sequence Alignment (MSA) is a wide-ranging problem. A lot of work is done around it and many different methods are proposed to deal with [6, 24, 27]. A subset of this problem, called multiple block alignment aims at aligning together highly conserved parts of the sequences before aligning the rest of the sequences. Here, the problem isn't considered as a whole. The focus is done on the underlining problem of MSA that is the alignment of blocks between the sequences.

3.1 Description of the Problem

Multiple alignment is an inescapable bioinformatic tool, essential to the daily work of the researchers in molecular biology. The simple versions come down to optimization problems which are for the majority NP-hard. However, multiple alignment is very strongly related to the question of the evolution of the organisms from which the sequences result. It is indeed allowed that the probability of having close sequences for two given organisms is all the more important as these organisms are phylogenetically close. One of the major difficulties of multiple alignment is to determine an alignment which, without considering explicitly evolutionary aspects, is biologically relevant.

Among the many ways followed to determine satisfactory alignments, one of them rests on the notion of block. A factor is a substring present in several sequences and should correspond to strongly preserved zones from an evolutionary point of view. This particular point makes this approach naturally relevant from the biological point of view. Building blocks consists in choosing and gathering identical factors common to several sequences in the more appropriate way. A block contains at most one factor per sequence and one given factor can only belong to one block. The construction of blocks is one step of the full process of multiple sequence alignment and the choice of factors for building blocks is the problem we address.

3.2 Proposed Solution

The general scheme for this kind of the method follows three steps:

1. Detection of common subsets (factors) in the set of sequences.
2. Determination of groups of common factors (blocks) without conflict between them.
3. Alignment of the rest of the sequences (parts of sequences that don't belong to the set of blocks).

The work done by the ant colony deals with the second step. The problem must be modeled as a graph. It will constitute the environment that the ants will be able to traverse. Solutions are observed in the graph as a set of relevant edges that link factors into blocks.

3.3 Graph Model

As only factors of sequences are manipulated, each DNA or protein sequence considered is reduced to the list of its common factors with the other sequences of alignment. Figure 3 illustrates such a conversion.

Commonly, a graph is used to figure out the environment in which ants are dropped off. A multiple alignment is often represented by laying out the sequences one under the other and by connecting the identical factors together. Factors and their relations in the alignment can be represented as graphs. The factors are the vertices and the edges between these factors are the edges of the graph.

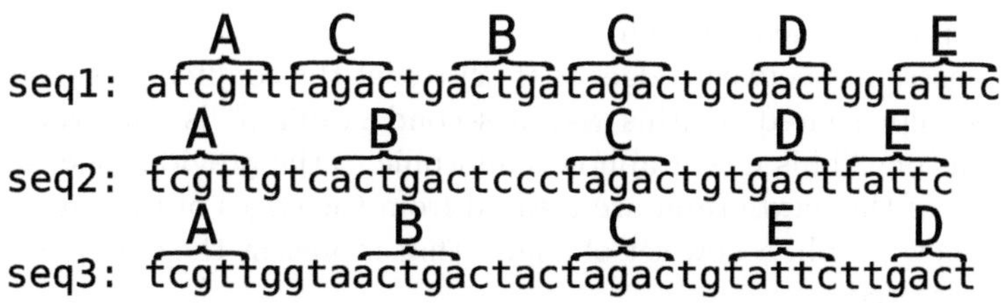

Fig. 3. Sequences conversion
This is an alignment composed of three sequences. Common subsequences of these sequences which are repeated are labeled. After the conversion, the alignment is described as sequence 1 = ACBCDE, sequence 2 = ABCDE and sequence 3 = ABCED.

A group of identical factors (with the same pattern) form a **factor graph**. The quantity of factor graphs is equal to the number of different factors. That is to say, given a factor 'A', the factor graph 'A' is a complete graph in which all the edges linking the factors belonging to the same sequence are removed.

A group of blocks is said to be **compatible** if all of these blocks can be present in the alignment with no cross and no overlap between them.

If two blocks cross each other, one of them will have to be excluded from the final alignment in order to respect the compatibility of the unit. In a graph representation, one can say that the crossings between the edges are to be avoided.

This representation makes it possible to locate the conflicts between the factors which will prevent later the formation of blocks. Indeed, as in Fig. 4, the factors 'D' and 'E' of sequence 2 are linked to the factors 'D' and 'E' of sequence 3. These edges cross each other, which translates a conflict between 2 potential blocks containing these factors.

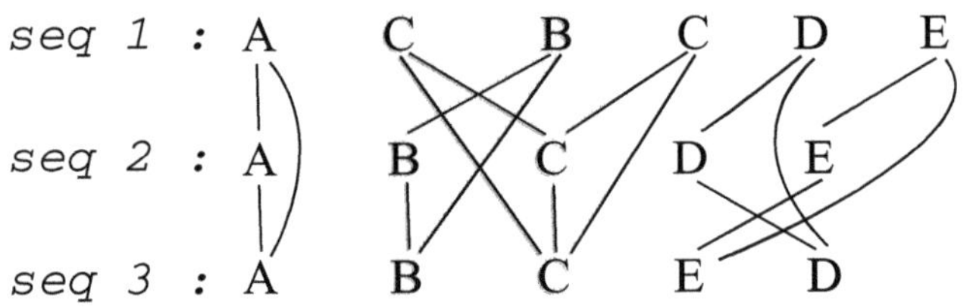

Fig. 4. The same alignment as in Fig. 3. Here, the factors and the edges between them are presented.

The alignment is finally represented as a set of factor graphs. This graph $G = (V, E)$ with V the set of vertices where each vertex is on factor of the alignment, and E the set of edges between vertices where an edge denotes the possibility for two common factors to be aligned together in the same block. The vertices of this graph are numbered. That is to say a factor X, appearing for the jth time in the ith sequence gives a vertex labeled $X_{i,j}$. Figure 5 represent such a graph.

The constraints between edges that represent the conflicts between factors have to be represented. In this way a second graph G' is constructed from the previous one. This new graph is isomorphic to the previous one, such that the vertices of this new graph are created from the edges of the previous one. From this new graph, a new set of edges, that represent the conflicts between factors, is created.

As it can be observed in Fig. 6 these new vertices have two kind of edges: one kind of edges when two vertices share a common factor, and one kind of edge to represent the conflicts.

In the following algorithm, the graph G is considered to be the environment the ant will use. G' is only used to determined conflicts.

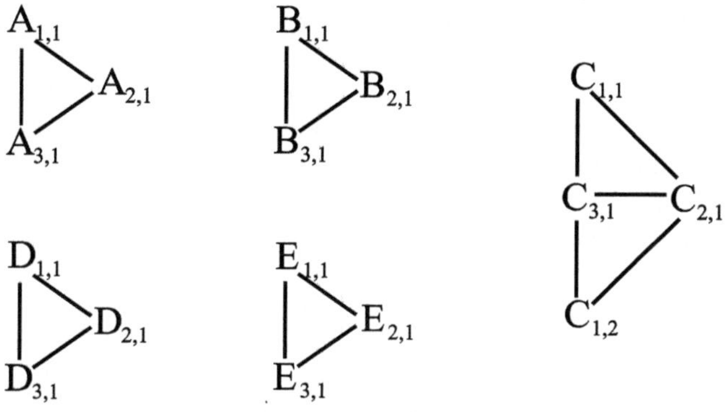

Fig. 5. The same alignment as in Fig. 3 and 4 represented with the graph model.

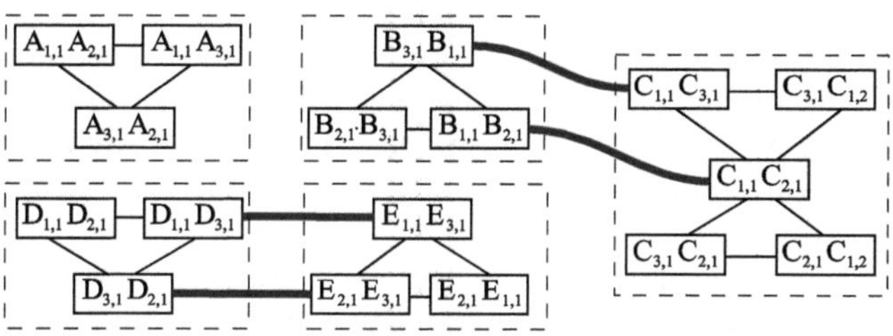

Fig. 6. This graph G' is isomorphic to G. It is called *conflicts graph*. It helps with the representation of conflicts between factors. The light edges represent the share of a common factor between two graphs. The Heavy edges represent the conflicts.

3.4 The Algorithm

The general scheme of the behavior of the ant colony based system follows these rules:

- Ants perform walks into the research graph.
- During these walks they lay pheromones down the edges they cross.
- They are attracted by the pheromone trails already laid down the environment.
- Constrained areas of the graph are repulsive for ants.

Like has been said in the Sect. 2, the first step of our approach is to characterize the solutions we are looking for.

Formulation of a Solution

This question is delicate because it returns to the nature of the problem. Indeed, multiple alignment is not only one fixed method. Many biological considerations come into play. Nothing guarantees besides that one is able to provide a coherent response without solids knowledge in molecular biology.

Moreover, biologist themselves do not agree whether or not a given alignment is a good one. Popular functions like the sum of pairs [3] are widely debated. In consequence, the formulation of an evaluation function seems doubtful. It doesn't seem to be so problematic since the method used doesn't evaluate solutions. Moreover ants don't construct solutions individually.

Nevertheless, it is known that final alignments should be composed of blocks without conflicts. The simplest and relatively natural idea is to maximize the number of blocks formed without conflicts. But this goal is not founded on any biological consideration. Such a goal can probably draw aside from the solutions biologically better than those required. An interesting property can be observed without scientific considerations. Indeed starting from the factors one can evaluate the relative distance between 2 blocks compared to the set the factors. This information is simple and gives an advantage to edges between close factors, "geographically" speaking.

Solutions in this model are carried out from the exploration of the system composed of ant colonies. High valuable edges in the environment end loaded with a large quantities of the pheromones. The most heavily loaded edges belong to the solution. These edges link vertices that represent parts of the blocks of the solution.

Since ants only perform walks in the environment, they do not consider any block formation nor try to minimize any conflict. From a general point of view, a non conflicted set of blocks rise from the graph.

Pheromone Lay Down

Let τ_{ij} be the quantity of pheromone present on the edge (i, j) of graph G which connects the vertices i and j. If $\tau_{ij}(t)$ is the quantity of pheromone present on the edge (i, j) at moment t, then $\Delta\tau_{ij}$ is the quantity of pheromone to be added to the total quantity on the edge at moment $t + 1$. So:

$$\tau_{ij}(t+1) = (1 - \rho).\tau_{ij}(t) + \Delta\tau_{ij} \tag{2}$$

ρ represents the evaporation rate of the pheromones. Indeed, the modeling of the evaporation (like natural pheromones) is useful because it makes it possible to control the importance of the produced effect. In practice, the control of this evaporation makes it possible to limit the risks of premature convergence toward a local minimum.

The quantity of pheromone $\delta\tau_{ij}$ added on the edge (i, j) is the sum of the pheromone deposited by all the ants crossing the edge (i, j) with the new step of time. If m ants use the edge (i, j), then:

$$\Delta\tau_{ij} = \sum_{k=1}^{m} \Delta\tau_{ij}^{k} \tag{3}$$

The volume of pheromone deposited on each passage of an ant is a constant value Q.

Feedback Loops

In a general scheme, negative feedback loops prevent the system from continually increasing or decreasing to critical limits. In our system, constraints between conflicted edges are repulsive for ants. These negative loops aim at maintaining bounded quantities of pheromones. Let consider an ant going over the edge (i, j) of the graph G. The information on the isomorphic graph G' returns that (i, j) is in conflict with edges (k, l) and (m, n). Then the quantities of pheromone will be modified consequently. $\delta\tau_{ij} = \delta\tau_{ij} + q$ for the normal deposit of pheromone and $\delta\tau_{kl} = \delta\tau_{kl} - q$ and $\delta\tau_{mn} = \delta\tau_{mn} - q$ to represent negative feedback on the edges in conflict with (i, j).

Transition Rule

When an ant is on vertex i, the choice of the next vertex to be visited must be carried out in a random way according to a definite probability rule.

According to the method classically proposed in ant algorithms, the choose of a next vertex to visit is influenced by 2 parameters. First, a local heuristic is made with the local information that is the relative distance between the factors. The other parameter is representative of the stygmergic behavior of the system. It is the quantity of pheromone deposited.

Note: Interaction between positive and negative pheromones can lead on some edges to an overall negative value of pheromone. Thus, pheromones quantities need normalization before the random draw is made. Let max be an upper bound value set to the largest quantity of pheromones on the neighborhood of the current vertex. The quantity of pheromone τ_{ij} between edge i and j is normalized as $\tau_{ij} = max - \tau_{ij}$.

The function $neighbors(i)$ returns the list of vertices next to i. The choice of the next vertex will be carried out in this list of successors. The probability for an ant being on vertex i, to go on j (j belonging to the list of successors) is:

$$P(ij) = \frac{[\frac{1}{max - \tau_{ij}}]^{\alpha} \times [\frac{1}{d_{ij}}]^{\beta}}{\sum_{s \in neighbors(i))} [\frac{1}{max - \tau_{is}}]^{\alpha} \times [\frac{1}{d_{is}}]^{\beta}} \tag{4}$$

In this equation, the parameters α and β make it possible more or less to attach importance to the quantities of pheromone compared to the relative distances between the factors.

3.5 Results

In the examples below, for a better readability, the factors are reduced to letters. A sequence is only represented with its factors. For example, `ABBC` is a sequence containing 4 factors, one factor labeled 'A', two labeled 'B' and 1 named 'C'.

Let's see some little alignments that require a particular attention. In the tables, the first column represents the original alignment. The other columns represent possible solutions. The last line of the table show the average number of time solutions are chosen over 10 trials.

Table 2.

AB	AB-	-AB	AB
AB	AB-	-AB	AB
BA	-BA	BA-	BA
	5	5	0

The choice, in the table 2 must be made between the 'A' factors and the 'B' factors to determine which block will be complete. The algorithm cannot make a clear choice between the 2 first solutions.

Table 3.

AB	AB-	--AB	--AB-
AA	A-A	A-A-	--A-A
ABA	ABA	ABA-	ABA--
	10	0	0

In the table 3, the choice of 3 blocks is wanted. The difficulty for the method is to discover the 'B' block inside all the 'A' factors.

Table 4.

DAZZZZ	DA-----ZZZZ	DAZZZZ-----	-----DAZZZZ
ACGSTD	-ACGSTD----	-A----CGSTD	ACGSTD-----
ACGSTD	-ACGSTD----	-A----CGSTD	ACGSTD-----
ACGSTD	-ACGSTD----	-A----CGSTD	ACGSTD-----
	0	10	0

In the table 4, the 'A' factor and the 'D' factor on the first sequence have the same conflicts. The choice will be made thanks to the relative distance between the factors. Here, the rest of he 'D' factor are far from the 'D' factor of the sequence 1, so it will not be chosen.

The goal, in table 5 tests the aggregation capacity of the method. It is the concept of "meta-block" that we want to highlight. Here, the first block of 'B' factors can be align with the first 'B' factor of the second sequence or with the second factor 'B' on the same sequence. The algorithm will chose the second

Table 5.

BABAB	B-ABA-B	-BABA-B	-BAB-AB
BBABAAB	BBABAAB	BBABAAB	BBABAAB
BABAB	B-ABA-B	-BABA-B	-BAB-AB
	0	5	5

option because there is a meta-block at this position, i.e. two or more blocks where each factor on each sequence follow in the same order.

4 Natural Language Processing

4.1 Description of the Problem

Understanding a text requires the comparison of the various meanings of any polysemous word with the context of its use. But, the context is precisely defined by the words themselves with all their meanings and their status in the text (verb, noun, adjective...). Given their status, there exist some grammatical relations between words. These relations can be represented by a tree-like structure called a morpho-syntactic analysis tree. Such a tree is represented on Fig. 7.

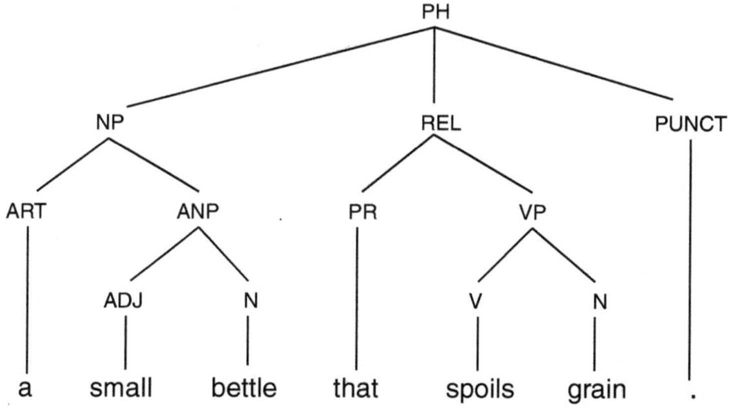

Fig. 7. Morphosyntactic tree obtained from a lexical and grammatical analysis of the text. ART denotes article, PUNCT the punctuation, V stands for verb, N for noun...

This structure represents the syntactical relations between words, but in no way their semantics relations. In order to represent this new relational structure, we consider words as the basic compounds of an interaction network which implicit dynamics reveals the pregnancy of each meaning associated to

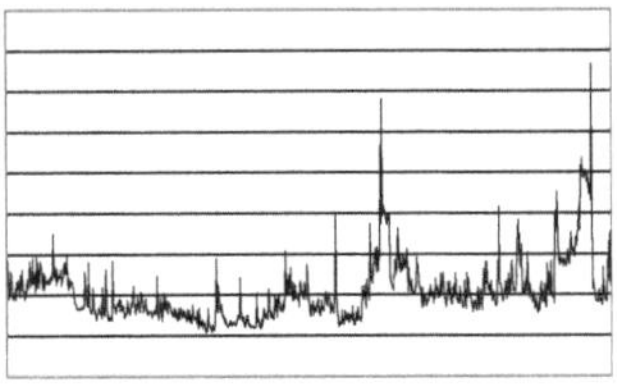

profit (profit).

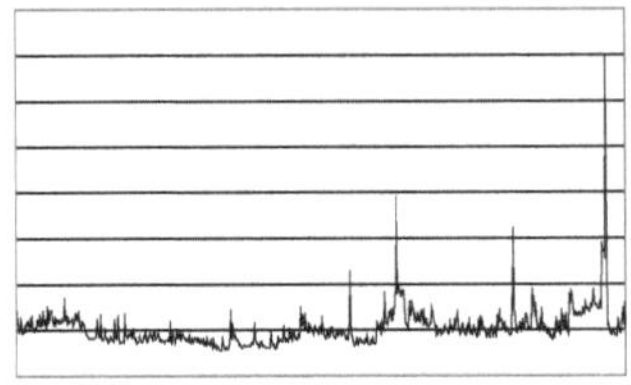

benefice (benefit).

Fig. 8. Conceptual vectors of two terms (profit and benefit). In French, we can observe that *profit* is more polysemous than *benefice.* Moreover, from the graphical representation, it appears that both terms are closely related.

any polysemous word. If we refer to the most commonly shared definition of a *complex system*, it states that it is a *network of entities which interactions lead to the emergence of structures that can be identified as high-level organizations.* The action of one entity may affect subsequent actions of other entities in the network, so that *the action of the whole is more than the simple sum of the actions of its parts.*[2]. The *actions* in our context correspond to the *meanings* of the words constituting the text, and the sum of the actions results in the global meaning of the text, which is, for sure, much more than the simple sum of the meanings of the words.

The addressed problem is twofold. On the one hand, we have to find a way of expressing words meanings interactions, and on the other hand, we have to conceive a system able to bring to the fore potential meanings for the whole text in order to help an expert for raising ambiguities due to polysemy. This problem is known as Word Sense Disambiguation (WSD). A more detailed description of the proposed solution can be found in [19].

4.2 Models and Tools

Thematic aspects of textual segments (documents, paragraphs, syntagms, etc.) can be represented by conceptual vectors. From a set of elementary notions, concepts, it is possible to build vectors (conceptual vectors) and to associate them to lexical items[3]. Polysemous words combine the different vectors corresponding to the different meanings (Fig. 8 gives an illustration of conceptual vectors).

This vector approach is based on well known mathematical properties, it is thus possible to undertake well founded formal manipulations attached to reasonable linguistic interpretations. Concepts are defined from a thesaurus (in the prototype applied to French, [21] has been chosen where 873 concepts are identified). The building of conceptual vectors is a research domain in

[2] [20] *Why do we need artificial life ?* page 305.

[3] Lexical items are words or expressions that constitute lexical entries. For instance, *car or* white ant are lexical items

itself and is out of the scope of this brief example, so in the following, we consider that each meaning of a word is attached to a conceptual vector.

In order to build the interaction network we need a way of deciding whether two terms are thematically related or not. For that purpose, we define two measures: $Sim(A, B)$ a *similarity measure* between two vectors A and B that can be expressed as the scalar product of both vectors divided by the product of their norm, and the *angular distance* D_A between two vectors A and B ($D_A(A, B) = \arccos(Sim(A, B))$) Intuitively, this latter function constitutes an evaluation of the *thematic proximity* and is the measure of the angle between the two vectors. We would generally consider that, for a distance $D_A(A, B) \leq \frac{\pi}{4}$, (i.e. less than 45 degrees) A and B are thematically close and share many concepts. For $D_A(A, B) \geq \frac{\pi}{4}$, the thematic proximity between A and B would be considered as loose. Around $\frac{\pi}{2}$, they have no relation.

The structures associated to solutions for our problem can be defined as a path or a set of paths. Each path correspond to a so-called *interpretation trail*. The fact that within a sentence several meanings may co-exist entails the possibility for several interpretation trails and thus of several distinct paths.

4.3 Ant-based Method

Our method for WSD relies on both kind of stigmergy. Sign-based stigmergy plays a role in ant behaviors. Sematectonic stigmergy is used for modifying nodes characteristics and for creating new paths between vertices. In the sequel, these new paths will be called bridges.

The "binary bridge" is an experiment developed by [23]. As reported in [10] *in this experiment, a food source is separated from the nest by a bridge with two equally long branches A and B.* Initially, both paths are visited and after some iterations, one path is selected by the ants, whereas the second, although as good as the first one, is deserted. This experiment interests us for two reasons. It first shows that ants have the ability of organizing themselves in order to determine a global solution from local interactions, thus, it is likely to obtain an emergent solution for a problem submitted to an ant-based method. This point is crucial for our approach, since we expect the emergence of a meaning for the analyzed text. But, the experiment also shows the inability of such method, in its classical formulation, to provide a set of simultaneous and distinct solutions instead of only one at a time. As these methods are based on the reinforcement of the current best solution, they are not directly suitable for our situation. Indeed, if several meanings are possible for a text, all these meanings should emerge. That was the main motivation for introducing color as a colony characteristic.

The computational problem is twofold. Indeed the meanings are not strictly speaking active entities. In order to ensure the interactions of the meanings of the whole text, an active framework made of "meaning transporters" must be supplied to the text. These "transporters" are intended to allow the interactions between meanings of text elements. They have to be both light (because of their possible large number) and independent (word meanings are intrinsic values). Moreover, when some meanings stemmed from different words are compatible (*engaged* with *job* for instance), the system has to keep a trace of this fact.

This was the original motivation for us to consider ant-based complex systems as described in Sect. 2. A similar idea already existed in [16] with the COPYCAT project, in which the environment by itself contributes to solution computation and is modified by an agent population where roles and motivations varies. We retain here some aspects mentioned in [28], that we consider as being crucial: (1) mutual information or semantic proximity is one key factor for lexical activation, (2) the syntactic structure of the text can be used to guide information propagation, (3) conceptual bridges can be dynamically constructed (or deleted) and could lead to *catastrophic events* (in the spirit of [26]). These bridges are the instrumental part allowing mutual-information exchange beyond locality horizons. Finally, as pointed by [16], biased randomization (which doesn't mean chaos) plays a major role in the model.

In order to build several structures corresponding to competing meanings, we consider several colonies with the color characteristic.

4.4 Environment

The underlying structure of the environment is the morphosyntactic analysis tree of the text to be analyzed. Each content word is a node. This node has as many children in the tree as senses. To each child associated to a sense corresponds a unique color: the conceptual vector of the sense. A child is also a *nest* and all children of a node associated to a content word are *competing nests*. An ant can walk through graph edges and, under some circumstances, can build new ones (called bridges). Each node contains the following attributes beside the morphosyntactic information computed by the analyzer: (1) a resource level R, and (2) a conceptual vector V. Each edge contains (1) a pheromone level. The main purpose of pheromone is to evaluate how popular a given edge is. The environment by itself is evolving in various aspects:

1. the conceptual vector of a node is slightly modified each time a new ant arrives. Only vectors of nests are invariant (they cannot be modified). A nest node is initialized with the conceptual vector of its word sense, other nodes with the null vector.
2. resources tend to be redistributed toward and between nests which *reinvest* them in ants production. Nodes have an initial amount of resources of 1.
3. the pheromone level of edges are modified by ant moves. The evaporation rate δ ensures that with time pheromone level tends to decrease toward zero if no ant are passing through. Only bridges (edges created by ants) would disappear if their pheromone level reaches zero.

The environment has an impact on an ant and in return ants continuously modify the environment. The results of a simulation run are decoded thanks to the pheromone level of bridges and the resource level of nests.

4.5 Nests, Ant Life and Death

A nest (word sense) has some resources which are used for producing new ants. At each cycle, among the set of nests having the same parent node (content word), only one is allowed to produced a new ant. The color of this ant is the one of the selected nest. In all generality, a content word has n children (nests), and the nest chosen for producing the next ant is probabilistically selected according to the level of resources. There is a cost ϵ for producing an ant, which is deducted from the nest resources. Resource levels of nests are modified by ants. The probability of producing an ant, is related to a sigmoid function applied to the resource level of the nest. The definition of this function ensures that a nest has always the possibility to produce a new ant although the chances are low when the node is inhibited (resources below zero). A nest can still borrow resources and thus a word meaning has still a chance to express itself even if the environment is very unfriendly.

The ant cost can be related to the ant life span λ which is the number of cycles the ant can forage before dying. When an ant dies, it gives back all the resources it carries plus its cost, to the currently visited node. This approach leads to a very important property of the system, that the total level of resources is constant. The resources can be unevenly distributed among nodes and ants and this distribution changes over time, sometimes leading to some stabilization and sometimes leading to periodic configurations. This is this *transfer of resources* that reflects the lexical selection, through word senses activation and inhibition.

The ant population (precisely the color distribution) is then evolving in a different way of classical approaches where ants are all similar and their number fixed in advance. However, at any time (greater than λ), the environment contains at most λ ants that have been produced by the nests of a given content word. It means that the global ant population size depends on the number of content words of the text to be analyzed, but not on the number of word meanings. To our views, this is a very strong point that reflects the fact some meanings will express more than others, and that, for a very polysemic word, the ant struggle will be intense. A monosemic word will often serve as a pivot to other meanings. Moreover, this characteristic allows us to evaluate the computing requirements needed for computing the analysis of a given text since the number of ants depends only on the number of words.

4.6 Ant Population

An ant has only one motivation: foraging and bringing back resources to its nest. To this purpose, an ant has two kinds of behavior (called modes), (1) searching and foraging and (2) returning resources back to the nest. An ant a has a resource storage capacity $R(a) \in [0, 1]$. At each cycle, the ant will decide between both modes as a linear function of its storage. For example, if the $R(a) = 0.75$, there is a 75% chance that this ant a is in *bringing back* mode.

Each time an ant visits a (non-nest) node, it modifies the node color by adding a small amount of its own color. This modification of the environment is one factor of the sematectonic stigmergy previously mentioned and is the means for an ant to find its way back home. The new value of the color is computed as follows: $C(N) = C(N) + \alpha C(a)$ with $0 < \alpha < 1$. In our application, colors are conceptual vectors and the "+" operation is a normalized vector addition ($V(N) = V(N) + \alpha V(a)$). We found heuristically, that $\alpha = 1/\lambda$ constitutes a good trade-off between a static and a versatile environment.

4.7 Searching Behavior

Given a node N_i. N_j is a neighbor of N_i if and only if there exists an edge E_{ij} linking both nodes. A node N_i is characterized by a resource level noted as $R(N_i)$. An edge E_{ij} is characterized by a pheromone level noted as $Ph(E_{ij})$. A searching ant will move according to the resource level of each adjacent node (its own nest excepted) and to the level of pheromones of the outgoing edges. More precisely an attraction value is computed for each neighbor. This value is proportional to the resource level and inversely proportional to the pheromone level.

$$\text{attract}_S(N_x) = \frac{\max(\text{R}(N_x), \eta)}{\text{Ph}(E_{ix}) + 1} \tag{5}$$

Where η is a tiny constant avoiding null values for attraction. The motivation for considering an attraction value proportional to the inverse of the pheromone level is to encourage ants to move to non visited parts of the graph. If an ant is at node N_i with p neighbors $N_k (k = 1 \cdots p)$, the probability $P_S(N_x)$ for this ant to choose node N_x in *searching* mode is:

$$P_S(N_x) = \frac{\text{attract}_S(N_x)}{\sum_{1 \le j \le p} \text{attract}_S(N_j)} \tag{6}$$

Then, if all neighbors of a node N_i have the same level of resources (including zero), then the probability for an ant visiting N_i to move to a neighbor N_x depends only on the pheromone level of the edge E_{ix}.

An ant is attracted by node with a large supply of resources, and will take as much as it can hold (possibly all node resources). A depleted node does not attract searching ants. The principle here, is a simple greedy algorithm.

4.8 Bringing Back Behavior

When an ant has found enough resources, it tends to bring them back to its nest. The ant will try to find its way back thanks to the color trail left back during previous moves. This trail could have been reinforced by ants of the same color, or inversely blurred by ants of other colors.

An ant a returning back and visiting N_i will move according to the color similarity of each neighboring node N_x with its own color and according to the level of pheromones of the outgoing edges. More precisely an attraction value is computed for each neighbor. This value is proportional to the similarity of colors and to the pheromone level:

$$\text{attract}_R(N_x) = \max(\text{sim}(\text{colorOf}(N_x), \text{colorOf}(a)), \eta) \times (\text{Ph}(E_{ix}) + 1)$$

Where η is a tiny constant avoiding null values for attraction.

If an ant is at node N_i with p neighbors $N_k(k = 1 \cdots p)$, the probability $P_B(N_x)$ for this ant to choose node N_x in *returning* mode is:

$$P_R(N_x) = \frac{\text{attract}_R(N_x)}{\sum_{1 \le j \le p} \text{attract}_R(N_j)} \tag{7}$$

All considered nodes are those connected by edges in the graph. Thus, the syntactic relations, projected into geometric neighborhood on the tree, dictate constraints on the ant possible moves. However, when an ant is at a friendly nest, it can create a shortcut (called a *bridge*) directly to its home nest. That way, the graph is modified and this new arc can be used by other ants. These arcs are evanescent and might disappear when the pheromone level becomes null.

From an ant point of view, there are two kinds of nests: friend and foe. Foe nests correspond to alternative word senses and ants stemmed from these nests are competing for resources. Friendly nests are all nests of other words. Friends can fool ants by inciting them to give resource. Foe nests instead are eligible as resource sources, that is to say an ant can steal resources from an enemy nest as soon as the resource level of the latter is positive.

4.9 Results

The evaluation of our model in terms of linguistic analysis is by itself challenging. To have a larger scale assessment of our system, we prefer to evaluate it through a Word Sense Disambiguation task (WSD).

A set of 100 small texts have been constituted and each term (noun, adjective, adverb and verb) has been manually tagged. A tag is a term that names one particular meaning. For example, the term *bank* could be annotated as *bank/river*, *bank/money institution* or *bank/building* assuming we restrict ourselves to three meanings. In the conceptual vector database, each word meaning is associated to at least one tag (in the spirit of [18]). Using tag is generally much easier than sense number especially for human annotators.

The basic procedure is quite straightforward. The unannotated text is submitted to the system which annotates each term with the guessed meaning. This output is compared to the human annotated text. For a given term, the annotation available to the human annotator are those provided by the conceptual vector lexicon (i.e. for bank the human annotator should choose between *bank/river*, *bank/money institution* or *bank/building*). It is allowed for the human annotator to add several tags, in case several meanings are equally acceptable. For instance, we can have *The frigate/{modern ship/ancient ship} sunk in the harbor.*, indicating that both meanings are acceptable, but excluding *frigate/bird.* Thereafter, we call *gold standard* the annotated text. We should note that only annotated words of the gold standard are target words and used for the scoring.

When the system annotates a text, it tags the term with all meanings which activation level is above 0. That is to say that inhibited meanings are ignored. The system associates to each tag the activation level in percent. Suppose, we have in the sentence *The frigate sunk in the harbor.* an activation level of respectively 1.5 , 1 and −0.2 for respectively *frigate/modern ship*, *frigate/ancient ship* and *frigate/bird.* Then, the output produced by the system is:

The frigate/{modern ship:0.6/ancient ship:0.4}.

Precisely, we have conducted two experiments with two different ranking methods.

A *Fine Grained* approach, for which only the first best meaning proposed by the system is chosen. If the meaning is one of the gold standard tag, the answer is considered as valid and the system scores 1. Otherwise, it is considered as erroneous and the system scores 0.
A *Coarse Grained* approach, more lenient, gives room to closely related meanings. If the first meaning is the good one, then the system scores 1. Otherwise, the system scores the percent value of a good answer if present. For example, say the system mixed up completely and produced:
The frigate/{bird:0.8/ancient ship:0.2}.
the system still gets a 0.2 score.

Table 6.

Scoring scheme	All terms	Nouns	Adjectives	Verbs	Adverbs
Fine Grain Scoring	0.68	0.76	0.78	0.61	0.85
Coarse Grain Scoring	0.85	0.88	0.9	0.72	0.94

These results compare quite favorably to other WSD systems as evaluated in SENSEVAL campaign [1]. However, our experiment is applied to French which figures are not available in Senseval-2 [2].

As expected, verbs are the most challenging as they are highly polysemous with quite often subtle meanings. Adverbs are on the contrary quite easy to figure when polysemous.

We have manually analyzed failure cases. Typically, in most cases the information that allows a proper meaning selection are not of thematic value. Other criteria are more prevalent. Lexical functions, like hyperonymy (is-a) or meronymy (part-of) quite often play a major role. Also, meaning frequency distribution can be relevant. Very frequent meanings can have a boost compared to rare ones (for example with a proportional distribution of initial resources). Only if the context is strong, then could rare meanings emerge.

All those criteria were not modeled and included in our experiments. However, the global architecture we propose is suitable to be extended to ants of other *caste* (following an idea developed by Bertelle et al. in [7]). In the prototype we have developed so far, only one caste of ants exists, dealing with thematic information under the form of conceptual vectors. Some early assessments seem to show that only with a semantic network restricted *part-of* and *is-a* relations, a 10% gain could be expected (roughly a gain of 12% and a lost of 2%).

5 Analysis

From the problems presented Sects. 3 and 4 the three main steps identified in Sect. 2 can be argued.

5.1 Identification of the Structures

The first necessity when using our approach is to clearly identify and define the shape of the desired structures in the environment graph. This definition must be clearly established when modeling the problem. In other words, the conception of a graph model for one particular problem must be done in consideration of the necessary structures. Finally, the generated model must be practicable by the system and must permit the production of such structure.

5.2 Choice of General Features to Achieve the Goal

Secondly, from the description of the wanted structures, a choice of general characteristics for nests, colonies and ants has to be made. These characteristics define general behaviors that lead the system to produce wanted structures. Let consider as an example two general groups of characteristics that lead to two different kind of structures:

- If a path or a set of paths is desired as solution to one problem, then, some nests must be defined in the graph. They are assigned a vertex of the search graph. Ants are given one or more destinations vertices so as to perform walks between a nest and a destination. Thus, the system will produce paths between two vertices. The pheromone deposit can be performed when going back to the nest. A tabu list is used, its size depends heavily on the shape of the graph but it is generally equal to the size of the whole tour of the ant. The tabu list prevents ants form making loops in their paths.
- If sub-graphs or sets of vertices/edges are wanted, there is not one special vertex that represents the nest, indeed, ants are displayed on every vertices of the graph. So it is considered that each vertex of the graph is a nest. More than one colony is used so as to define competitive behaviors between ants. Ants are assigned a colony specific color and deposit colored pheromones at each edge cross. A tabu list is used but its purpose is to prevent ants from stagnating on a same couple of vertices, so it has a relatively small size.

This general settings define a global behavior of the system for common kinds of solution structures but are not enough to lead the system in a precise problem.

5.3 Relevant Parameters Tuning

Thirdly, after the definition of a structure and a general set of characteristics to solve one problem, a set of relevant parameters need to be identified. Modifications of these parameters bring behavioral changes to the system. In effect, once the general process defined, these relevant parameters permit the adaptation to the precise problem and to the precise environment. Finally the tuning of those parameters sounds quiet difficult and require attention. As said by Theraulaz in [25], one of the *signatures* of self-organizing processes is their sensitivity to the values of some critical parameters. It has been observed from experiments that some parameters are probably linked there is probably a sort of ratio that is to be found between them. Another idea that these parameters are strongly related to the nature of the environment that is considered. One simple example of this correlation is the relative dependency between the number of ants and the number of elements (vertices and edges) of the graph. It is true that a bigger graph will require a bigger colony to explore it, but how much ant for how much element is an hard question and probably depends to other values like the average degree of each vertex (its connectivity).

5.4 An Example of this Analysis

The quite simple shortest path problem has been modeled. In this problem, the shortest path between two points of an environment is to be found. The environment is modeled with a graph where the start point and the end point are two particular vertices of this graph.

The first step that is the definition of the desired structures is straightforward. One solution is one path in the graph starting from one of the two previously defined points and stopping at the other one.

In the second step, the general characteristics are defined so as to create this path. One ant colony is used. One nest is located on the start vertex. Ants start their exploration from the nest and are looking for the end vertex. When the destination is found, they go back to the nest by laying pheromone down the edges of the paths they found. They use a tabu list who's size is of the order of the constructed path's length.

In the third step, experiments allowed the identification of four relevant parameters:

- The number of ants, that define the total number of ant agent in the colony. It seams to be highly related to the distribution of the edges in the graph.
- The evaporation rate for the pheromone quantities. If a high evaporation value is set, it is difficult for ants influence each other with pheromones. They will explore new parts of the graph better than exploiting already visited areas. If the evaporation is set to a lower value, pheromone trails evaporate slowly, they are stronger and bring ants to already visited paths.
- The tabu list size defines the number of already visited edges that cannot be crossed again. This parameter prevents ants from making loops. It would normally be set to the length of the constructed path, in other words, ants are not allowed to visit twice the same vertex. Actually, this parameter needs sometimes to be set to a lower value depending on the shape of the graph.
- The exploration threshold defines the sensitivity of the ants to the pheromone trails. It is usually used to find a balance between the intensification of pheromone trails and the diversification of the search. For one ant situated on vertex i, this parameter (q_0) influences the calculation of the transition rule. q_0 is the probability ($0 \leq q_0 \leq 1$) of choosing the most heavily loaded edge in pheromones among outgoing edges of the local vertex i. The normal transition rule (1) is chosen with probability $1 - q_0$. When ants are too much attracted by pheromones they follow already defined trails and don't try to visit unexplored parts of the graph. On the contrary, if they don't pay enough attention to the pheromones, no stigmergy is available.

As an example, let's consider a torus graph composed of 375 nodes and 750 edges. The shortest path in this graph has to be determined between two defined vertices. The shortest path between those vertices is 20 edges long. A torus has no border. Each vertex has exactly 4 outgoing edges. Among experiments carried out the following parameter set gives good results for his graph:

- Number of ants = 100.
- Evaporation rate = 0.03. Note: the evaporation process is performed at each step. One step represents one vertex move for each ant of the colony.
- Tabu list size = size of the length of the path.
- Exploration threshold = 0.6.

Figure 9 shows 4 different states of the graph when ants are running on it. This figure illustrates the way a path emerges from the activity of the whole

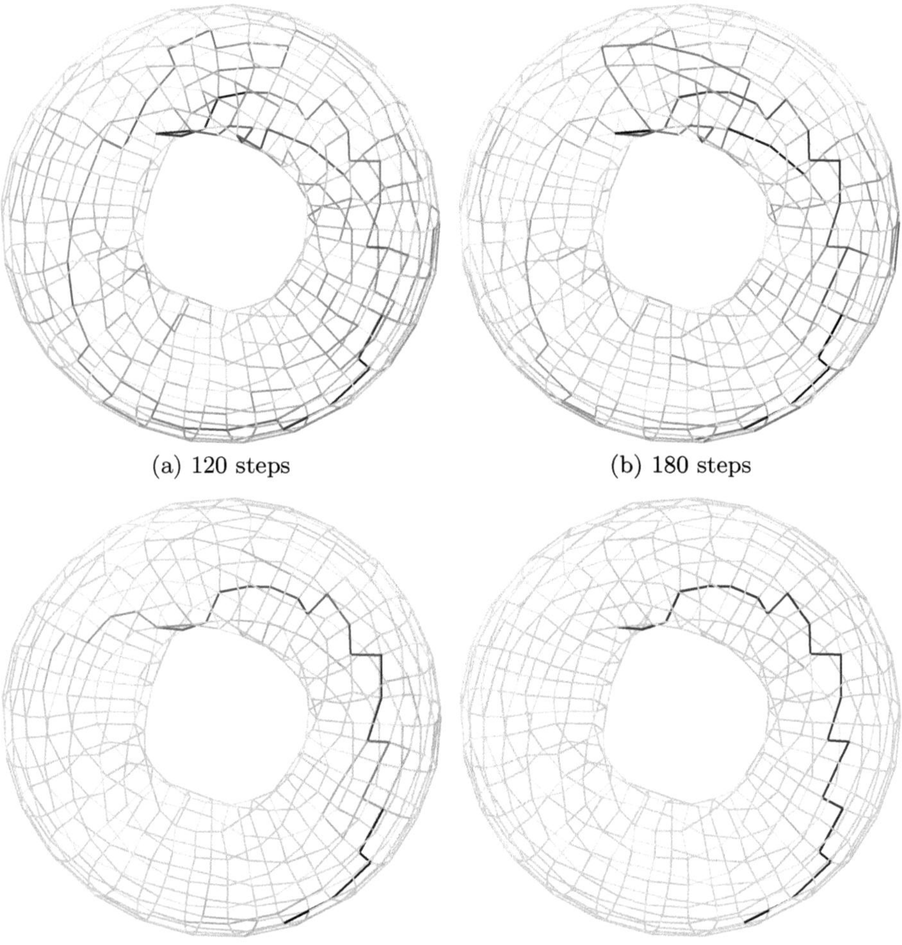

(a) 120 steps (b) 180 steps

(c) 230 steps (d) 360 steps

Fig. 9. The search of the shortest path between two points of a torus graph. The darker the arcs of the graph, the higher the quantity of pheromone. Figure (a) is a picture of the graph at step 120. A *step* is a one vertex move for each ant of the colony. Figure (b) is a picture of the same graph after 180 steps and so on.

colony. At first ants perform random walks. When ants discover the destination vertex, they go back to the nest and lay down the graph a pheromone trail. That why in the early steps of the run, pheromones are lay approximatively everywhere on the graph. When ants performing shorter paths they are faster and can lay down more pheromone trails. Finally a shortest emerges in the pheromone trails.

This is true that the values of different parameters are highly related to the structure of the graph

Between two vertices of this graph, the quantity of possible shortest paths can be important. In this graph, ants can easily make a lot of loops without finding the destination vertex. The use of a tabu list becomes useful in this kind of graph. In effect, the tabu list prevents ants from making loops in the graph. Experiments show that the use of a tabu list that represent the size of the whole path of the ant gives good results. It is to say that ants are never allowed to cross already visited nodes.

If it's true that the use of a tabu list is efficient in the case of a torus graph, but it is not in all circumstances. Let us consider the graph Fig. 10. In this graph there is no cycle, it is a tree. If a tabu list is used, ants will loose them self in leafs of the tree without been able to go back. If no tabu list is use at all, then strange phenomena occur where some ant go from vertex A to vertex B, then go back to A and so on. Finally the best solution in this case is to use a tabu list to prevent stagnation and to allow ants to go back when no more vertices are available. A relatively few ants are necessary for this graph.

5.5 Questions Hold by this Approach

The first question is about the system's ability to produce such structures. How can one be sure that the system will produce structures that have not been defined anywhere in the system? No proof can be given that such structure will rise. The only evaluation is the user's point of view when observing the graph.

The method we propose here, as well as the majority of the metaheuristics, is composed of a set of control parameters that need to be tuned, more or less according to the problem considered. Values taken by these parameters represent a multidimensional space. One given set of parameter will produce one solution. If another set of parameters is given to the system, another solution may appear. The question asked deals with the existence of particular conditions bounded to the values of parameters. Is it possible to make a parallel between the different sets of parameters and the one dimensional cellular automata classification made by Stefen Wolfram [31] where different classes of solutions exist:

- Class I: Fixed configurations. All the automaton cells have the same state.
- Class II: Simple structures. Repetitions can be observed.

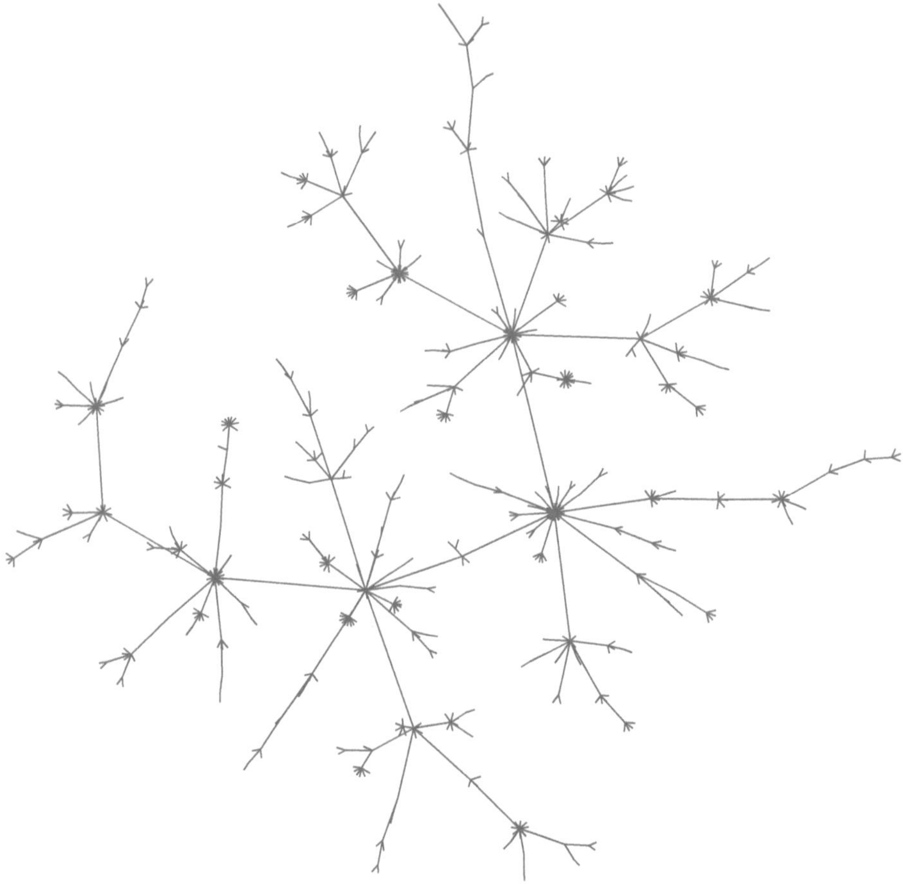

Fig. 10. A tree with 750 edges and 751 nodes.

- Class III: Chaotic solutions. This kind of automaton usually produce fractal structures.
- Class IV: In this class, complex organizations appear, many different structures are observed.

Jean-Claude Heudin [15] proposed to compare Wolfram's classification with experiments made on two dimensional automata where one parameter can take different values. In his model, the different possible values of the β parameter produce different solutions that can be classified with the above scheme. The *beta* parameter defines the number of active neighbors around one particular cell necessary to maintain it in its state. For one particular value $\beta = 2$, the automaton produced becomes a "game of life" witch is part of the IV class. Back to the model presented here, is it possible to make such a parallel so as to classify the produced solutions. In other words, are there particular areas in the parameter space where the system produces nothing, where it produces incomprehensible solutions (many different structures, continually changing) and where it produces understandable solutions (a structure with the shape of what we are looking for). Finally if such a comparison can be done, let's call systems with good sets of parameters, class IV systems.

In this case of emerging class IV structures:

- Are the produced structures stable when a little variation occur in the parameter set ? Is the system robust to few changes in the parameter space ?
- What are the characteristics that can be carried out from the solutions ?
- Do all the structures that belong to the class IV share a common (or closed) area in the multidimensional space of parameters or are there islands of parameters ?
- In the case of islands, in one particular island, does a little change in the parameter lead to a similar solution with little differences ?

6 Conclusion

This paper presented an implicit building solution approach using emergent properties of ant-based complex systems. Ant Systems are particularly adapted to the observation of emerging structures in an environment as solutions to a given problem. This approach is relevant especially in case of problems where no global evaluation function can be clearly defined and/or when the environment changes.

The according modeling and solving of two problems illustrated the use of this model. It raised the heavy importance that must be given to the tuning of some critical parameters. This set of parameters is seen as a multidimensional space of possible values. In this space some islands of values may lead to produce wanted structures. If these structures are considered as class IV structures, like Wolfram's automata, then one part of the future work will consist in the identification and the understanding of the areas of parameters.

An analyze of the method was started, it introduced the basis of the future work to be done. The final aim in this project is to construct a kind of "grammar" of the different possible structures and the different systems running on it.

References

1. Senseval (2000) URL: http://www.itri.brighton.ac.uk/events/senseval/
2. Senseval-2 (2001) URL: http://www.sle.sharp.co.uk/senseval2/
3. S. F. Altschul (1989) Gap costs for multiple sequence alignment. *Journal of Theoretical Biology*, 138:297–309
4. P. Amar, J.-P. Cornet, F. Képès and V. Norris, editors (2004) *Proceedings of the Evry Spring School on "Modelling and Simulation of Biological Processes in the Context of Genomics"*

5. P. Bak (1996) *How Nature works: the science of self-organized criticality.* Springer-Verlag
6. G. J. Barton and M. J. E. Sternberg (1987) A strategy for the rapid multiple alignment of protein sequences: Confidence levels from tertiary structure comparisons. *J. Mol. Biol.*, 198:327–337
7. C. Bertelle, A. Dutot, F. Guinand and D. Olivier (2002) Dna sequencing by hybridization based on multi-castes ant system. In *proceedings of the 1st AAMAS Workshop on Agents in Bioinformatics (Bixmas)*, Bologna (Italy)
8. F. G. C. Bertelle, A. Dutot and D. Olivier (2004) Colored ants for distributed simulations. In *proceedings of* ANTS 2004. *Brussels (Belgium), September 5-8*, volume 3172 of *LNCS*, pages 326–333
9. H. L. E. Ben-Jacob (2004) Des fleurs de bactéries. *Pour la Science – Les formes de la vie*, HS-44:78–83
10. E. Bonabeau, M. Dorigo and G. Theraulaz (1999) *Swarm Intelligence: from natural to artificial systems.* Oxford University Press
11. T. Giraud, J. Pedersen and L. Keller (2002) Evolution of Supercolonies: the Argentine ants of southern Europe. *PNAS*, 99(9):6075–6079
12. P.-P. Grassé (1959) La reconstruction du nid et les coordinations inter-individuelles chez belicositermes natalensis et cubitermes s.p. la théorie de la stigmergie : essai d'interprétation du comportement des termites constructeurs. *Insectes sociaux*, 6:41–80
13. P.-P. Grassé (1984) *Termitologia, Tome II.* Fondation des sociétés. Construction. Masson. Paris
14. G. G. H. Chaté (2004) La forme des groupements animaux. *Pour la Science – Les formes de la vie*, HS-44:57–61
15. J. C. Heudin (1998) *L'Evolution au bord du chaos.* Hermes
16. D. Hofstadter (1995) *Fluid Concepts and Creative Analogies: Computer Models of the Fundamental Mechanisms of Thought (together with the Fluid Analogies Research Group).* Basic Books. NY
17. R. C. E. J. Kennedy (2001) *Swarm Intelligence.* Morgan Kaufmann Publishers
18. Jalabert and M. Lafourcade (2002) From sense naming to vocabulary augmentation in papillon. In *Proceedings of PAPILLON-2003*, page 12, Sapporo (Japan)
19. M. Lafourcade and F. Guinand (2006) Artificial ants for natural language processing. *to appear in: International Journal of Computational Intelligence Research*
20. C. G. Langton, editor (1996) *Artificial Life: an overview.* MIT. 2nd edition
21. Larousse, editor (1992) *Thésaurus Larousse – des idées aux mots, des mots aux idées.* Larousse
22. T. S. M. Dorigo (2004) *Ant Colony Optimization.* MIT Press
23. J. Pasteels, J. Deneubourg, S. Goss, D. Fresneau and J. Lachaud (1987) Self-organization mechanisms in ant societies (ii): learning in foraging and division of labour. *Experientia Supplementa*, 54:177–196
24. W. Taylor (1988) A flexible method to align large numbers of biological sequences. *J Molec Evol*, 28:161–169
25. G. Theraulaz (1997) *Auto-organisation et comportement*, chapter Auto-organisation et comportements collectifs dans les sociétés animales : introduction, pages 79–83. Collection Systèmes Complexes. Hermès
26. R. Thom (1972) *Stabilité structurelle et morphogénèse.* InterEditions (Paris)

27. J. D. Thompson, D. G. Higgins and T. J. Gibson (1994) Clustal w: improving the sensitivity of progressive multiple sequence alignment through sequence weighting, position specific gap penalties and weight matrix choice. *Nucleic Acids Research*, 22:4673–4680
28. K. W. C. W. Gale and D. Yarowsky (1992) A method for disambiguating word senses in a large corpus. *Computers and the Humanities*, 26:415–439
29. H. Walter (1994) *L'aventure des langues en occident.* Robert Laffont
30. S. Wolfram (1994) Universality and complexity in cellular automata. *Physica D*, 10:91–125
31. S. Wolfram (1984) Universality and complexity in cellular automata. *Physica D*, 10:1–35
32. H. Zwirn (2003) La complexité, science du xxie siècle ? *Pour la Science – La complexité*, 314:28–29

Changing Levels of Description in a Fluid Flow Simulation

Pierrick Tranouez, Cyrille Bertelle, and Damien Olivier

LITIS – University of Le Havre
25 rue Philippe Lebon, BP 540
76058 Le Havre Cedex, France
[FirstName.Surname]@univ-lehavre.fr

Summary. We describe here our perception of complex systems, of how we feel the different layers of description are an important part of a correct complex system simulation. We describe a rough models categorization between rules based and law based, of how these categories handled the levels of descriptions or scales. We then describe our fluid flow simulation, which combines different fineness of grain in a mixed approach of these categories. This simulation is built keeping in mind an ulterior use inside a more general aquatic ecosystem.

Key words: complex system, simulation, multi-scale, cooperative problem solving, heterogeneous multiagent system

1 Introduction

Not everyone agrees on what a complex system is, but many authors agree on some properties of these objects, among them the difficulty of their simulation. We will describe here some of the causes we perceive of this difficulty, notably the intricate levels of description necessary to tackle their computer simulation. We will then explain how this notion guided our simulation, a fluid flow computation with solid obstacles, meant as a step toward a broader simulation of an aquatic ecosystem.

2 Simulation of Complex Systems

2.1 Complexity vs Reductionism

As explained in [1], to study how mass works in a material system, dividing this system into smaller parts is a good method. Indeed, each of his subsystems is massive, and therefore the study, the reductionism, can continue.

This is not so with living systems. If you divide a living system into smaller parts, the odds are good that all you reap is a heap of dead things. That's because the life question of a live system is *complex*. This means that what is important is not so much the parts of the systems, nor the parts of these parts, but the functioning relations that exist between them.

This is one of the two main reasons why one may want to integrate the multiple possible scales of description into a simulation. When you enquire about a complex question in a system, you need to choose carefully the needed levels of description, as you can't simplify them. Furthermore, these needed levels may change during the simulation, and it would be a fine thing if the simulation could adapt to these variations.
Thus changing the scales of description during the simulation could be useful for the accuracy of the answers to complex questions regarding the system the simulation may provide.

Then, there is the understanding of these answers.

2.2 Clarity of the Simulation

Users of a simulation *question* it. Final users ponder about the future of the thing simulated in various circumstances, developers try to ascertain the validity of their model and of its implementation, but all use it with a purpose in mind.

Choosing the right level of description is then important to give a useful answer. If the simulation is able to adapt its descriptions to what is needed by its user, lowering the noise and strengthening the signal, by choosing the right level(s) of description, it will be a better tool. For example in our application, a simulation of a fluid flow in an ecosystem, this help takes the form of hiding tiny perturbation and putting forward the main structures of the flow that emerged during the simulation.

3 Methods for Changing the Scale in a Simulation

3.1 Law-Based vs. Rule-Based Models

Classifying the various ways science can tackle problems is an arduous task. We will nonetheless distinguish two rough categories of models.

Law-based models are the most used in science, most notably in physics. They are often continuous, especially in their handling of time and space, and based on a differential formulation whose resolution, ideally formal but often numerical, computes the values of state variables that describe the studied domain. Those methods are sometimes also called *global* or *analytical*.

In rule-based models , the studied domain is discretized in a number of entities whose variations are computed with rules. There is therefore no longer a global description of the domain, nor is there a priori continuity. Cellular automata fall in this category of course, and so do objects/actors/agents.

Those models have had a strong influence on game theory, and from there directly on social models, and later on other domain through computer science for instance, at least by way of metaphors. Those models have other names depending on the domain where they are used, ranging from *micro-analytical* in sociology, to *individual-based* in life sciences or just simply *local.*

Both kinds of models can be deterministic or stochastic. Finally, to blur the distinctions a bit more, models may include sub-parts falling in any of these categories. This is often the case with ecosystems for instance.

3.2 Changing the Scale in Law-Based Models

Accessing different levels of description in these models is often done through integration. Indeed, as said before, state function in these models are often continuous, and can therefore be integrated. New state functions are then valued or even built, on another domain and based on different phenomenological equations. For example, A. Bourgeat [4] describes fluid flows in porous milieus, where, from Navier-Stokes equations, through integration and the addition of an extra parameter, he builds a Darcy law. These changes of equations description from one level to another alter sometimes drastically the linearity of the models and may lead to the introduction of new parameters that act as a memory of the local domain inside the global one.

In a similar way to this example, the change of level of description in analytical models is often performed *a priori*, at the building of the model.

3.3 Changing the Scale in Rule-Based Models

Models based on rules offer a wider variety of ways of changing the levels of description. Indeed, local approaches are better designed to integrate particularities of very different entities and their mutual influence, as is the case when entities of various scales interact.

Cellular automata

The first individual based computer science structures may have been cellular automata. If they were created by Stanislas Ulam, Von Neumann self-replicating automata may have been the foundation of their success [18]. Ulam himself already noticed that complex geometric shape could appear starting with only simple basic blocks. Von Neumann then Langton [10] expanded this work with self-replicating automata.

If shapes and structures did appear in the course of these programs, it must be emphasized that it were users, and not the programs themselves, that perceive them. Crutchfield [6] aimed at correcting that trend, by automating the detection of emergent structures.

Detecting structures has therefore been tried, but reifying these structures, meaning automatically creating entities in the program that represent the detected structures has not been tackled yet, as far as cellular automata are concerned. It could be that the constraint on its geometry and the inherent isotropy of the cellular automata are in this case a weakness.

Ecology

Since the beginning of the use of individual based models in ecology, the problem of handling the interactions between individuals and populations occurred [7]. The information transfers between individual was handled either statistically [5] or through the computing of action potential [14].

DAI uses

Most software architectures designed to handle multiple levels of description are themselves hierarchical. They often have two levels, one fine grained and the other coarse grained. Communication between these two levels could be called decomposition and recomposition, as in [13].

In 1998, members of the RIVAGE project remarked that it was necessary in multi-agent simulations, to handle the emergent organizations, by associating them with behaviors computed by the simulation [15]. Before that, were handled only border interactions between entities and groups [9].

This led in D. Servat PhD thesis to a hydrodynamic model incorporating in part these notions. In his Rivage application, water bowls individuals are able to aggregate in pools and rivulets. The individuals still exist in the bigger entities. The pros are that it enables their easily leaving the groups, the cons that it doesn't lighten in any way the burden of computing. Furthermore, these groups do not have any impact on the trajectories of the water bowls.

4 Application to a Fluid Flow

4.1 Ontological Summary

The fluid flows that constitute the ocean currents on the planet are the result of an important number of vortexes of different scales. Turbulent movement can also be decomposed into vortexes, on scales going down to the near molecular. Viscosity then dissipates kinetic energy thus stopping the downward fractal aspect of these vortexes [12]. There are qualitatively important transfers of energy between these various scales of so different characteristic length. Representing these is a problem in classic modeling approaches.

In classic, law based models, turbulent flows are described as a sum of a deterministic mean flow and of a fluctuating, probabilistic flow. These equations (Navier-Stokes) are not linear, and space-time correlation terms must be introduced to compensate for that. These terms prevent any follow up of the turbulent terms, and thus of the energy they transmit from one level to another.

A pure law based approach is therefore not capable of a qualitative analysis of the transfer of energy between the different scales of a turbulent flow. A multi-level model, where multiple scales of vortexes would exist, and where they would be able to interact, would be a step in this qualitative direction.

4.2 Treatment of Multiple Scales

Fluid mechanic model and its structures

There are a number of models used to describe fluid flows. The set we use here are based on a discretisation of the flow, and are called vortex methods [11].

In vortex methods, the flow is separated in a number of abstract particles, each being a local descriptor of the flow. These particles indicate the speed, vorticity etc ... of the flow where they are located.
These particles are not fixed: they are conveyed by the fluid they describe.
We find this model interesting as it is a local model, hence better able to deal with local heterogeneities. The values of the properties the particles describe are computed through the interactions between the particles, most notably through Biot-Savart formula. More details on this computation can be found in [3].

The vortex method we use is of $O(n^2)$ complexity. Finding ways of lightening this calculus is therefore important. One lead is through making our model multi-scale, and only computing entities at the scale we need them. This is our second motivation for our using different levels of description. In order to have different levels of description, we will have to use an adapted description of the simulation entities.
These entities come and go during the simulation, and thus we need a method to change the level of their description *during* the simulation, and not beforehand the way it is usually done.

In our fluid flow, the main entities as we explained are vortexes. Not only do we therefore need to detect emerging vortexes by monitoring lower level vortexes particles, but also, as these vortexes aggregate among themselves to form even bigger vortexes, make this detection process iterative. Detecting the structures is not enough: we also need to create them in the simulation once they are detected. We must make these new entities live in the simulation, interacting with its various inhabitants (most notably particles, vortexes). They must evolve, whether it is growing or decaying to its possible disintegration.

Let us now describe our recursive detection-creation/evolution-destruction cycle.

Detecting emergent vortexes among the vortex particles

Structures are detected as clusters of particles sharing some properties. For vortexes these properties are spatial coordinates and rotation sense. As described in the following figure (figure 1), the process is:

1. Delaunay triangulation of the particles
2. Computation of a minimal spanning tree of this triangulation
3. Edges that are too much longer than the average length of edges leading to the particles are removed. So are edges linking particles of opposite rotational.
4. The convex hull of the remaining trees is computed
5. An ellipse approximates the hull through a least square method. Geometric constraints on the proportions of the ellipse must be satisfied or the vortex is not created.

Further details on this process can be found in [16].

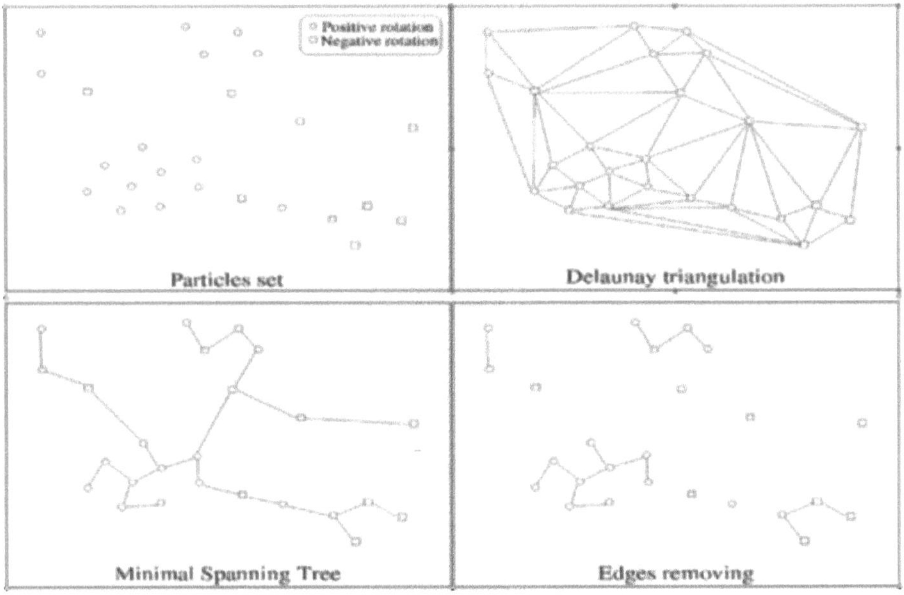

Fig. 1. Basic detection scheme

Scale transfer : making simulation entities of the detected structures

Detected structures are created in the simulation where they take the place of the particles whose interactions gave them birth.
The vortex structures are implemented through multiplicity automata [2]. These automata handle both the relations between higher level vortexes and the relations between them and the basic particles.

Part of the relations between vortexes and their environment are handled through a method based on the eco-resolution model [8], in which entities are described through a perception and combat metaphor.

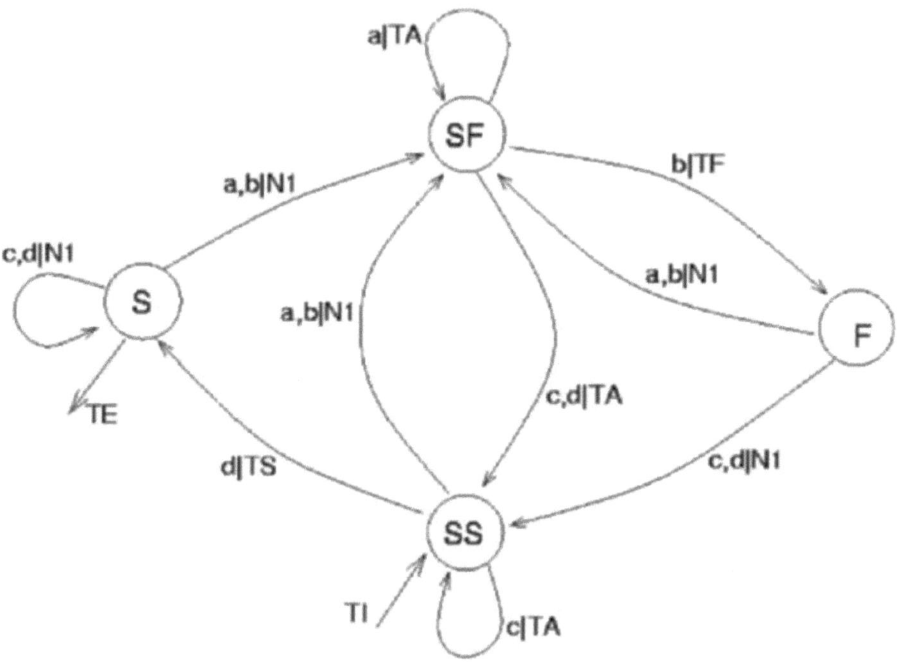

Fig. 2. Transducer managing the stability of a vortex

The associated perceptions and actions are:

1. Perceiving an intruder means being on a collision course with another vortex. Figure 3 sums up the various possibilities of interception by vortexes of opposed rotation and how each is translated in a perception.
2. Attacking another vortex means sending it a message.
3. Being attacked means receiving such a message.
4. Fleeing means being destabilized: the vortex structure shrinks and creates particles on its border. (Figure 4). Too much flight can lead to the death of the structure, which is then decomposed in its basic particles.

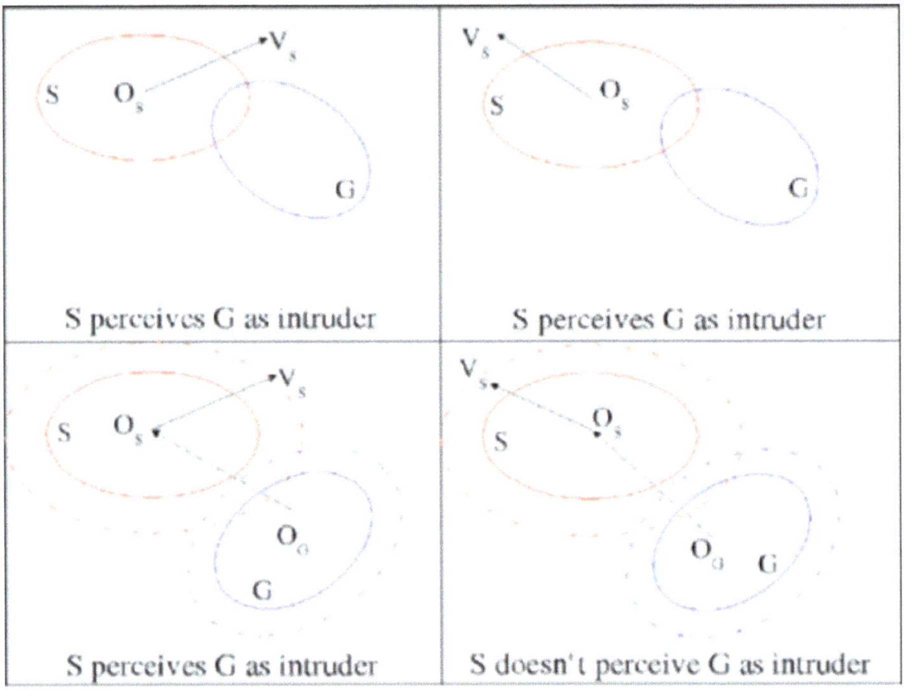

Fig. 3. Perception of an intruder

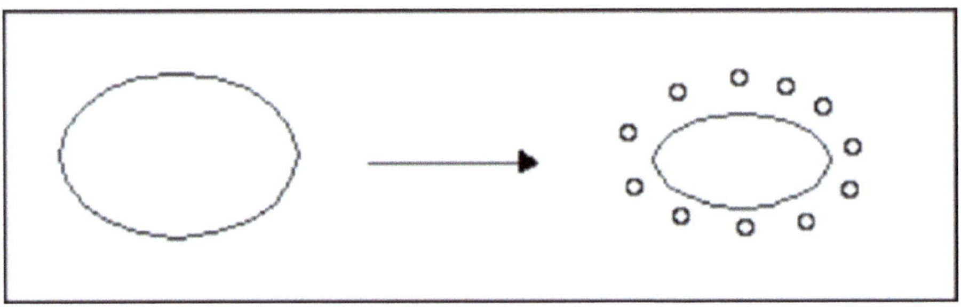

Fig. 4. Destabilization

5. Getting satisfaction can mean two things. One is aggregating surrounding particles of compatible vorticity. This calculation is done through a method close to the initial structure detection: Delaunay triangulation, spanning tree, removal of edges. Compacity criteria are then used to estimate whether the tree should be added to the vortex and thus a new ellipse be computed or not. For instance in Figure 5, the particles on the lower left will be aggregated while those on top will not. The other is fusing with a nearby vortex of the same rotation sense, if the resultant vortex satisfies compacity geometric constraints (it mustn't be too stretched, as real fluid flow vortexes are not).

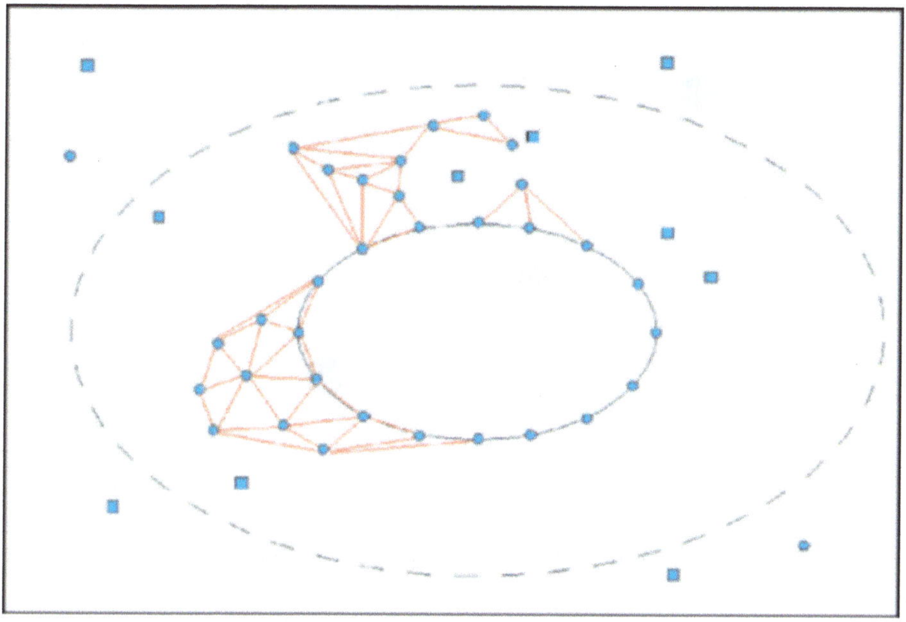

Fig. 5. To aggregate or not to aggregate

The Eco-agent manages the stability of the structures. Its trajectory behavior depends on adapted Biot-Savart formula.

The described process is then iterated. New structures are detected and implemented, while others grow, shrink or disappear altogether. They move according to fluid mechanics laws. What remains to be seen is: how do they interact with solids?

Further details on this process can be found in [16].

4.3 Interaction With Solids

Solids in a fluid flow simulation are a necessary evil, acting as obstacles and borders. Their presence requires a special treatment, but their absence would prevent the simulation of most interesting cases. They bring strong constraints in their interactions with the fluid, and as such introduce the energy that gives birth to the structures we're interested in.

We manage structures using virtual particles. A solid close enough to real particles generates these particles symmetrically to its border (cf. Figure 6 and Figure 7). The real particles perceive the virtual particles as particles in the Biot-Savart formula. The virtual particles on the other hand do not move. They are generated at each step if necessary, at the right place to repel the particles the obstacle perceived.

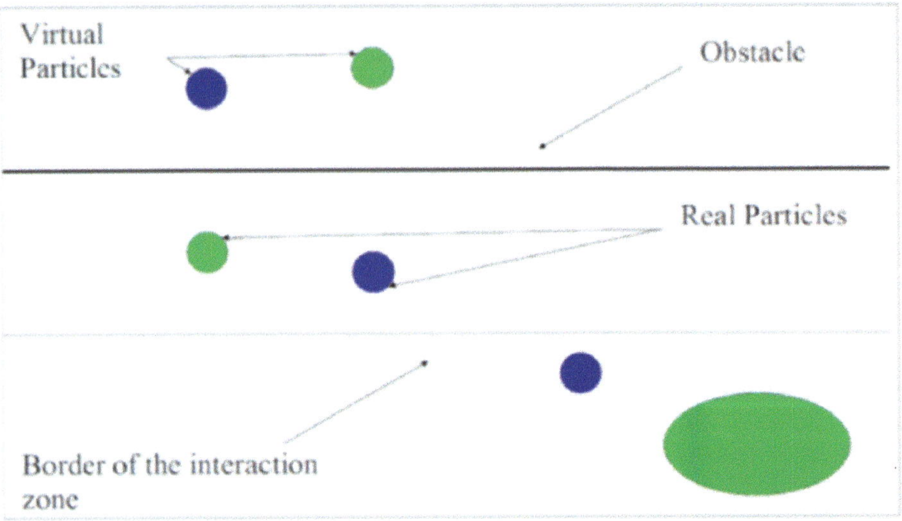

Fig. 6. Plane obstacle (e.g. border)

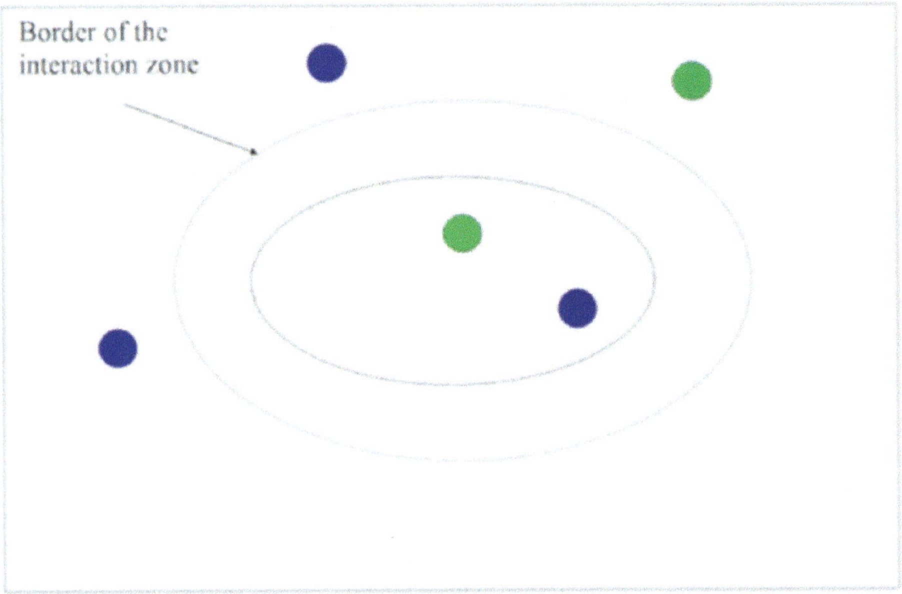

Fig. 7. Elliptic obstacle

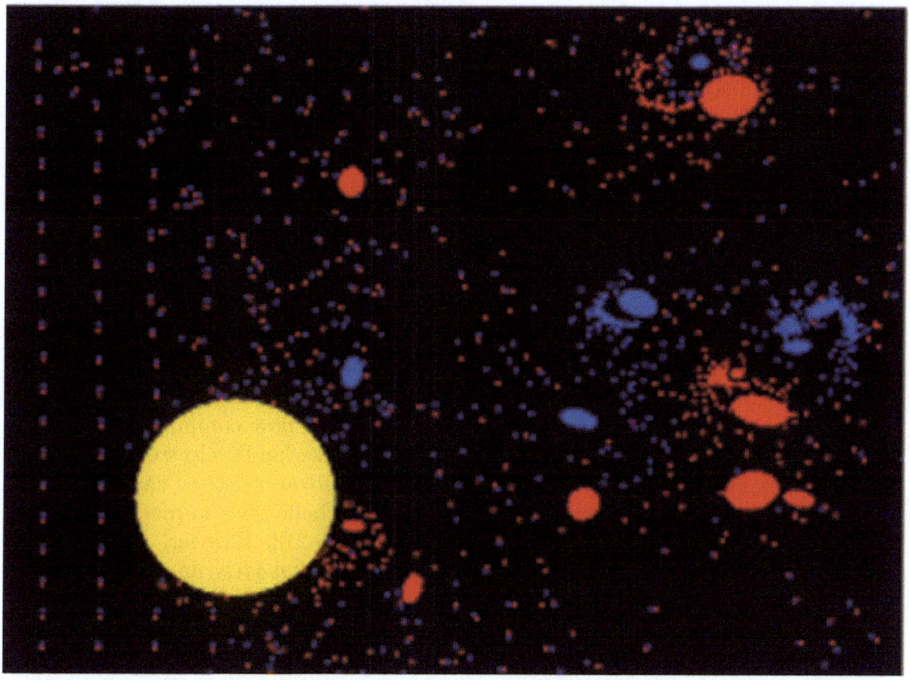

Fig. 8. Simulation screenshot Red and blue : particles and vortexes Yellow disc : Obstacle In case of shades of grey, the disc is at the bottom left, surrounded by the fluid

Obstacles also generate virtual particles to try to repel structures. If it fails, obstacles aggress the vortex, thus making it shrink so as to deal with the interaction solid/fluid in a finer grain.

5 Conclusion

We described here our perception of complex systems, of how we feel the different layers of description are an important part of a correct complex system simulation. We described a rough models categorization between rules based and law based, of how these categories handled the levels of descriptions or scales. We then described our fluid flow simulation, which combines different fineness of grain in a mixed approach of these categories. This simulation is built keeping in mind an ulterior use inside a more general aquatic ecosystem.

Our result show so far an adequate real time handling of the interactions between fluids and solids. Other methods can statically simulate in better details this interaction, but it often requires a knowledge beforehand of the placement of the solids. We can deal with solids moving at random, or more interestingly computed by the simulation itself, and not known before the simulation begins. We hope it will enable us to simulate for example animals or dynamic obstacles, thus integrating our more global works on ecosystem simulation [17].

References

1. L. Von Bertalanffy (1968) *General System Theory: Foundations, Development, Applications.* George Braziller Inc, New York
2. C. Bertelle, M. Flouret, V. Jay, D. Olivier and J.L-Ponty (2001) Automata with multiplicities as behaviour model in multi-agent simulations. In *SCI'2001*
3. C. Bertelle, D. Olivier, V. Jay, P. Tranouez and A. Cardon (2000) A multi-agent system integrating vortex methods for fluid flow computation. In *16th IMACS Congress*, volume 122-3, Lausanne (Switzerland), electronic edition.
4. A. Bourgeat (1997) *Tendances nouvelles en modélisation pour l'environnement*, chapter Quelques problèmes de changement d'échelle pour la modèlisation des ï¿½écoulements souterrains complexes, pages 207–213. Elsevier,
5. H. Caswell and A. John (1992) *Individual-based models and approaches in ecology*, chapter From the individual to the population in demographic models, 36–66. Chapman et Hall
6. J. Crutchfield (1992) *Nonlinear Dynamics of Ocean Waves*, chapter Discovering Coherent Structures in Nonlinear Spatial Systems, 190–216. World Scientific., Singapore
7. D. L. DeAngelis and L. J. Gross, editors (1992) *Individual-Based Models and Approaches in Ecology: Populations, Communities and Ecosystems.* Chapman and Hall
8. A. Drogoul and C. Dubreuil (1992) *Decentralized Artificial Intelligence III*, chapter Eco-Problem-Solving : results of the N-Puzzle, 283–295. North Holland
9. L. Gasser (1992) *Decentralized A.I*, chapter Boundaries, Identity and Agggregation : Pluralities issues in Multi-Agent Systems. Elsevier
10. C. Langton (1986) Studying artificial life with cellular automata. *Physica D*, 22
11. A. Leonard (1980) Vortex methods for flow simulation. *Journal of Computational Physics*, 37:289–335
12. M. Lesieur (1987) *Turbulence in fluids.* Martinus Nijhoff Publishers
13. P. Marcenac (1997) Modélisation de systèmes complexes par agents. *Techniques et Sciences Informatiques*
14. J. Palmer (1992) *Individual-based models and approaches in ecology*, chapter Hierarchical and concurrent individual-based modelling, 36–66. Chapman et Hall
15. D. Servat, E. Perrier, J.-P. Treuil and A. Drogoul (1998) When agents emerge from agents : Introducing multi-scale viewpoints in multi-agent simulations. In *MABS*, 183–198
16. P. Tranouez, S. Lerebourg, C. Bertelle and D. Olivier (2003) Changing the level of description in aquatic ecosystem models: an overview. In *ESMc' 2003*, Naples, Italy

17. P. Tranouez, G. Préost, C. Bertelle and D. Olivier (2005) Simulation of a compartmental multiscale model of predator-prey interactions. *Dynamics of Continuous, Discrete and Impulsive Systems Journal*
18. J. VonNeuman and A. Burks (1966) Theory of self-reproduction automata. University of Illinois Press

DNA Supramolecular Self Assemblies as a Biomimetic Complex System

Thierry A.R.[1], Durand D.[2], Schmutz M.[3], and Lebleu B.[1]

[1] Laboratoire des Défenses Antivirales et Antitumorales,
UMR 5124, Univ. Montpellier 2
34095 Montpellier, France
thierrya@univ-montp2.fr

[2] Laboratoire de Physique des Solides
CNRS UMR 8502, Bat 510, Centre universitaire Paris-Sud
91405 Orsay, France

[3] Institut Charles Sadron
6 rue Boussingault
67083 Strasbourg Cedex, France

Summary. The structure of complexes made from DNA and suitable lipids (Lx), and designed for gene transfer was examined. Cryo Electron Microscopy, Small angle X-ray scattering and Dynamic Light Scattering showed that Lx form monodisperse and spherical multilamellar particles with a distinct concentric ring-like pattern. The same concentric and lamellar structure with different packing regimes was also observed when using linear dsDNA, ssDNA, oligodeoxynucleotides (ODN) and RNA. Lx ultrastructure is of highly ordered crystalline nature exhibiting lamellar and/or hexagonal phase. We have demonstrated structural similarities between this synthetic supramolecular auto-organization and that found in some viruses. Our data point towards the possible existence of a ubiquitous organization of genetic materials.

Key words: DNA complex structures, Cryo Electron Microscopy, lipoplexes, supramolecular auto-organization

1 Introduction

Synthetic gene-transfer vectors have been subject to intense investigation since this strategy appears to be clinically safe. Potential methods of gene delivery that could be employed include DNA/polymer complexes [1] or DNA/cationic lipid complexes (Lipoplex, Lx, [2]) [3–6]. The genetic material to be delivered to target cells by these methods is plasmids. Plasmids (pDNA) are circular DNA which can be modified to contain a promoter and the gene coding for the protein of interest. Such plasmids can be expressed in the nucleus of transfected cells in a transient manner. In rare events, the plasmids may be integrated or partly integrated in the cell host genome and might therefore be stably expressed. Plasmids have a promising potential considering the fact that they may be applied in combination with a synthetic vector as carrier and that gene therapy by this means may be safe, durable, and used as drug-like therapy. Plasmid preparation is simple, quick, safe, and inexpensive representing important advantages over retroviral vector strategy. The successful use of this genetic tool for "in vivo" approaches to gene therapy will rely on the development of an efficient cell delivery system [7].

In general, the transport of nucleic acids (NA) is limited by their major biodegradability. pDNA in particular must be delivered totally intact to the nucleus of the target cell to enable expression of the transgene. The pDNA pathway after systemic administration involves many stages, each of which is a potential barrier to transgene expression. The following are the characteristics of an optimal synthetic transport system for gene transfer: (i), DNA compaction within micro particles of homogeneous size; (ii), protection of DNA against nucleases; (iii), transport of DNA in the target cell; (iv), intracellular separation of pDNA from synthetic vectors, and (v), nuclear penetration.

2 Lipoplexes as Vectors for Nucleic Acids

DNA/cationic lipid complexes were first designed in 1987 [3]. Although they are numerous commercially available reagents of high ability in transfecting cells in in vitro cell culture models [3, 4, 8], only a few Lx system were successfully applied for in vivo gene transfer, especially following systemic administration [5, 6, 8].

We have contributed and accumulated knowledge in delivering DNA by using various synthetic vectors. Our first priority was at that time to obtain pharmaceutically suitable vectors regarding stability and reproducibility. We have designed and developed an efficient lipidic vector termed as DLS [6]. Intracellular distribution and uptake was studied in numerous cell culture models with various ODN types (modified or unmodified) [9–11]. Our observations suggested that complete release of the DNA from the endocytic vesicles can be achieved and support the notion of the complete or partial release of the DNA from the lipidic carrier [12].

We were the first to show that systemic administration of plasmid DNA led to widespread and longlasting reporter gene expression [6]. We further demonstrated increased DNA plasma half life and efficient uptake in blood cells following intravenous administration in mice [12]. The DLS system was developed for transgene expression [12, 16] and applied in various experimental therapeutic models for gene transfer such as human MDR1 in vivo expression in mouse bone marrow progenitor cells [13], and glucocerobrosidase gene transfer [14].

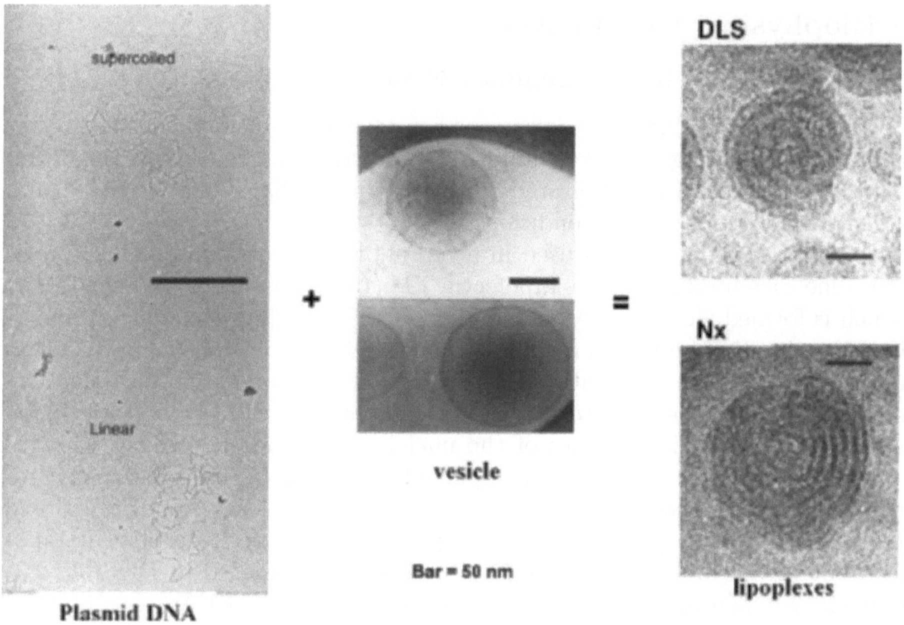

Fig. 1. Cryo-EM observation of Lx formation

ODN delivery by the DLS system was studied in various in vitro cell cultured models such as inhibition of HIV [9, 10]. We recently reported the use of antisense ODN directed against VEGF RNA for treating AIDS Kaposi's Sarcoma in vitro and in vivo setting [11]. Significant activity (39% inhibition) was observed at nanomolar range dose (0.010 μM). This result confirms those previously obtained in HIV culture models showing activity even at subnanomolar concentration. This level of ODN activity is unprecedented illustrating the potential of the DLS system for ODN delivery. Daily intratumoral administrations of VEGF-ODN conduce to a marked change in tumor growth, in cell proliferation and in the number of mitotic figures as observed in thin tissue sections [11].

In order to enhance stability in serum following systemic administration we have designed the first globally anionic charged lipoplex termed as Neutraplex [15]. Neutraplex delivery system improved significantly the bioavailability and the ODN pharmacokinetics profile using whole body Positron Emission Tomography (PET) and an enzyme-based competitive hybridization assay [16]. The anionic vector appears as a promising delivery system for in vivo administration of therapeutic ODN. We are currently applying these vectors for the delivery of ON aimed at correcting splicing alteration and of siRNAs (small interfering RNA).

3 Biophysical Examination

3.1 Components Before Complex Formation

In order to obtain stable, homogeneous complex assemblies, we decided to set a method by adding the two components in a stable and homogeneous form. The cationic lipid was integrated in Small Unilamellar Vesicles (SUV) and pDNA as solute in precise conditions.

Several lipids have been used in attempts to prepare liposome-like particles. One such lipid mixture is Lipofectin TM (Invitrogen Corp, Carlsbad, CA) which is formed with the cationic lipid DOTMA, N[1-(2,3-dioleyloxy)propyl]-N,N,N- trimethyl-ammonium chloride, and DOPE, dioleylphosphatidyl ethanolamine at a 1:1 molar ratio. The lipidic particles prepared with this formulation spontaneously interact with DNA through the electrostatic interaction of the negative charges of the nucleic acids and the positive charges at the surface of the cationic lipidic particles. This DNA/liposome- like complex fuses with tissue culture cells and facilitates the delivery of functional DNA into the cells [3]. Cationic lipid particles have been developed: LipofectamineTM. (Invitrogen Corp, Carlsbad, CA), composed of DOSPA, 2,3-dioleyloxy-N[2(sperminecarboxamido)ethyl]-N,N-dimethyl-1-propanaminium trifluoracetate and DOPE at a 1:1 molar ratio. Lipofectace.TM. (Invitrogen Corp, Carlsbad, CA) composed of DDAB, dimethyidioctadecylammonium chloride and DOPE at a 1:1 molar ratio. DOTAP.TM. (Boehringer Mannheim, Ind.) is 1-2-dioleoyloxy-3 (trimethyl ammonia) propane. Behr et al. [17] have reported the use of a lipopolyamine (DOGS, Spermine-5-carboxy- glycinediotade-cylamide) to transfer DNA to cultured cells. Lipopolyamines are synthesized from a natural polyamine spermine chemically linked to a lipid. For example, DOGS is made from spermine and dioctadecylamidoglycine [17].

DLS SUV are composed of DOGS and the neutral lipid DOPE. Briefly, they are formed following injection of water in excess to an ethanol solution of the lipids. Neutraplex SUV are composed as well of DOGS, DOPE, and a anionic phospholipid (cardiolipin) which interact with DNA under a meticulous formulation process allowing formation of globally negative charged particles [15, 16].

Neutraplex SUV are carefully prepared to obtain a highly homogeneous population in regard to size (mean size: 198 nm, and width: 54 nm) as measured by dynamic light scattering size analysis and structure (>95% unilamellar vesicles) as observed by cryoEM (Figure 1). The bilayer measured 5.0 nm.

The genetic material to be delivered to target cells by these methods is plasmids. Plasmids are autonomous extra chromosomal circular DNA. They can be modified to contain a promoter and the gene coding for the protein of interest. Such plasmids can be expressed in the nucleus of transfected cells in a transient manner. In rare events, the plasmids may be integrated or partly integrated in the cell host genome and might therefore be stably expressed. Plasmids have a promising potential considering the fact that they may be applied in combination with a synthetic vector as carrier and that gene therapy by this means may be safe, durable, and used as drug-like therapy. Plasmid preparation is simple, quick, safe, and inexpensive representing important advantages over retroviral vector strategy. For the Lx preparation pDNA was purified in the supercoiled form (Figure 1) and free (<1%) of endotoxin (bacterial protein contaminating pDNA extraction following the recombinant production). pDNA is presented in a low salt solution (50mM NaCl) of pDNA.

3.2 Lipoplexes

Lx preparation is spontaneously formed by adding DNA to cationic small unilamellar vesicles (SUV) [6, 15].

The addition of pDNA to highly homogenous SUV, in precise experimental conditions, resulted in the formation of stable lipid-pDNA complexes. Complex formation of DNA with cationic lipids leads to the respective condensation of both entities by electrostatic interactions. As a consequence, control of the thermodynamic parameters of complex formation is crucial to obtain homogeneous and reproducible particles.

DLS (Fig. 2) and Neutraplex (Fig. 3) particles appear spherical, monodisperse, homogenous in size and remarkably structured (mostly lamellar, rolled and condensed). In contrast to DOGS/pDNA complex particles do not exhibit concentric winding but rather aggregates of planar packed DNA chain (Fig. 2). DOGS allows for the side-by-side alignment of DNA in a liquid crystalline phase (Figure 2A). DOGS spontaneously condense DNA on a cationic lipid layer and result in the formation of nucleolipidic particles. This lipospermine-coated DNA shows some instability but high transfection efficiency [17].

We have extensively examined preparation of Lx for elucidating their ultrastructure. As shown in Fig. 3A, cryoEM examination of Neutraplex Lx prepared with a pDNA of 10,4 kbp reveals spherical particles exhibiting two kind of structures: mostly (>90% of total particles) a multilamellar organization of spherulite-like pattern and rarely (<10%) a punctuated pattern.

Based on the cryoEM images, the thickness of the striated layers has a mean value of 5.3 nm with a periodicity of 7.5 nm. The outer striated layer has a thickness of 4.1 nm. Perpendicular to these layers, faint striations can be seen mainly on the edge and also inside the particles. These striations are 2.0 nm thick with a periodicity of 3.4 nm (Figure 3). The particles with punctate array [15] are confined by an outer shell. This outer shell bears the same thin striations than those found on multilamellar structures. When these striations can be detected in a cryoEM image they are observed in most of the Lx particles. In addition, the punctuated structures are occasionally associated with these multilamellar structures.

Analysis by SAXS of our Lx sample (Figure 3) revealed three diffraction peaks: one indicating a repetitive spacing of $2\pi/0.094=6.9+/-1$ nm and the second and fourth order indicating the multi-layer structure ($2\pi/0.18$) [15].

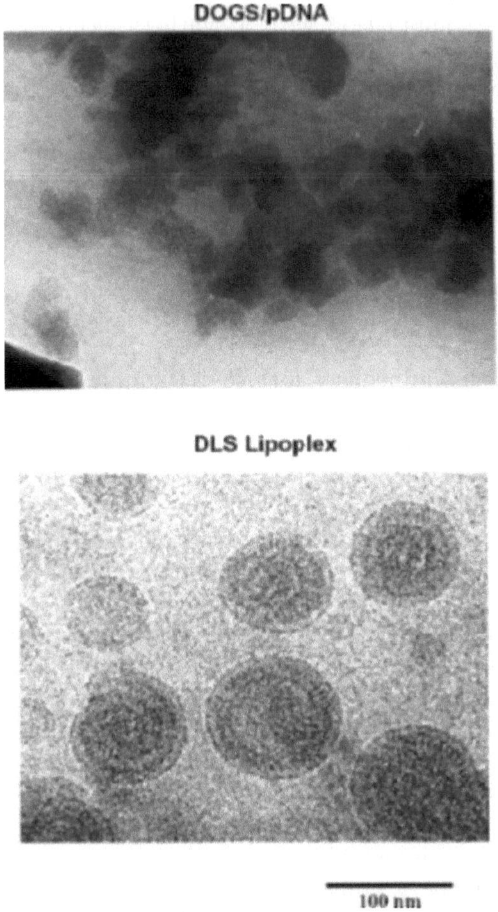

Fig. 2. Cryo-EM observation of DOGS/pDNA and DLS lipoplex

The relative width of X-ray diffraction suggested highly ordered spherulite-like particles made of at least a dozen layers. Thus SAXS clearly confirmed the multilamellar structure of Lx and the periodicity observed with cryoEM [15].

The lamellar symmetry in Lx structure has been previously demonstrated by means of Small Angle X-ray Scattering (SAXS) [18, 19]. Radler et al [18] and Lasic et al [19] showed the multilamellar nature of Lx and a periodicity of 6.5 nm. We could not reliably detect a peak corresponding to the DNA sandwiched between lipid bilayers, as previously observed [18, 19]. This could be due to the beam intensity that did not allow for the observation of the weak DNA-DNA correlation peak; or it could also suggest that the DNA chains are not arranged in a parallel array between lipid layers as assumed by these authors.

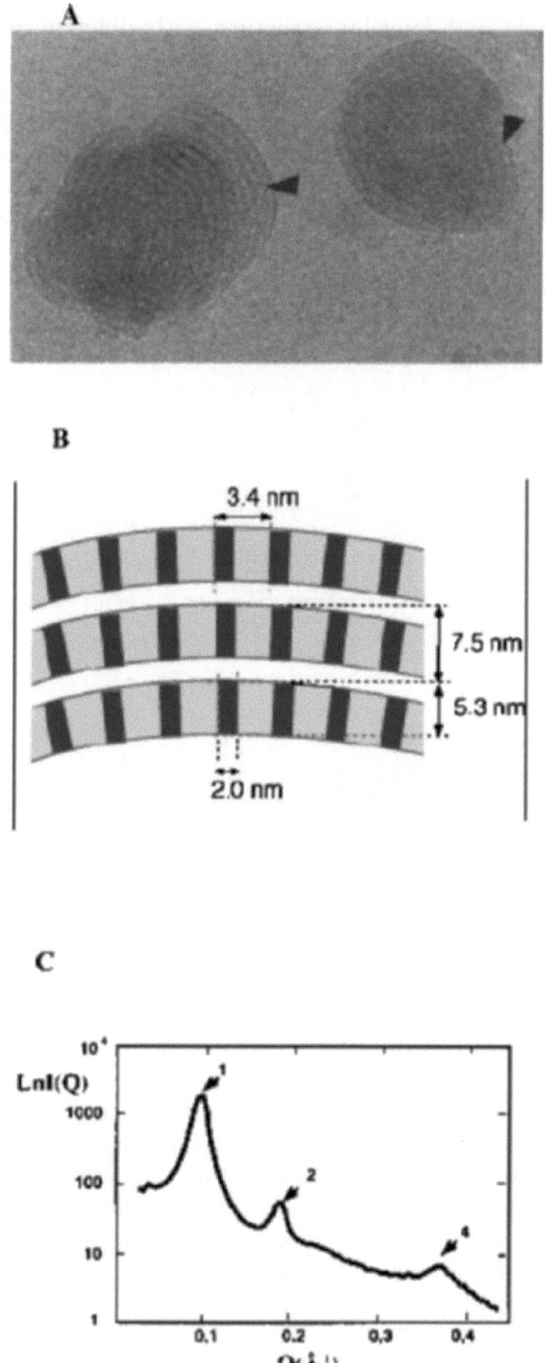

Fig. 3. Analysis of the ultrastructure of Lx as determined by cryo-EM (A,B) and SAXS (C) analysis [15]

This concentric and lamellar structure with different packing regimes was also observed by cryoEM when using linear dsDNA, ssDNA, RNA and oligodeoxynucleotides (Figure 4) [15] .We can obtain a similar structural morphology in Lx formulated with all the nucleic acids tested despite of highly different structures and sizes as with pDNA (circular supercoiled DNA of 10.4 kbp, PM 6,870,600) and ODN (linear single strand DNA of 30 bases, PM=9900). SAXS analysis confirmed the crystalline phase nature of the complexes. However, there are some discrepancies in the details of these structures especially in the periodicity number order (data not shown).

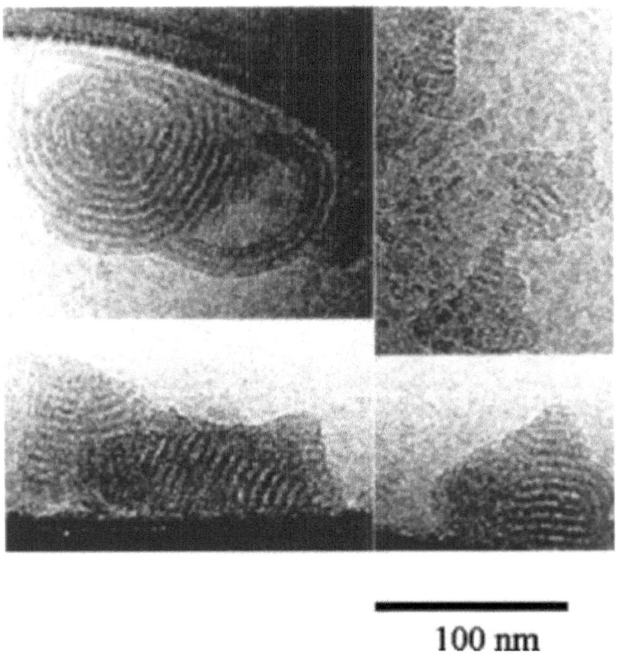

Fig. 4. Cryo-EM analysis of Lx formed with oligodeoxynucleotides [15]

Dynamic light scattering examination [20] indicated a pDNA/Lx mean size of 254 nm (width: 108 nm) and exposed a monomodal population with a polydispersity of 0.193 [15]. As presented in Table 1, Lx formed with other DNA types, presented the same particles characteristics. Mean size may significantly vary but not to a high extent (184 to 254 nm). Polydispersity values are < 0.2 and attest for every preparation of the monodisperse nature of the colloidal suspension (Table 1).

Table 1. Particle Size analysis by dynamic light scattering: Long linear double strand DNA is the T4 phage DNA (169,372 base pairs). Linear single strand DNA (M13rmp8, 7,229 bases) is the DNA from M13 bacteriophage derivatives. Linear double strand DNA (M13mp8; 7229 base pairs) is the intracellular replicative form of the M13mp8 phage produced by chronic infection in E. Coli (Sigma, St Quentin, France). RNA (ribonucleotide) is the 18S and 28S ribosomal RNA from calf liver (2000 and 5300 bases). ODN is a 27 bases oligodeoxynucleotides. (Xba I, Promega, Wis.).

Nucleic acid	Size (polydispersity)
Circular double strand DNA	254 nm (0.193)
Linear double strand DNA	221 nm (0.048)
Long linear double strand DNA	224 nm (0.198)
Linear single strand DNA	261 nm (0.113)
RNA	188 nm (0.172)
Oligodeoxynucleotides	184 nm (0.045)

CryoEM examination of T4 phage DNA packed either in T4 capsides or in lipid particles showed similar patterns (Figure 5 and 6). SAXS suggested a hexagonal phase in Lx-T4 DNA. Most of T4 DNA/Lx particles exhibit a high majority of punctuate array (Figure 5). Conversely, spherulite/concentric motif are the majority specie in Lx-pDNA cryoEM images corroborating with the lamellarity detected by SAXS. As punctate and multilamellar structure were associated in some Lx particles it might be possible that both liquid cristalline and hexagonal coexist in the same particle. T4 DNA is more than 16 times longer than the pDNA used in this study. Our results indicate that both lamellar and hexagonal phases may coexist in the same Lx preparation or particle and that transition between both phases may depend upon equilibrium influenced by type and length of the DNA used.

Lx is a multicomponent system governed by a combination of interactions caused by charge neutralization [22]. At critical concentrations and charge densities, liposome-induced DNA collapse and DNA-dependent liposome fusion are initiated28. We suggest that these phenomena induce thermodynamic forces towards compaction of the DNA macromolecule in the complex through a concentric winding where DNA is adsorbed on to the cationic head groups of the lipid bilayers.

4 Discussion and Perspectives

In light of these observations we suggest that those Lx supramoleculaire assemblies are useful tool to basic research for studying: (i), DNA condensation/decondensation phenomena in cell; (ii), DNA prebiotic assembling; and for applied research regarding development of DNA synthetic vectors.

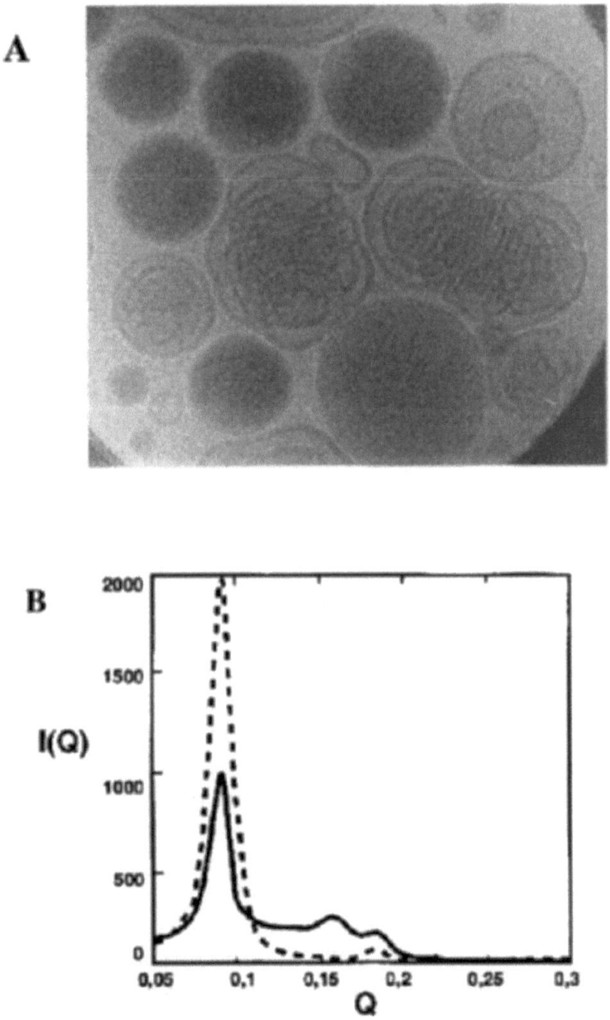

Fig. 5. Cryo-EM and SAXS analysis of Lx formed with T4 phage virus [15]

Lx are developed for therapeutic gene transfer and requirements regarding homogeneity and reproducibility are high when considering clinical use. Elucidating their ultrastructure and the mechanisms involved will provide crucial information towards better controlling manufacturing procedure.

Lx specific supramolecular organization is the result of thermodynamic forces, which cause compaction to occur through concentric winding of DNA in a liquid crystalline phase. Lipid fusion and cooperative DNA collapse processes are initiated at critical concentrations and charge densities [22]. Those parameters must be clearly delineated to define an optimized preparation. In addition, the more the supramolecular assemblies are homogeneous in liquid

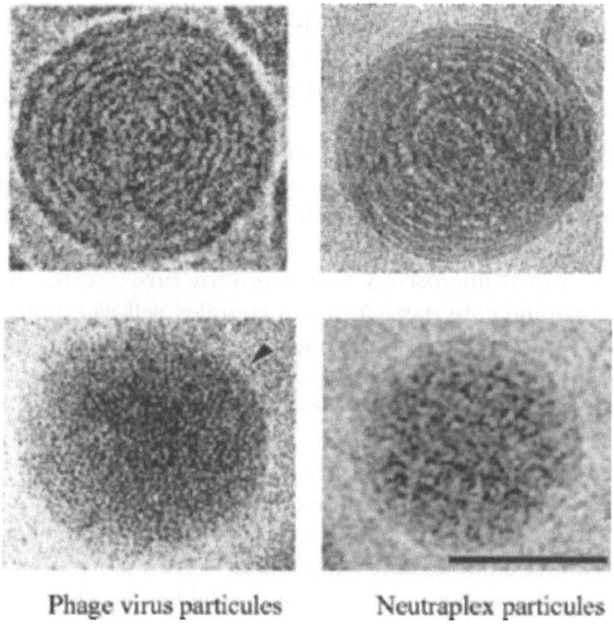

Fig. 6. Biomimetism of Lx particles and transition between multilamellar (upper images) and hexagonal (bottom images) phases. Bar=50 nm

crystal structure, the more stable and efficient for gene transfer they are (data not shown). Results suggest that transition between lamellar and hexagonal phase in Lx particles may exist. To date, we do not know whether particles of one of this phase leads to a better stability or gene transfer efficacy. This specific question is under investigation in our laboratory.

In the Lx formed here neutralization of DNA charges is due to the presence of cationic spermidine containing lipids (DOGS). The side-by-side alignment of DNA in a liquid crystalline phase may be further promoted by spermine [23]. Tight supercoiling structures can be obtained in vivo by polycationic molecules or proteins such as spermine, spermidine, histones or histone-like proteins [20, 24].

Polyamine spermine or spermidine are antique and ubiquitous compounds which stabilize the DNA double helix in vivo and which, consequently, were intensively studied as a valuable model system for studying DNA organization in biological structures [25].

DNA such as pDNA used in this study is recombinant, thus purely "natural". Consequently, DNA/cationic lipids complexes might be considered as hemi-synthetic or hemi-natural systems. It was established that DNA has an identical reactivity along his molecule chain. Ghirlando et al [26] indicated that DNA segments as short as 20 bp were able to condense in micellar aggregates and, in general, condensation of DNA by multivalent cations seems due to mechanisms independent from the length of the individual DNA molecules [27]. Thus, use of short chain or sequence might be representative in some conditions to the micro-(de)condensation occurring during replication, transcription or splicing.

In the nucleus DNA is wound up onto successive nucleosome core particles, forming a "beads on a string" complex. It has been an open question whether nuclear structure is affected by these same packing considerations. In a technically challenging work, Livolant and Leforestier [28] employed a combination of optical microscopy and freeze-fracture electron microscopy to show that nucleosome core particles form complex self-assembled structures.

Our results confirm that DNA condensation in liquid crystal complexes was insensitive to DNA size when using SUV, as previously found when using micelles [26, 29]. As a consequence, elucidation of the ultrastructure of the Lx formed here might provide information in the DNA condensation/decondensation phenomena in cell.

The cryoEM images of punctuated Lx particles formed with T4 DNA that we observe are strikingly similar to the images of complete tail- deletion mutant of T4 [30] (Figure 4) and T7 [31] phages. T4 phage mutants produce a normal mature head but no tail, which enables a better observation of DNA organization33,34. These T4-related phage particles showed a spherical shape and an average diameter of 80 nm33. Recent investigations performed by using tailless mutants, indicated DNA packing domains in viral particles. It is noteworthy that T7 tail-deletion mutants exhibit in cryoEM images a concentric ring motif as well as a punctuated motif as observed for Lx-pDNA [31].

Cerritelli et al30 observed T7 tail-deletion mutants by CryoEM which revealed a concentric ring motif "along the axis via which DNA is packaged and a punctuated pattern in side views". They suggest that T7 DNA is spooled around this axis in coaxial shells in a quasi-crystalline packing predicting the viability of such a model for other phage heads.

Interestingly, another report demonstrated thatT5 phage appears condensed in lamellar and concentric winding form when released in proteoliposomes following virus attachment [32]. DNA of bacteria under stress self organize in crystalline phase by forming complexes with cationic peptides [33]. Three-dimensional liquid-crystalline arrangements of biomolecules have been known since the pioneering work of Bouligand [34,35]. Self organization leads to the possible phase transition in live cells.

Furthermore, organization of such nucleotidic supramolecular assemblies is relevant for prebiotic chemistry [36]. As postulated by Lipowsky [37], cellular life might begin with a membranar vesicle containing just the right mixture of polymers. In light of the numerous observations made on DNA packaging in a natural setting or by various organic or inorganic condensing agents, our data corroborate the notion that a parallel between natural and synthetic DNA compaction can be drawn. We demonstrated for the first time a structural similarity between a synthetic supramolecular organization and viruses.

Synthetic gene delivery particles must obey to reversible DNA condensation just as viruses do. Ultrastructure of Lx appears as a complex system and could constitute a valuable model system for studying nucleotidic organization in biological structures.

We strongly believe that the Lx complexes correspond to a ubiquitous supramolecular self organization of DNA and could be considered as an emergent complex system [38]. Self-organization in general, refers to the various mechanisms by which pattern, structure and order emerge spontaneously in complex systems.

Self-organization is a process in which pattern emerges at the global (collective) level by means of interactions among components of the system at the individual level. [39]. What makes a system self-organized is that the collective patterns and structures arise without: (i), the guidance of well-informed leaders; (ii), any set of predetermined blueprints; (iii), recipes or templates to explicitly specify the pattern [39]. Instead, structure is as an emergent property of the dynamic interactions among components in the system.

Self-organization appears to be an important mechanism useful for explaining pattern and structure in physical, chemical and biological systems. Lx appear as one DNA supramolecular self-organization model and elucidation of their ultrastructure is under investigation.

References

1. Cristiano RJ, Smith LC, and Woo SL (1993) Hepatic gene therapy: adenovirus enhancement of receptor-mediated gene delivery and expression in primary hepatocytes Proc Natl Acad Scí U S A 90: 2122-2126
2. Felgner PL, Barenholz Y, Behr JP, Cheng SH, Cullis P, Huang L, Jessee JA, Seymour L, Szoka F, Thierry AR, Wagner E, and Wu G (1997) Nomenclature for synthetic gene delivery systems Hum Gene Ther 8: 511-512
3. Felgner PL, Gadek TR, Holm M, Roman R, Chan HW, Wenz M, Northrop JP, Ringold GM (1987) and Danielsen M Lipofection : a highly efficient, lipid-mediated DNA-transfection procedure Proc Natl Acad Sci U S A 84: 7413-7417
4. Nabel EG, Plautz G, and Nabel GJ (1990) Site-specific gene expression in vivo by direct gene transfer into the arterial wall Science 249: 1285-1288
5. Zhu N, Liggitt D, Liu Y, and Debs R. (1993) Systemic gene expression after intravenous DNA delivery into adult mice Science 261: 209-211
6. Thierry AR, Lunardi-Iskandar Y, Bryant JL, Rabinovich P, Gallo RC, and Mahan LC (1995) Systemic gene therapy: biodistribution and long-term expression of a transgene in mice Proc Natl Acad Sci U S A 92: 9742-9746
7. Thierry A.R. and Mahan,LC (1998) Advanced Gene Delivery, A.Rolland, Harwood Press, Amsterdam
8. Thierry AR, Vives E, Richard JP, Prevot P, Martinand-Mari C, Robbins I, and Lebleu B (2003) Cellular uptake and intracellular fate of antisense oligonucleotides. Curr Opin Mol Ther 5: 133-138
9. Lavigne C and Thierry AR (1997) Enhanced antisense inhibition of human immunodeficiency virus type 1 in cell cultures by DLS delivery system. Biochem Biophys Res Commun 237: 566-571

10. Lavigne C, Yelle J, Sauve G, and Thierry AR (2002) Is antisense an appropriate nomenclature or design for oligodeoxynucleotides aimed at the inhibition of HIV-1 replication? AAPS PharmSci 4: E9
11. Lavigne C, Lunardi-Iskandar Y, Lebleu B, and Thierry AR (2004) Cationic liposomes/lipids for oligonucleotide delivery: application to the inhibition of tumorigenicity of Kaposi's sarcoma by vascular endothelial growth factor antisense oligodeoxynucleotides. Methods Enzymol 387: 189-210
12. Thierry AR, Rabinovich P, Peng B, Mahan LC, Bryant JL, and Gallo RC (1997) Characterization of liposome-mediated gene delivery: expression, stability and pharmacokinetics of plasmid DNA. Gene Ther 4: 226-237
13. Aksentijevich I, Pastan I, Lunardi-Iskandar Y, Gallo RC, Gottesman MM, and Thierry AR (1996) In vitro and in vivo liposome-mediated gene transfer leads to human MDR1 expression in mouse bone marrow progenitor cells. Hum Gene Ther 7: 1111-1122
14. Baudard M, Flotte TR, Aran JM, Thierry AR, Pastan I, Pang MG, Kearns WG, and Gottesman MM (1996) Expression of the human multidrug resistance and glucocerebrosidase cDNAs from adeno-associated vectors: efficient promoter activity of AAV sequences and in vivo delivery via liposomes. Hum Gene Ther 7: 1309-1322
15. Schmutz M, Durand D, Debin A, Palvadeau Y, Etienne A, and Thierry AR (1999) DNA packing in stable lipid complexes designed for gene transfer imitates DNA compaction in bacteriophage. Proc Natl Acad Sci U S A 96: 12293-12298
16. Tavitian B, Marzabal S, Boutet V, Kuhnast B, Terrazzino S, Moynier M, Dolle F, Deverre JR, and Thierry AR (2002) Characterization of a synthetic anionic vector for oligonucleotide delivery using in vivo whole body dynamic imaging. Pharm Res 19: 367-376
17. Behr JP, Demeneix B, Loeffler JP, and Perez-Mutul J (1989) Efficient gene transfer into mammalian primary endocrine cells with lipopolyamine-coated DNA. Proc Natl Acad Sci U S A 86: 6982-6986
18. Radler JO, Koltover I, Salditt T, and Safinya CR (1997) Structure of DNA-cationic liposome complexes: DNA intercalation in multilamellar membranes in distinct interhelical packing regimes. Science 275: 810-814
19. Lasic DD (1997) Colloid chemistry. Liposomes within liposomes Nature 387: 26-27
20. Luger K, Mader AW, Richmond RK, Sargent DF, and Richmond TJ (1997) Crystal structure of the nucleosome core particle at 2.8 A resolution. Nature 389: 251-260
21. Xu Y and Szoka FC Jr (1996) Mechanism of DNA release from cationic liposome/DNA complexes used in cell transfection. Biochemistry 35: 5616-5623
22. Safinya CR (2001) Structures of lipid-DNA complexes: supramolecular assembly and gene delivery. Curr Opin Struct Biol 11: 440-448
23. Bloomfield VA (1991) Condensation of DNA by multivalent cations: considerations on mechanism. Biopolymers 31: 1471-1481
24. Marx KA and Ruben GC (1983) Evidence for hydrated spermidine-calf thymus DNA toruses organized by circumferential DNA wrapping. Nucleic Acids Res 11: 1839-1854
25. Sikorav JL, Pelta J, and Livolant F (1994) A liquid crystalline phase in spermidine-condensed DNA. Biophys J 67: 1387-1392

26. Ghirlando R, Wachtel EJ, Arad T, and Minsky A (1992) DNA packaging induced by micellar aggregates: a novel in vitro DNA condensation system. Biochemistry 31: 7110-7119
27. Reich Z, Ghirlando R, and Minsky A (1991) Secondary conformational polymorphism of nucleic acids as a possible functional link between cellular parameters and DNA packaging processes. Biochemistry 30: 7828-7836
28. Livolant F and Leforestier A. Chiral (2000) discotic columnar germs of nucleosome core particles. Biophys J 78: 2716-2729
29. Gershon H, Ghirlando R, Guttman SB, and Minsky A (1993) Mode of formation and structural features of DNA-cationic liposome complexes used for transfection. Biochemistry 32: 7143-7151
30. Lepault J, Dubochet J, Baschong W, and Kellenberger E (1987) Organization of double-stranded DNA in bacteriophages: a study by cryo-electron microscopy of vitrified samples. Embo J 6: 1507-1512
31. Cerritelli ME, Cheng N, Rosenberg AH, McPherson CE, Booy FP, and Steven AC (1997) Encapsidated conformation of bacteriophage T7 DNA. Cell 91: 271-280
32. Lambert O, Letellier L, Gelbart WM, and Rigaud JL (2000) DNA delivery by phage as a strategy for encapsulating toroidal condensates of arbitrary size into liposomes. Proc Natl Acad Sci U S A 97: 7248-7253
33. Wolf SG, Frenkiel D, Arad T, Finkel SE, Kolter R, and Minsky A (1999) DNA protection by stress-induced biocrystallization Nature 400: 83-85
34. Bouligand Y and Norris V (2001) Chromosome separation and segregation in dinoflagellates and bacteria may depend on liquid crystalline states. Biochimie 83: 187-192
35. Norris V, Alexandre S, Bouligand Y, Cellier D, Demarty M, Grehan G, Gouesbet G, Guespin J, Insinna E, Le Sceller L, Maheu B, Monnier C, Grant N, Onoda T, Orange N, Oshima A, Picton L, Polaert H, Ripoll C, Thellier M, Valleton JM, Verdus MC, Vincent JC, White G, and Wiggins P (1999) Hypothesis: hyperstructures regulate bacterial structure and the cell cycle. Biochimie 81: 915-920
36. Baeza I, Ibanez M, Wong C, Chavez P, Gariglio P, and Oro J (1991) Possible prebiotic significance of polyamines in the condensation, protection, encapsulation, and biological properties of DNA. Orig Life Evol Biosph 21: 225-242
37. Lipowsky R (1991) The conformation of membranes. Nature 349: 475-481
38. Van Regenmortel MH (2004) Biological complexity emerges from the ashes of genetic reductionism. J Mol Recognit 17: 145-148
39. Carl Anderson, Scott Camazine:
http://www.scottcamazine.com/personal/research/index.htm

Part III

Dynamic Systems & Synchronization

Slow Manifold of a Neuronal Bursting Model

Jean-Marc Ginoux and Bruno Rossetto

PROTEE Laboratory, Université du Sud
B.P. 20132
83957, La Garde Cedex, France
ginoux@univ-tln.fr, rossetto@univ-tln.fr

Summary. Comparing neuronal bursting models (NBM) with slow-fast autonomous dynamical systems (S-FADS), it appears that the specific features of a (NBM) do not allow a determination of the analytical slow manifold equation with the singular approximation method. So, a new approach based on Differential Geometry, generally used for (S-FADS), is proposed. Adapted to (NBM), this new method provides three equivalent manners of determination of the analytical slow manifold equation. Application is made for the three-variables model of neuronal bursting elaborated by Hindmarsh and Rose which is one of the most used mathematical representation of the widespread phenomenon of oscillatory burst discharges that occur in real neuronal cells.

Key words: differential geometry, curvature, torsion, slow-fast dynamics, neuronal bursting models.

1 Slow-Fast Autonomous Dynamical Systems, Neuronal Bursting Models

1.1 Dynamical Systems

In the following we consider a system of differential equations defined in a compact E included in $\mathbb{R}$:

$$\frac{d\mathbf{X}}{dt} = \Im(\mathbf{X}) \tag{1}$$

with

$$\mathbf{X} = [x_1, x_2, ..., x_n]^t \in E \subset \mathbb{R}^n$$

and

$$\Im(\mathbf{X}) = [f_1(\mathbf{X}), f_2(\mathbf{X}), ..., f_n(\mathbf{X})]^t \in E \subset \mathbb{R}^n$$

The vector $\Im(\mathbf{X})$ defines a velocity vector field in E whose components f_i which are supposed to be continuous and infinitely differentiable with respect to all x_i and t, i.e., are C^∞ functions in E and with values included in $\mathbb{R}$, satisfy the assumptions of the Cauchy-Lipschitz theorem. For more details, see for example [2]. A solution of this system is an integral curve $\mathbf{X}(t)$ tangent to $\Im$ whose values define the *states* of the *dynamical system* described by the Eq. (1). Since none of the components f_i of the velocity vector field depends here explicitly on time, the system is said to be *autonomous*.

1.2 Slow-Fast Autonomous Dynamical System (S-FADS)

A (S-FADS) is a *dynamical system* defined under the same conditions as previously but comprising a small multiplicative parameter ε in one or several components of its velocity vector field:

$$\frac{\mathrm{d}\mathbf{X}}{\mathrm{d}t} = \Im_\varepsilon(\mathbf{X}) \tag{2}$$

with

$$\Im_\varepsilon(\mathbf{X}) = \left[\frac{1}{\varepsilon}f_1(\mathbf{X}), f_2(\mathbf{X}), ..., f_n(\mathbf{X})\right]^t \in E \subset \mathbb{R}^n$$

$$0 < \varepsilon \ll 1$$

The functional jacobian of a (S-FADS) defined by (2) has an eigenvalue called "fast", i.e., great on a large domain of the phase space. Thus, a "fast" eigenvalue is expressed like a polynomial of valuation -1 in ε and the eigenmode which is associated with this "fast" eigenvalue is said:

- "evanescent" if it is negative,
- "dominant" if it is positive.

The other eigenvalues called "slow" are expressed like a polynomial of valuation 0 in ε.

1.3 Neuronal Bursting Models (NBM)

A (NBM) is a *dynamical system* defined under the same conditions as previously but comprising a large multiplicative parameter ε^{-1} in one component of its velocity vector field:

$$\frac{\mathrm{d}\mathbf{X}}{\mathrm{d}t} = \Im_\varepsilon(\mathbf{X}) \tag{3}$$

with

$$\Im_{\varepsilon}(\mathbf{X}) = [f_1(\mathbf{X}), f_2(\mathbf{X}), ..., \varepsilon f_n(\mathbf{X})]^t \in E \subset \mathbb{R}^n$$

$$0 < \varepsilon \ll 1$$

The presence of the multiplicative parameter ε^{-1} in one of the components of the velocity vector field makes it possible to consider the system (3) as a kind of *slow-fast* autonomous dynamical system (S-FADS). So, it possesses a *slow manifold*, the equation of which may be determined. But, paradoxically, this model is not *slow-fast* in the sense defined previously. A comparison between three-dimensional (S-FADS) and (NBM) presented in Table (1) emphasizes their differences. The dot $(\cdot)$ represents the derivative with respect to time and $\varepsilon \ll 1$.

Table 1. Comparison between (S-FADS) and (NBM)

(S-FADS) vs (NBM)	
$\frac{d\mathbf{X}}{dt}\begin{pmatrix}\dot{x}\\ \dot{y}\\ \dot{z}\end{pmatrix} = \Im_{\varepsilon}\begin{pmatrix}\frac{1}{\varepsilon}f(x,y,z)\\ g(x,y,z)\\ h(x,y,z)\end{pmatrix}$	$\frac{d\mathbf{X}}{dt}\begin{pmatrix}\dot{x}\\ \dot{y}\\ \dot{z}\end{pmatrix} = \Im_{\varepsilon}\begin{pmatrix}f(x,y,z)\\ g(x,y,z)\\ \varepsilon h(x,y,z)\end{pmatrix}$
$\frac{d\mathbf{X}}{dt}\begin{pmatrix}\dot{x}\\ \dot{y}\\ \dot{z}\end{pmatrix} = \Im_{\varepsilon}\begin{pmatrix}fast\\ slow\\ slow\end{pmatrix}$	$\frac{d\mathbf{X}}{dt}\begin{pmatrix}\dot{x}\\ \dot{y}\\ \dot{z}\end{pmatrix} = \Im_{\varepsilon}\begin{pmatrix}fast\\ fast\\ slow\end{pmatrix}$

2 Analytical Slow Manifold Equation

There are many methods of determination of the analytical equation of the slow manifold. The classical one based on the singular perturbations theory [1] is the so-called *singular approximation method.* But, in the specific case of a (NBM), one of the hypothesis of the Tihonov's theorem is not checked since the *fast dynamics* of the *singular approximation* has a periodic solution. Thus, another approach developed in [4] which consist in using *Differential Geometry* formalism may be used.

2.1 Singular Approximation Method

The *singular approximation* of the *fast* dynamics constitutes a quite good approach since the third component of the velocity is very weak and so, z is nearly constant along the periodic solution. In dimension three the system (3) can be written as a system of differential equations defined in a compact E included in $\mathbb{R}$:

$$\frac{d\mathbf{X}}{dt} = \begin{pmatrix} \frac{dx}{dt} \\ \frac{dy}{dt} \\ \frac{dz}{dt} \end{pmatrix} = \Im_\varepsilon \begin{pmatrix} f(x,y,z) \\ g(x,y,z) \\ \varepsilon h(x,y,z) \end{pmatrix}$$

On the one hand, since the system (3) can be considered as a (S-FADS), the *slow* dynamics of the *singular approximation* is given by:

$$\begin{cases} f(x,y,z) = 0 \\ g(x,y,z) = 0 \end{cases} \tag{4}$$

The resolution of this reduced system composed of the two first equations of the right hand side of (3) provides a one-dimensional *singular manifold*, called *singular curve*. This curve doesn't play any role in the construction of the periodic solution. But we will see that there exists all the more a *slow dynamics*. On the other hands, it presents a *fast* dynamics which can be given while posing the following change:

$$\tau = \varepsilon t \Leftrightarrow \frac{d}{dt} = \varepsilon \frac{d}{d\tau}$$

The system (3) may be re-written as:

$$\frac{d\mathbf{X}}{d\tau} = \begin{pmatrix} \frac{dx}{d\tau} \\ \frac{dy}{d\tau} \\ \frac{dz}{d\tau} \end{pmatrix} = \Im_\varepsilon \begin{pmatrix} \varepsilon^{-1} f(x,y,z) \\ \varepsilon^{-1} g(x,y,z) \\ h(x,y,z) \end{pmatrix} \tag{5}$$

So, the *fast* dynamics of the *singular approximation* is provided by the study of the reduced system composed of the two first equations of the right hand side of (5).

$$\left.\frac{d\mathbf{X}}{d\tau}\right|_{fast} = \begin{pmatrix} \frac{dx}{d\tau} \\ \frac{dy}{d\tau} \end{pmatrix} = \Im_\varepsilon \begin{pmatrix} \varepsilon^{-1} f(x,y,z^*) \\ \varepsilon^{-1} g(x,y,z^*) \end{pmatrix} \tag{6}$$

Each point of the *singular curve* is a singular point of the *singular approximation* of the *fast* dynamics. For the z value for which there is a periodic solution, the *singular approximation* exhibits an unstable focus, attractive with respect to the *slow* eigendirection.

2.2 Differential Geometry Formalism

Now let us consider a three-dimensional system defined by (3) and let's define the instantaneous acceleration vector of the *trajectory curve* $\mathbf{X}(t)$. Since the functions f_i are supposed to be C^∞ functions in a compact E included in

$\mathbb{R}$, it is possible to calculate the total derivative of the vector field $\Im_\varepsilon$. As the instantaneous vector function $\mathbf{V}(t)$ of the scalar variable t represents the velocity vector of the mobile M at the instant t, the total derivative of $\mathbf{V}(t)$ is the vector function $\gamma(t)$ of the scalar variable t which represents the instantaneous acceleration vector of the mobile M at the instant t. It is noted:

$$\gamma(t) = \frac{\mathrm{d}\mathbf{V}(t)}{\mathrm{d}t} \tag{7}$$

Even if neuronal bursting models are not exactly slow-fast autonomous dynamical systems, the new approach of determining the *slow manifold* equation developed in [4] may still be applied. This method is using *Differential Geometry* properties such as *curvature* and *torsion* of the *trajectory curve* $\mathbf{X}(t)$, integral of *dynamical systems* to provide their *slow manifold* equation.

Proposition 1. *The location of the points where the local torsion of the trajectory curves integral of a dynamical system defined by (3) vanishes, provides the analytical equation of the slow manifold associated with this system.*

$$\frac{1}{\Im} = -\frac{\dot{\gamma}\cdot(\gamma\times\mathbf{V})}{\|\gamma\times\mathbf{V}\|^2} = 0 \Leftrightarrow \dot{\gamma}\cdot(\gamma\times\mathbf{V}) = 0 \tag{8}$$

Thus, this equation represents the slow manifold of a neuronal bursting model defined by (3).

The particular features of neuronal bursting models (3) will lead to a simplification of this proposition (1). Due to the presence of the small multiplicative parameter ε in the third components of its velocity vector field, instantaneous velocity vector $\mathbf{V}(t)$ and instantaneous acceleration vector $\gamma(t)$ of the model (3) may be written:

$$\mathbf{V}\begin{pmatrix}\dot{x}\\ \dot{y}\\ \dot{z}\end{pmatrix} = \Im_\varepsilon\begin{pmatrix}O(\varepsilon^0)\\ O(\varepsilon^0)\\ O(\varepsilon^1)\end{pmatrix} \tag{9}$$

and

$$\gamma\begin{pmatrix}\ddot{x}\\ \ddot{y}\\ \ddot{z}\end{pmatrix} = \frac{\mathrm{d}\Im_\varepsilon}{\mathrm{d}t}\begin{pmatrix}O(\varepsilon^1)\\ O(\varepsilon^1)\\ O(\varepsilon^2)\end{pmatrix} \tag{10}$$

where $O(\varepsilon^n)$ is a polynomial of n degree in ε

Then, it is possible to express the vector product $\mathbf{V}\times\gamma$ as:

$$\mathbf{V}\times\gamma = \begin{pmatrix}\dot{y}\ddot{z}-\ddot{y}\dot{z}\\ \ddot{x}\dot{z}-\dot{x}\ddot{z}\\ \dot{x}\ddot{y}-\ddot{x}\dot{y}\end{pmatrix} \tag{11}$$

Taking into account what precedes (9, 10), it follows that:

$$\mathbf{V} \times \gamma = \begin{pmatrix} O\left(\varepsilon^2\right) \\ O\left(\varepsilon^2\right) \\ O\left(\varepsilon^1\right) \end{pmatrix} \tag{12}$$

So, it is obvious that since ε is a small parameter, this vector product may be written:

$$\mathbf{V} \times \gamma \approx \begin{pmatrix} 0 \\ 0 \\ O\left(\varepsilon^1\right) \end{pmatrix} \tag{13}$$

Then, it appears that if the third component of this vector product vanishes when both instantaneous velocity vector $\mathbf{V}(t)$ and instantaneous acceleration vector $\gamma(t)$ are collinear. This result is particular to this kind of model which presents a small multiplicative parameter in one of the right-hand-side component of the velocity vector field and makes it possible to simplify the previous proposition (1).

Proposition 2. *The location of the points where the instantaneous velocity vector* $\mathbf{V}(t)$ *and instantaneous acceleration vector* $\gamma(t)$ *of a neuronal bursting model defined by (3) are collinear provides the analytical equation of the slow manifold associated with this dynamical system.*

$$\mathbf{V} \times \gamma = \mathbf{0} \Leftrightarrow \dot{x}\ddot{y} - \ddot{x}\dot{y} = 0 \tag{14}$$

Another method of determining the *slow manifold* equation proposed in [14] consists in considering the so-called *tangent linear system approximation*. Then, a coplanarity condition between the instantaneous velocity vector $\mathbf{V}(t)$ and the *slow* eigenvectors of the *tangent linear system* gives the *slow manifold* equation.

$$\mathbf{V}.\left(\mathbf{Y}_{\lambda_2} \times \mathbf{Y}_{\lambda_3}\right) = 0 \tag{15}$$

where $\mathbf{Y}_{\lambda_i}$ represent the *slow* eigenvectors of the *tangent linear system*. But, if these eigenvectors are complex the *slow manifold* plot may be interrupted. So, in order to avoid such inconvenience, this equation has been multiplied by two conjugate equations obtained by circular permutations.

$$\left[\mathbf{V} \cdot \left(\mathbf{Y}_{\lambda_2} \times \mathbf{Y}_{\lambda_3}\right)\right] \cdot \left[\mathbf{V} \cdot \left(\mathbf{Y}_{\lambda_1} \times \mathbf{Y}_{\lambda_2}\right)\right] \cdot \left[\mathbf{V} \cdot \left(\mathbf{Y}_{\lambda_1} \times \mathbf{Y}_{\lambda_3}\right)\right] = 0$$

It has been established in [4] that this real analytical *slow manifold* equation can be written:

$$(J^2\mathbf{V}) \cdot (\gamma \times \mathbf{V}) = 0 \tag{16}$$

since the the *tangent linear system approximation* method implies to suppose that the functional jacobian matrix is stationary. That is to say

$$\frac{\mathrm{d}J}{\mathrm{d}t} = 0$$

and so,

$$\dot{\gamma} = J\frac{\mathrm{d}\mathbf{V}}{\mathrm{d}t} + \frac{\mathrm{d}J}{\mathrm{d}t}\mathbf{V} = J\gamma + \frac{\mathrm{d}J}{\mathrm{d}t}\mathbf{V} = J^2\mathbf{V} + \frac{\mathrm{d}J}{\mathrm{d}t}\mathbf{V} \approx J^2\mathbf{V}$$

Proposition 3. *The coplanarity condition (15) between the instantaneous velocity vector and the slow eigenvectors of the tangent linear system transformed into the real analytical equation (16) provides the slow manifold equation of a neuronal bursting model defined by (3).*

3 Application to a Neuronal Bursting Model

The transmission of nervous impulse is secured in the brain by action potentials. Their generation and their rhythmic behaviour are linked to the opening and closing of selected classes of ionic channels. The membrane potential of neurons can be modified by acting on a combination of different ionic mechanisms. Starting from the seminal works of Hodgkin-Huxley [7, 11] and FitzHugh-Nagumo [3, 12], the Hindmarsh-Rose [6, 13] model consists of three variables: x, the membrane potential, y, an intrinsic current and z, a *slow* adaptation current.

3.1 Hindmarsh-Rose Model of Bursting Neurons

$$\begin{cases} \frac{\mathrm{d}x}{\mathrm{d}t} = y - f(x) - z + I \\ \frac{\mathrm{d}y}{\mathrm{d}t} = g(x) - y \\ \frac{\mathrm{d}z}{\mathrm{d}t} = \varepsilon\,(h(x) - z) \end{cases} \tag{17}$$

I represents the applied current, $f(x) = ax^3 - bx^2$ and $g(x) = c - dx^2$ are respectively cubic and quadratic functions which have been experimentally deduced [5]. ε is the time scale of the slow adaptation current and $h(x) = x - x^*$ is the scale of the influence of the *slow* dynamics, which determines whether the neuron fires in a tonic or in a burst mode when it is exposed to a sustained current input and where (x^*, y^*) are the co-ordinates of the leftmost equilibrium point of the model (1) without adaptation, i.e., $I = 0$.

Parameters used for numerical simulations are:
$a = 1$, $b = 3$, $c = 1$, $d = 5$, $\varepsilon = 0.005$, $s = 4$, $x^* = \frac{-1-\sqrt{5}}{2}$ and $I = 3.25$.

While using the method proposed in the section 2 (2) it is possible to determine the analytical *slow manifold* equation of the Hindmarsh-Rose 84'model [6].

3.2 Slow Manifold of the Hindmarsh-Rose 84'model

Figure 1 presents the *slow manifold* of the Hindmarsh-Rose 84'model determined with proposition (1).

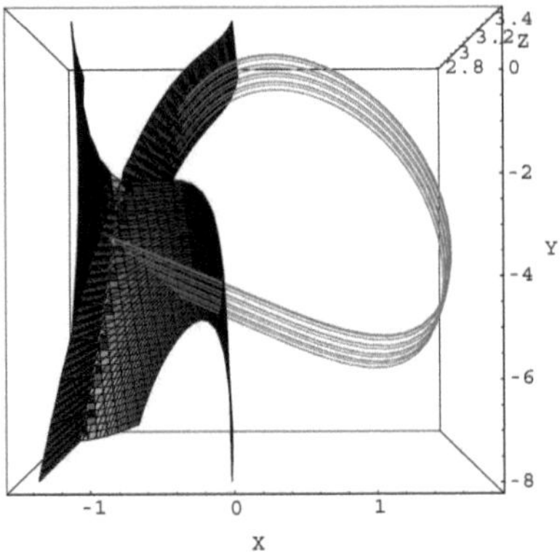

Fig. 1. *Slow manifold* of the Hindmarsh-Rose 84'model with the Proposition (1).

The *slow manifold* provided with the use of the *collinearity condition* between both instantaneous velocity vector and instantaneous acceleration vector, i.e., while using the Proposition (2) is presented in Fig. 2.

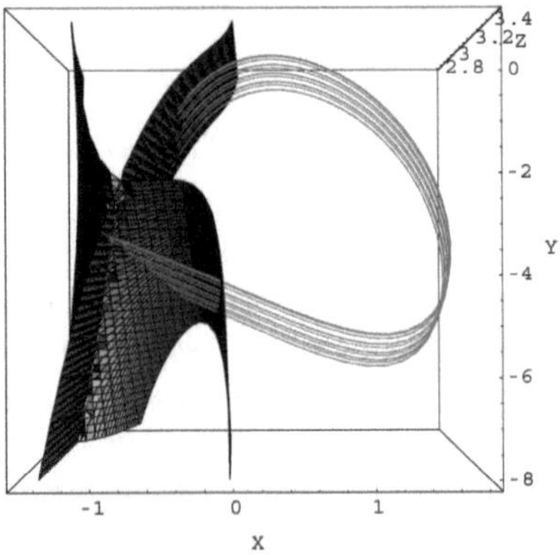

Fig. 2. *Slow manifold* of the Hindmarsh-Rose 84'model with the Proposition (2).

Figure 3 presents the *slow manifold* of the Hindmarsh-Rose 84'model obtained with the *tangent linear system approximation*, i.e., with the use of Proposition (3).

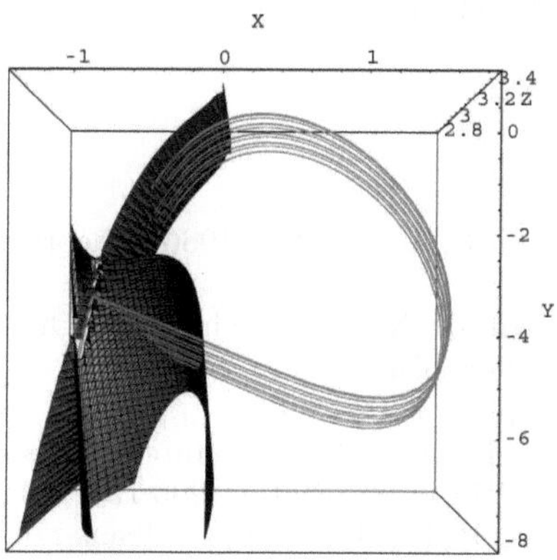

Fig. 3. *Slow manifold* of the Hindmarsh-Rose 84'model with the Proposition (3).

4 Discussion

Since in the case of neuronal bursting model (NBM) one of the Tihonov's hypothesis is not checked, the classical *singular approximation method* can not be used to determine the analytical *slow manifold* equation. In this work the application of the *Differential Geometry* formalism provides new alternative methods of determination of the *slow manifold* equation of a neuronal bursting model (NBM).

- the *torsion method*, i.e., the location of the points where the local *torsion* of the *trajectory curve*, integral of *dynamical systems* vanishes,
- the *collinearity condition* between the instantaneous velocity vector $\vec{V}$, the instantaneous acceleration vector $\vec{\gamma}$,
- the *tangent linear system approximation*, i.e., the coplanarity condition between the instantaneous velocity vector eigenvectors transformed into a real analytical equation.

The striking similarity of all figures due to the smallness of the parameter ε highlights the equivalence between all the propositions. Moreover, even if the presence of this small parameter ε in one of the right-hand-side component of the instantaneous velocity vector field of a (NBM) prevents from using the *singular approximation method*, it clarifies the Proposition (1) and transforms it into a *collinearity condition* in dimension three, i.e., Proposition (2). Comparing (S-FADS) and (NBM) in Table (1) it can be noted that in a (S-FADS) there is one fast component and two fast while in a (NBM) the situation is exactly reversed. Two fast components and one slow. So, considering (NBM) as a particular class of (S-FADS) we suggest to call (NBM) *fast-slow* instead of *slow-fast* in order to avoid any confusion. Further research should highlight other specific features of (NBM).

References

1. Andronov AA, Khaikin SE, & Vitt AA (1966) Theory of oscillators, Pergamon Press, Oxford
2. Coddington EA & Levinson N, (1955) Theory of Ordinary Differential Equations, Mac Graw Hill, New York
3. Fitzhugh R (1961) Biophys. J 1:445–466
4. Ginoux JM & Rossetto B (2006) Int. J. Bifurcations and Chaos, (in press)
5. Hindmarsh JL & Rose RM (1982) Nature 296:162–164
6. Hindmarsh JL & Rose RM (1984) Philos. Trans. Roy. Soc. London Ser. B 221:87–102
7. Hodgkin AL & Huxley AF (1952) J. Physiol. (Lond.) 116:473–96
8. Hodgkin AL & Huxley AF (1952) J. Physiol. (Lond.) 116: 449–72
9. Hodgkin AL & Huxley AF (1952) J. Physiol. (Lond.) 116: 497–506
10. Hodgkin AL & Huxley AF (1952) J. Physiol. (Lond.) 117: 500–44
11. Hodgkin AL, Huxley AF & Katz B (1952) B. Katz J. Physiol. (Lond.) 116: 424–48
12. Nagumo JS, Arimoto S & Yoshizawa S (1962) Proc. Inst. Radio Engineers 50:2061–2070
13. Rose RM & Hindmarsh JL (1985) Proc. R. Soc. Ser. B 225:161–193
14. Rossetto B, Lenzini T, Suchey G & Ramdani S (1998) Int. J. Bifurcation and Chaos, vol. 8 (11):2135-2145

Complex Emergent Properties and Chaos (De)synchronization

Aziz-Alaoui M.A.[1]

Laboratoire de Mathématiques Appliquées,
Université du Havre, BP 540, 76058 Le Havre Cedex, France
aziz.alaoui@univ-lehavre.fr

Summary. Emergent properties are typically novel and unanticipated. In this paper, using chaos synchronization tools and ideas, we demonstrate, via two examples of three-dimensional autonomous differential systems, that, by simple uni- or bi-directional coupling, regular (resp. chaotic) behaviour can emerge from chaotic (resp. regular) behaviour.

Key words: complex systems, emergence, chaos, synchronization, Lorenz-type system, predator-prey.

1 Introduction : Complexity and Emergent Properties

First of all, let us start with the citation below.

> "I think the next century (21th) will be the century of complexity", (Stephen Hawking).

But what is complexity? An extremely difficult "I know it when I see it" concept to define, see [1]. However, intuitively, complexity is usually greater in systems whose components are arranged in some intricate difficult-to-understand pattern or, in the case of a dynamical system, when the outcome of a process is difficult to predict from its initial state (sensitive dependence on initial conditions, see below). A complex system is an animate or inanimate system composed of many interacting components whose behaviour or structure is difficult to understand. Sometimes a system may be structurally complex, like a mechanical clock, but behave very simply.

While several measures of complexity have been proposed in the research literature, they all fall into two general classes:

(1) Static Complexity which addresses the question of how an object or system is put together (i.e. only pure structural informational aspects of an object).

(2) Dynamic Complexity which addresses the question of how much dynamical or computational effort is required to describe the informational content of an object or state of a system.

These two measures are clearly not equivalent. In this paper we embrace the following definition.

Complexity is a scientific theory that asserts that some systems display behavioural phenomena completely inexplicable by any conventional analysis of the systems' constituent parts.

Besides, **emergence** refers to the appearance of higher-level properties and behaviours of a system that while obviously originating from the collective dynamics of that system's components -are neither to be found in nor are directly deductable from the lower-level properties of that system. Emergent properties are properties of the 'whole' that are not possessed by any of the individual parts making up that whole. For example, an air molecule is not a cyclone, an isolated species doesn't form a food chain and an 'isolated' neuron is not conscious: emergent behaviours are typically novel and unanticipated.

Moreover, it is becoming a commonplace that, if the 20th was the century of physics, the 21st will be the century of biology, and, more specifically, mathematical biology, see [8]. We will concentrate our attention in demonstrating the emergence of complex (chaotic) behaviour in coupled (non chaotic) systems. That is to show that for uni- or bi-directionally coupled non chaotic systems, chaos can appear even for large values of coupling parameter. This discussion is based on continuous autonomous differential systems, firstly of Lorenz-type illustrating identical chaos synchronization and regular behaviour emerging from chaotic one, and, secondly, systems modeling predator-prey food-chain showing chaotic behaviours emerging from regular ones.

2 Synchronization and Desynchronization

Synchronization is a ubiquitous phenomenon characteristic of many processes in natural systems and (nonlinear) science. It has permanently remained an object of intensive research and is today considered as one of the basic nonlinear phenomena studied in mathematics, physics, engineering or life science. Synchronization of two dynamical systems generally means that one system somehow traces the motion of another. Indeed, it is well known that many coupled oscillators have the ability to adjust some common relation that they have between them due to weak interaction, which yields to a situation in which a synchronization-like phenomenon takes place, see [2].

Since this discovery, periodic synchronization has found numerous applications in various fields, for instance in biological systems and living nature where synchronization is encountered on differents levels. Examples range from the modeling of the heart to the investigation of the circardian rhythm, phase locking of respiration with a mechanical ventilator, synchronization of oscillations of human insulin secretion and glucose infusion, neuronal information processing within a brain area and communication between different brain areas. Synchronization also plays an important role in several neurological diseases such as epilepsies and pathological tremors, or in different forms of cooperative behaviour of insects, animals or humans. For more details, see [10]. This process may also be encountered in other areas, celestical mechanics or radio engineering and acoustics.

But, even though original notion and theory of synchronization implies periodicity of oscillators, during the last decades, the notion of synchronization has been generalized to the case of interacting chaotic oscillators.

Roughly speaking, a system is **chaotic** if it is deterministic, has a long-term aperiodic behaviour, and shows sensitive dependence on initial conditions on a closed invariant set.

Chaotic oscillators are found in many dynamical systems of various origins, the behaviour of such systems is characterized by instability and, as a result, limited predictability in time.

Despite this, in the two last decades, the research for synchronization moved to chaotic systems. A lot of research has been done and, as a result, researchers showed that two chaotic systems could be synchronized by coupling them : synchronization of chaos is actual and chaos could then be exploitable, see [9], and for a review see [2]. Ever since, many researchers have discussed the theory, the design or applications of synchronized motion in coupled chaotic systems. A broad variety of applications have emerged, for example to increase the power of lasers, to synchronize the output of electronic circuits, to control oscillations in chemical reactions or to encode electronic messages to secure communications. Moreover, in the topics of coupled chaotic systems, many different phenomena, which are usually referred to as *synchronization*, exist and have been studied for more than a decade.

2.1 Synchronization and Stability : Definitions

For the basic *master-slave* configuration where an autonomous chaotic system (the master) :

$$\frac{dX}{dt} = F(X), \quad X \in \mathbb{R}^n \tag{1}$$

drives another system (the slave):

$$\frac{dY}{dt} = G(X,Y), \quad Y \in \mathbb{R}^m, \tag{2}$$

synchronization takes place when Y asymptotically copies, in a certain manner, a subset X_p of X. That is to say, it exists a relation between the two coupled systems, which could be a smooth invertible function ψ, the last carries trajectories on the attractor of a first system on the attractor of a second system. In other words, if we know, after a transient regime, the state of the first system, it allows us to predict the state of the second: $Y(t) = \psi(X(t))$. Generally, it is assumed $n \geq m$, however, for the sake of easy readability, we will reduce, even if it is not a necessary restriction, to the case $n = m$, and thus $X_p = X$. Henceforth, if we denote the difference $Y - \psi(X)$ by $X_\perp$, in order to reach at a synchronized motion, one expects to have:

$$||X_\perp|| \longrightarrow 0, \ \ as \ t \longrightarrow +\infty. \tag{3}$$

If ψ is the identity function, the process is called *identical synchronization* (IS hereafter).

Definition of IS. System (2) synchronizes with system (1), if the set $M = \{(X,Y) \in \mathbb{R}^n \times \mathbb{R}^n, Y = X\}$ is an attracting set with a basin of attraction B ($M \subset B$) such that $\lim_{t \to} \infty ||X(t) - Y(t)|| = 0$, forall $(X(0), Y(0)) \in B$.

Thus, this regime corresponds to the situation where all the variables of two (or more) coupled chaotic systems converge. If ψ is not the identity function, the phenomenon is more general and is referred to as *generalized synchronization* (GS).

Definition of GS. System (2) synchronizes with system (1), in the generalized sense, if it exists a transformation $\psi \ : \ \mathbb{R}^n \longrightarrow \mathbb{R}^m$, a manifold $M = \{(X,Y) \in \mathbb{R}^{n+m}, Y = \psi(X)\}$ and a subset B ($M \subset B$), such that for all $(X_o, Y_o) \in B$, the trajectory based on the initial conditions (X_o, Y_o) approaches M as time goes to infinity.

Henceforth, in the case of identical synchronization, equation (3) above means that a certain hyperplane M, called *synchronization manifold*, within $\mathbb{R}^{2n}$, is asymptotically stable. Consequently, for the sake of synchrony motion, we have to prove that the origin of the transverse system $X_\perp = Y - X$ is asymptotically stable. That is, to prove that the motion transversal to the synchronization manifold dies out.

The Lyapunov exponents associated with the variational equation corresponding to the transverse system $X_\perp$:

$$\frac{dX_\perp}{dt} = DF(X)X_\perp \tag{4}$$

where $DF(X)$ is the Jacobian of the vector field evaluated onto the driving trajectory X, are referred to as transverse or conditional Lyapunov exponents (CLE hereafter).

In the case of IS it appears that the condition $L^\perp_{max} < 0$ is sufficient to insure synchronization, where $L^\perp_{max}$ is the largest CLE. Indeed, Equation (4) gives the dynamics of the motion transverse to the synchronization manifold, therefore CLE will tell us if this motion die out or not and hence, whether the synchronization state is stable or not. Consequently, if $L^\perp_{max}$ is negative, it will insure the stability of the synchronized state. This will be better explained using the two examples below.

2.2 Identical Synchronization

The simplest form of chaos synchronization and the best way to explain it is *identical synchronization* (IS), also referred to as *Conventional* or *Complete synchronization*, see [4]. It is also the most typical form of chaotic synchronization often observable in two identical systems.

There are various processes leading to synchronization, depending on the used particular coupling configuration they could be very different. Thus, one has to distinguish the following two main situations, even if they are, in some sense, similar: the **uni-directional** and the **bi-directional** coupling . Indeed, synchronization of chaotic systems is often studied for schemes of the form:

$$\begin{aligned} \frac{dX}{dt} &= F(X) + kN(X - Y) \\ \frac{dY}{dt} &= G(Y) + kM(X - Y) \end{aligned} \tag{5}$$

where F and G act in $\mathbb{R}^n$, $(X, Y) \in (\mathbb{R}^n)^2$, k is a scalar and M and N are coupling matrices belonging to $\mathbb{R}^{n \times n}$. If $F = G$ the two subsystems X and Y are identical. Moreover, when both matrices are nonzero then the coupling is called *bi-directional*, while it is referred to as *uni-directional* if one is the zero matrix, and the other being nonzero.

Other names were given in the literature of this type of synchronization, such as *one-way diffusive* coupling, *drive-response* coupling, *master-slave* coupling or *negative feedback control.*

System (5) above with $F = G$ and $N = 0$ becomes uni-directionlly coupled, and reads:

$$\begin{aligned} \frac{dX}{dt} &= F(X) \\ \frac{dY}{dt} &= F(Y) + kM(X - Y) \end{aligned} \tag{6}$$

M is then a matrix that determines the linear combination of X components that will be used in the difference, and k determines the strength of the coupling.

In uni-directional synchronization, the evolution of the first system (the drive) is unaltered by the coupling, the second system (the response) is then constrained to copy the dynamics of the first.

By contrast to the uni-directional coupling, for the bi-directionally coupling (also called *mutual* or *two-way*), both drive and response systems are connected in such a way that they mutually influence each other's behaviour. Many biological or physical systems consist in bi-directionally interacting elements or components, examples range from cardiac and respiratory systems to coupled lasers with feedback.

3 Emergence of Regular Properties: A Lorenz-Type Example

3.1 Uni- and Bi-Directional Identical Synchronization for a Lorenz-Type System

Let us give an example, and for the sake of simplicity, let us develop the idea on the following 3-dimensional simple autonomous system, which belongs to the class of dynamical systems called *generalized Lorenz systems*, see [7] and references therein:

$$\begin{cases} \dot{x} = -9x - 9y \\ \dot{y} = -17x - y - xz \\ \dot{z} = -z + xy \ . \end{cases} \tag{7}$$

The signs used differentiate system (7) from the well-known Lorenz system:

$$\dot{x} = -10x + 10y, \ \dot{y} = 28x - y - xz, \ \dot{z} = -\frac{8}{3}z + xy.$$

From previous observations, it has been shown that system (7) oscillates chaotically, its Lyapunov exponents are +0.601, 0.000 and −16.470, it shows the chaotic attractor of Figure 1, with a 3D feature very similar to that of Lorenz attractor.

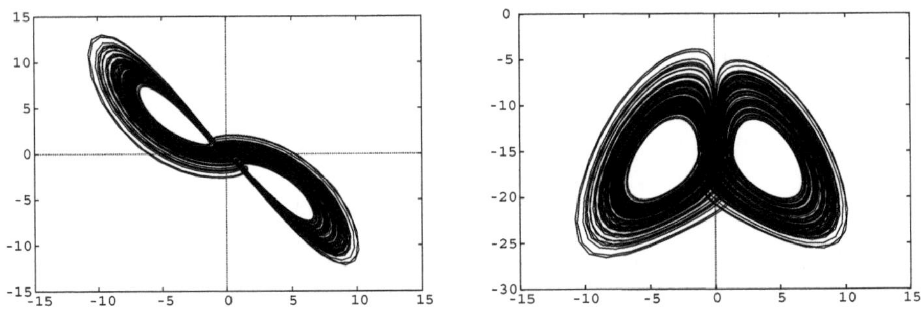

Fig. 1. The chaotic attractor of system (7) : xy and xz-plane projections.

Uni-directional coupling
Let us consider an example with two copies of system (7), and for

$$M = \begin{pmatrix} 1\,0\,0 \\ 0\,0\,0 \\ 0\,0\,0 \end{pmatrix} \tag{8}$$

that is, by adding a damping term to the first equation of the response system, we get a following uni-directionally coupled system, coupled through a linear term $k > 0$ according to variables $x_{1,2}$:

$$\begin{cases} \dot{x_1} = -9x_1 - 9y_1 \\ \dot{y_1} = -17x_1 - y_1 - x_1 z_1 \\ \dot{z_1} = -z_1 + x_1 y_1 \\ \dot{x_2} = -9x_2 - 9y_2 - k(x_2 - x_1) \\ \dot{y_2} = -17x_2 - y_2 - x_2 z_2 \\ \dot{z_2} = -z_2 + x_2 y_2 \end{cases} \tag{9}$$

For $k = 0$ the two subsystems are uncoupled, for $k > 0$ both subsystems are uni-directionally coupled. Our numerical computations yield the optimal value $\tilde{k}$ for the synchronization , we found that for $k \geq \tilde{k} = 4.999$ both subsystems of (9) synchronize. That is to say, starting from random initial conditions, and after some transient time, system (9) generates the same attractor as for system (7), see Figure 1. Consequently, all the variables of the coupled chaotic subsystems converge, that are x_2 converges to x_1, y_2 to y_1 and z_2 to z_1, see Figure 2. Thus, the second system (the response) is locked to the first one (the drive). One could also give correlation plots, that are the amplitudes

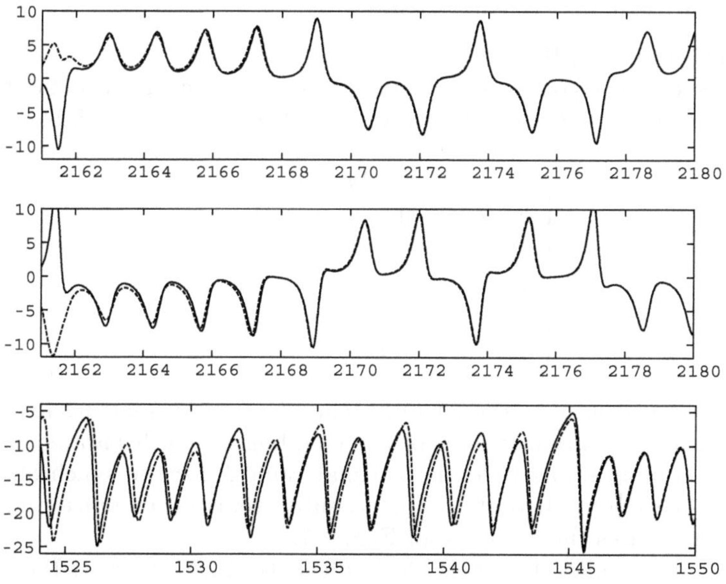

Fig. 2. Time series for $x_i(t)$, $y_i(t)$ and $z_i(t)$ in system (9), $(i = 1, 2)$, for the coupling constant k = 5.0, that is beyond the threshold necessary for synchronization. After transients die down, the two subsystems synchronize perfectly: Regular behaviour emerges from chaotic behaviours.

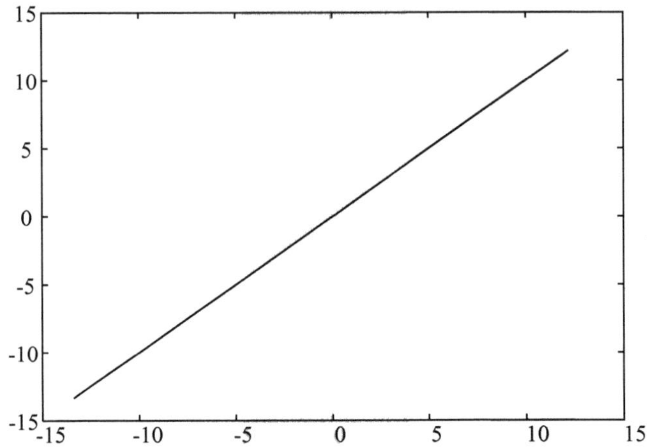

Fig. 3. The plot of amplitudes y_1 against y_2, after transients die down, shows a diagonal line, which also indicates that the receiver and the transmitter are maintaining synchronization. The plot of z_1 against z_2 shows a similar figure: Regular behaviour emerges from chaotic behaviours.

x_1 against x_2, y_1 against y_2 and z_1 against z_2, and observe diagonal lines, meaning also that the system synchronizes.

Bi-directional coupling

Let us then take two copies of the same system (7) as given above, but two-way coupled through a linear constant term $k > 0$ according to variables $x_{1,2}$:

$$\begin{cases} \dot{x_1} = -9x_1 - 9y_1 - k(x_1 - x_2) \\ \dot{y_1} = -17x_1 - y_1 - x_1 z_1 \\ \dot{z_1} = -z_1 + x_1 y_1 \\ \dot{x_2} = -9x_2 - 9y_2 - k(x_2 - x_1) \\ \dot{y_2} = -17x_2 - y_2 - x_2 z_2 \\ \dot{z_2} = -z_2 + x_2 y_2 \end{cases} \tag{10}$$

We can get an idea of the onset of synchronization by plotting for example x_1 against x_2 for various values of the coupling strength parameter k. Our numerical computations yield the optimal value $\tilde{k}$ for the synchronization : $\tilde{k} \simeq 2.50$, see Figure 4, both (x_i, y_i, z_i)-subsystems synchronize and system (10) also generates the attractor of Figure 1.

These results also show that, for sufficiently lage values of the coupling parameter k, simple uni- or bi-directional coupling of two chaotic systems does not increase the chaoticity of the new system, unlike what one might expect. Thus, in some sense (see synchronization manifold below), regular behaviour emerges from chaotic behaviour (the motion is confined in some manifold).

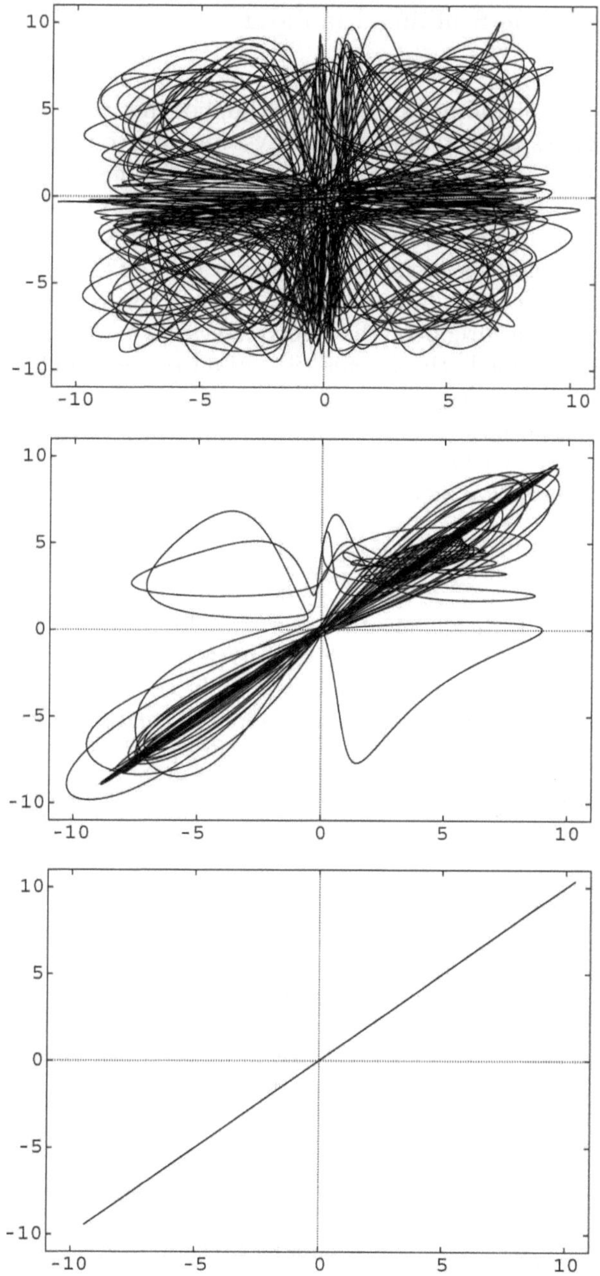

Fig. 4. Illustration of the synchronization onset of system (10). (a), (b) and (c) plot the amplitudes x_1 against x_2 for values of the coupling parameter $k = 0.5$, $k = 1.5$ and $k = 2.8$ respectively. The system synchronizes for $k \geq 2.5$.

3.2 Remark on the Stability Manifold

Geometrically, the fact that systems (9) and (10) beyond synchronization generate the same attractor as system (7), implies that the attractors of these combined drive-response 6-dimensional systems are confined to a 3-dimensional hyperplane (the *synchronization manifold*) defined by $Y = X$. This hyperplane is stable since small perturbations which take the trajectory off the synchronization manifold will decay in time. Indeed, as we said before, conditional Lyapunov exponents of the linearization of the system around the synchronous state could determine the stability of the synchronized solution. This means that the origin of the transverse system $X_\perp$ is asymptotically stable. To see this, for both systems (9) and (10), we switch to the new set of coordinates, $X_\perp = Y - X$, that is $x_\perp = x_2 - x_1$, $y_\perp = y_2 - y_1$ and $z_\perp = z_2 - z_1$. The origin $(0, 0, 0)$ is obviously a fixed point for this transverse system, within the synchronization manifold. Therefore, for small deviations from the synchronization manifold, this system reduces to a typical variational equation:

$$\frac{dX_\perp}{dt} = DF(X)X_\perp \tag{11}$$

where $DF(X)$ is the Jacobian of the vector field evaluated onto the driving trajectory X. Previous analysis, see [2], shows that $L^\perp_{max}$ becomes negative as k increases, for both uni- or bi-directionally coupling, which insures the stability of the synchronized state for systems (9) and (10), Figure 5.

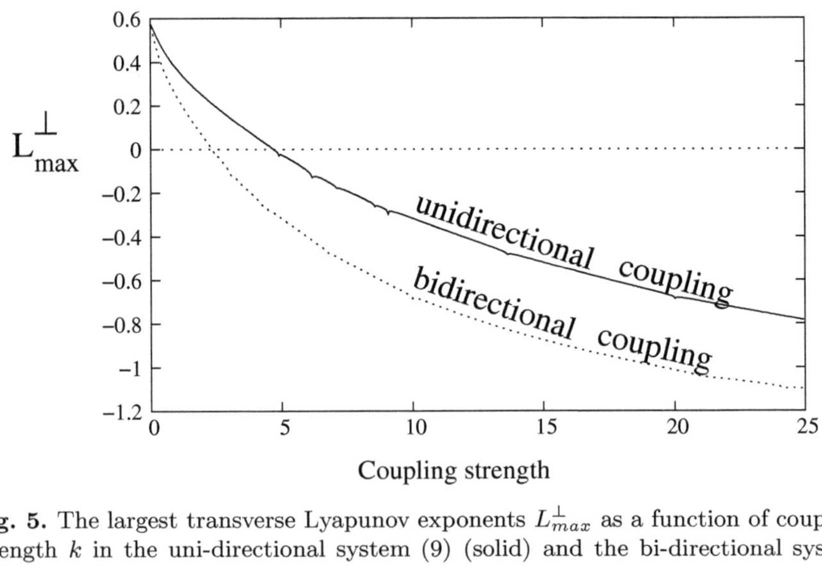

Fig. 5. The largest transverse Lyapunov exponents $L^\perp_{max}$ as a function of coupling strength k in the uni-directional system (9) (solid) and the bi-directional system (10) (dotted).

Let us note that this can also be proved analytically, as done in [7], by using a suitable Lyapunov function, and using a new extended version of LaSalle invariance principle.

Desynchronization motion. Synchronization depends on the coupling strength, but also on the vector field and the coupling function. For a choice of these quantities, synchronization may occur only within a finite range $[k_1, k_2]$ of coupling strength, in such a case a *desynchronization* phenomenon occurs: thus, increasing k beyond the critical value k_2 yields loss of the synchronized motion ($L^{\perp}_{max}$ becomes positive).

4 Emergence of Chaotic Properties : A Predator-Prey Example

This example, contrary to the first, shows a situation where the larger is the coupling coefficient the weaker is the synchronization.

4.1 The Model

As we said in the introduction, it is becoming a commonplace that, if the 20th was the century of physics, the 21st will be the century of biology and more specifically, mathematical biology, ecology, ... and in general, nonlinear dynamics and complexity in life sciences.

In this last part, we will hence focus ourselves on emergent (chaotic or regular) properties that arise when we couple uni- or bi-directionally two 3-dimensional autonomous differential systems, modeling predator-prey food-chains.

Let us then consider a continuous time dynamical system, model for a tritrophic food chain, based on a modified version of the Leslie-Gower scheme, see [3, 6], for which the rate equations of the three components of the chain population can be written as follows:

$$\begin{cases} \dfrac{dX}{dT} = a_o X - b_o X^2 - \dfrac{v_o XY}{d_o + X} \\ \dfrac{dY}{dT} = -a_1 Y + \dfrac{v_1 XY}{d_1 + X} - \dfrac{v_2 YZ}{d_2 + Y} \\ \dfrac{dZ}{dT} = c_3 Z - \dfrac{v_3 Z^2}{d_3 + Y}, \end{cases} \tag{12}$$

with $X(0) \geq 0$, $Y(0) \geq 0$ and $Z(0) \geq 0$, where X, Y and Z represent the population densities at time T ; a_0, b_0, v_0, d_0, a_1, v_1, d_1, v_2, d_2, c_3, v_3 and d_3 are model parameters assuming only positive values and defined as follows: a_o

is the growth rate of prey X, b_o measures the strength of competition among individuals of species X, v_o is the maximum value which *per capita* reduction rate of X can attain, d_o measures the extent to which environment provides protection to prey X, a_1 represents the rate at which Y will die out when there is no X, v_1 and d_1 have a similar meaning to v_0 and d_o, v_2 and v_3 have a similar biological connotation as that of v_o and v_1, d_2 is the value of Y at which the *per capita* removal rate of Y becomes $v_2/2$, c_3 describes the growth rate of Z, assuming that the number of males and females is equal, d_3 represents the residual loss in species Z due to severe scarcity of its favourite food Y ; the second term on the right hand side in the third equation of (12) depicts the loss in the predator population.

For the origin of this system and for some theoretical results, boundedness of solutions, existence of an attracting set, existence and local or global stability of equilibria, etc ..., see [3, 6]. In these works, using intensive numerical qualitative analysis, it has been demonstrated that the model could show periodic solutions, Figure (6), and quasi-periodic or chaotic dynamics, Figure (7), for the following parameters and state values:

$$\begin{cases} b_o = 0.06, \quad v_o = 1.0, \quad d_0 = d_1 = d_2 = 10.0, \\ a_1 = 1.0, \; v_1 = 2.0, \; v_2 = 0.9, \\ c_3 = 0.02, \; v_3 = 0.01, \; d_3 = 0.3. \end{cases} \tag{13}$$

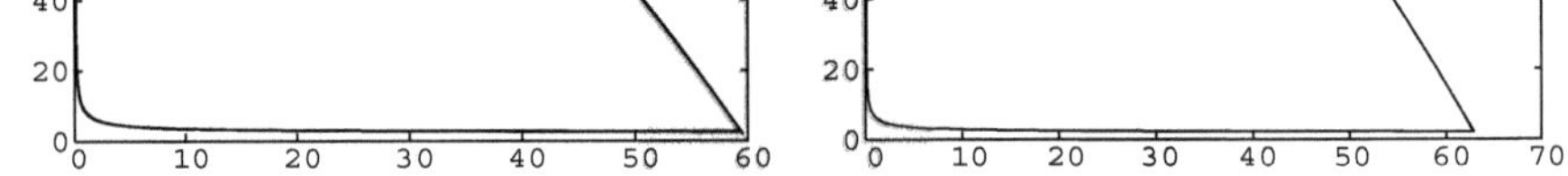

Fig. 6. Limit cycles of period one and two, for $a_0 = 3.6$ and $a_0 = 3.8$ respectively, found for system (12) and for parameters given by (13).

We will set, for the rest of the paper, the system parameters as given in (13) in order that system (13) oscillates in a regular way around a stable limit cycle of period one, Figure 6(b).

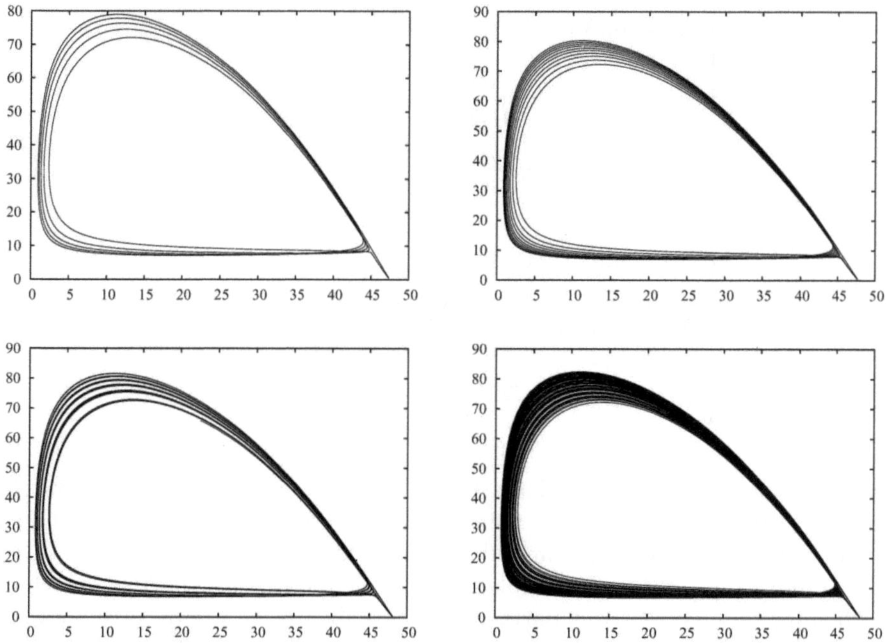

Fig. 7. Transition to chaotic (or quasi-periodic) behaviour found for system (12), it is established via period doubling bifurcation, respectively for $a_0 = 2.85$, $a_0 = 2.87$, $a_0 = 2.89$ and $a_0 = 2.90$, with parameters given by (13).

4.2 Uni-Directional Desynchronization : Predator-Prey System

Usually, as we have seen in the first example, section 2, in uni-directional synchronization, while the evolution of the first system (the drive) is unaltered by the coupling, the second system (the response) is constrained to copy the dynamics of the first.

However, it is not the case for our example below, for which, as we will see, while both subsystems evolve periodically (limit cycles of Figure 6(b)), the coupled system behaviour is extremely complex.

Let us then consider two copies of system (12). By adding a damping term to the first equation of the response system, we get a following uni-directionally coupled system, coupled through a linear term $k > 0$ according to variables $x_{1,2}$:

$$\begin{cases} \dot{X}_1 = a_o X_1 - b_o X_1^2 - \dfrac{v_o X_1 Y_1}{d_o + X_1} \\ \dot{Y}_1 = -a_1 Y_1 + \dfrac{v_1 X_1 Y_1}{d_1 + X_1} - \dfrac{v_2 Y_1 Z_1}{d_2 + Y_1} \\ \dot{Z}_1 = c_3 Z_1 - \dfrac{v_3 Z_1^2}{d_3 + Y_1} \\ \\ \dot{X}_2 = a_o X_2 - b_o X_2^2 - \dfrac{v_o X_2 Y_2}{d_o + X_2} - k(X_2 - X_1) \\ \dot{Y}_2 = -a_1 Y_2 + \dfrac{v_1 X_2 Y_2}{d_1 + X_2} - \dfrac{v_2 Y_2 Z_2}{d_2 + Y_2} \\ \dot{Z}_2 = c_3 Z_2 - \dfrac{v_3 Z_2^2}{d_3 + Y_2} \end{cases} \tag{14}$$

For $k = 0$ the two subsystems are uncoupled, for $k > 0$ both subsystems are uni-directionally coupled, and for $k \longrightarrow +\infty$ one can expect to obtain the same results as those obtained for the previous example in section 2, that is strong synchronization and two subsystems which evolve identically.

We have chosen, for the coupled system, a range of parameters for which both subsystem constituent parts evolve periodically, as Figure 6(b) shows.

However, our numerical computations show that both subsystems of (14) never synchronize nor identically neither *generally*, unless the coupling parameter k is very small. In such a case a certain generalized synchronization form takes place, see Figure 8(a). That is, starting from random initial conditions, and after some transient time, system (14) generates an attractor different from those showed by system (12) in Figure 6(b). Consequently, all the variables of the coupled limit cycle subsystems surprisingly do not converge, as, at first sight, one may intuitively expect, see Figure 8.

These results show that uni-directional coupling of these two non-chaotic systems (that are the subsystem constituents of system (14)) increases the bahaviour complexity, and transforms a periodic situation into a chaotic one.

Emergent chaotic properties are typically novel and unanticipated, for this example.

In fact, this phenomenon corresponds to the classical cascade of periodic-doubling bifurcation processus, with a sequence of order and disorder windows.

4.3 Bi-Directional Desynchronization : Predator-Prey System

As many biological or physical systems consist in bi-directional interacting elements or components, let us use a bi-directionally (*mutual*) coupling, in order that both drive and response subsystems are connected in such a way that they mutually influence each other's behaviour. Let us then take two copies of the same system (12) given above, but two-way coupled through a linear constant term $k > 0$ according to variables $x_{1,2}$:

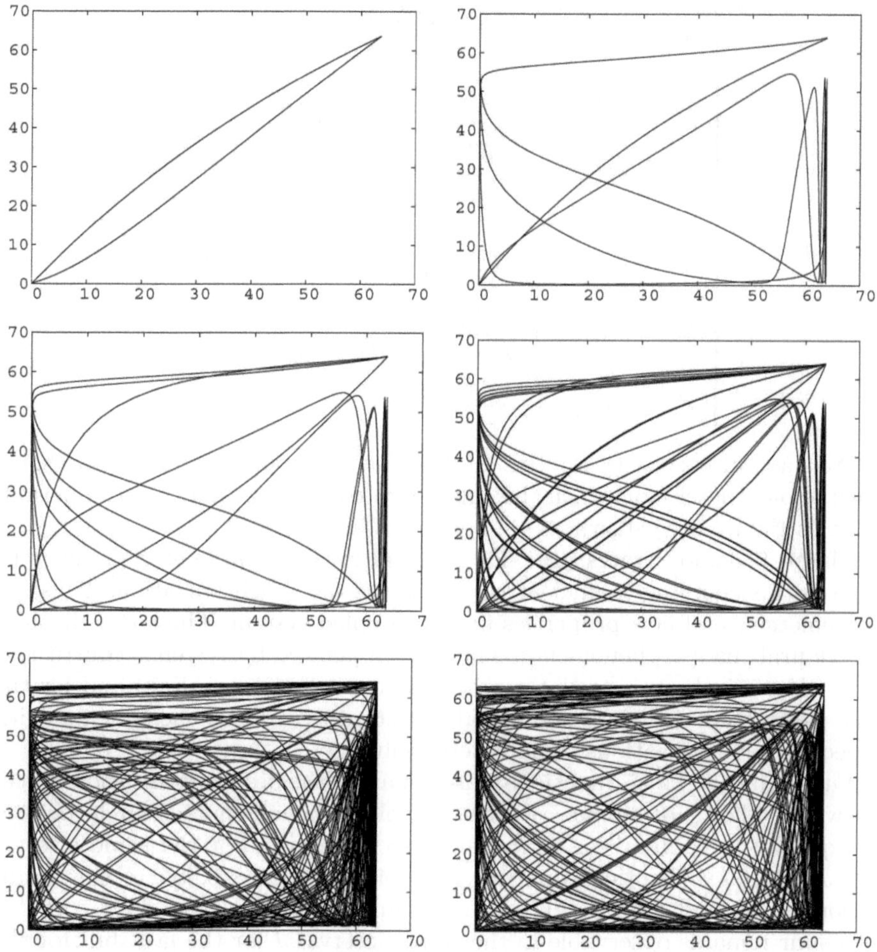

Fig. 8. Illustration of the desynchronization onset of the unidirectional coupled system (14). Figures (a), (b), ... (f) are done from left to right and up to down, and plot the amplitudes x_1 against x_2 for values of the coupling parameter (a) $k = 0.01$, (b) $k = 0.055$, (c) $k = 0.056$, (d) $k = 0.0565$, (e) $k = 0.057$ and (f) $k = 0.1$. Figure (a) shows a generalized synchronization phenomenon: the system synchronizes (in the generalized sense) for very small values of k. But a desynchronization processus quickly arises by increasing k, Figures (b,c,d,e,f): in some interval for k, the larger is the coupling coefficient the weaker is the synchronization. Hence, we have the emergence of chaotic properties: the coupled system displays behavioural chaotic phenomena which are not showed by the systems' constituent parts (that are the two predator-prey systems without coupling) which point out the limit-cycle of Figure 6(b), for the same parameters and the same initial conditions. This phenomenon is robust with respect to small parameters variations.

$$
\begin{cases}
\dot{X}_1 = a_o X_1 - b_o X_1^2 - \dfrac{v_o X_1 Y_1}{d_o + X_1} - k(X_1 - X_2) \\
\dot{Y}_1 = -a_1 Y_1 + \dfrac{v_1 X_1 Y_1}{d_1 + X_1} - \dfrac{v_2 Y_1 Z_1}{d_2 + Y_1} \\
\dot{Z}_1 = c_3 Z_1 - \dfrac{v_3 Z_1^2}{d_3 + Y_1} \\
\\
\dot{X}_2 = a_o X_2 - b_o X_2^2 - \dfrac{v_o X_2 Y_2}{d_o + X_2} - k(X_2 - X_1) \\
\dot{Y}_2 = -a_1 Y_2 + \dfrac{v_1 X_2 Y_2}{d_1 + X_2} - \dfrac{v_2 Y_2 Z_2}{d_2 + Y_2} \\
\dot{Z}_2 = c_3 Z_2 - \dfrac{v_3 Z_2^2}{d_3 + Y_2}
\end{cases}
\tag{15}
$$

We have also chosen, for this bi-directionally coupled system, the same range of parameters for which the subsystem constituent parts evolve periodically, as Figure 6(b) shows.

Figure 9 demonstrates also, for some interval of parameter k, that the larger is this coupling coefficient the weaker is the synchronization. Thus, we have emergence of new properties for the coupled system. The latter displays behavioural chaotic phenomenon which is not showed by systems' constituent parts (that are the two predator-prey systems without coupling) and for the same parameters and the same initial conditions. A robust phenomenon with respect to small variations of parameter values.

Furthermore, the bi-directional case enhances the desynchronization processus which allows occurence of new complex phenomenon. The latter occurs quickly in comparison to the uni-directinal case. For the same interval $k \in J =]0, 0.1]$, chaotic properites take place for $k = 0.055$ in the unidirectional case and for $k = 0.057$ in the uni-directional case. This complex behaviour remains observable in the whole interval J for the last, but for the first, it disappears after $k = 0.056$ -some regular generalized synchronization takes place- and appears again for $k \in]0.057, 0.1]$

Thus, we can conclude that, the larger is the coupling coefficient k the weaker is the synchronization (within some interval for k).

All these numerical results show that the whole predator-prey food chain in 6-dimensional space, displays behavioural phenomena which are completely inexplicable by any conventional analysis of the 3-dimensional systems' constituent parts, which have for the same ranges of parameters a one-peridoic solutions. They have to be compared to the results obtained in the previous section, in which it has been shown that the larger is the coupling coefficient the stronger is the synchronization

Therefore, our predator-prey system is an example pointing out new emergent properties, which are properties of the "whole" 6-dimensional system, being not possessed by any of the individual parts (which are the two 3-dimensional subsystems).

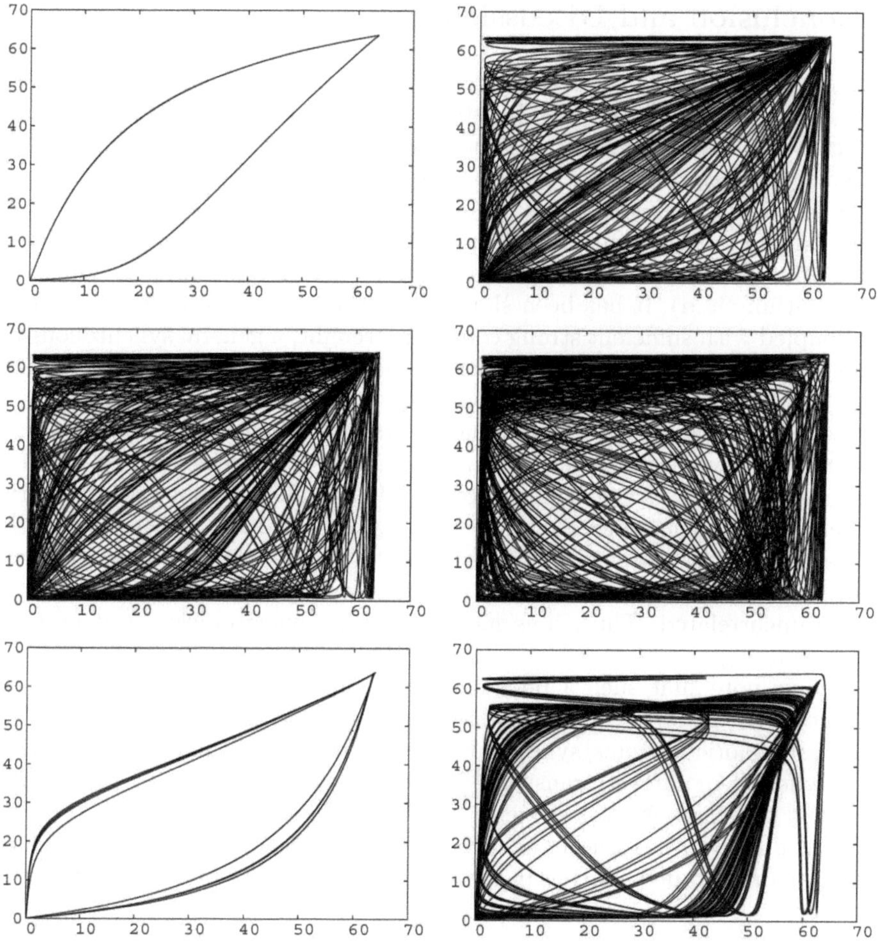

Fig. 9. Bi-directional coupling. Figures are done from left to right and up to down and plot amplitudes x_1 against x_2 for the same values as done in the previous figure, respectively for $k = 0.01$, $k = 0.055$ and $k = 0.056$ $k = 0.0565$, $k = 0.057$ and $k = 0.1$. These figures, Illustrate a window of generalized synchronization and desynchronization of system (12). (a), (b) and (c) plot The system synchronizes (in the generalized sense) for $k \leq 0.01$, as it has been shown in the uinidirectional case. But the desynchronization processus arises by increasing k, quickly in comparison with the unidirectional case.

5 Conclusion and Discussion

Identical chaotic systems synchronize by following the same chaotic trajectory (IS). However, in the real world systems are in general not identical. For instance, when the parameters of two-coupled identical systems do not match, or when these coupled systems belong to different classes, complete IS is not to be expected, because it does not exist such an invariant manifold $Y = X$, as for identical synchronisation. For non-identical systems, the possibility of some other types of synchronization has been investigated (see [2] and references within cited). It has been showed [11] that when two different systems are coupled with sufficient strong coupling strenght, a general synchronous relation between their states could exist and be expressed by a smooth invertible function, $Y(t) = \psi(X(t))$, as we have done in the previous section.

But, for coupled non-identical chaotic systems, other types of synchronization exist. For example *phase synchronization* (PS hereafter) which is rather a weak degree of synchronization, see [10]. It is a hidden phenomenon, in the sense that the synchronous motion is not visible. Indeed, in case of PS, the phases of chaotic systems are locked, that is to say that it exists a certain relation between them, whereas the amplitudes vary chaotically and are practically uncorrelated. Thus, it is mostly close to synchronization of periodic oscillators.

Let us note that such a phenomenon occurs when a zero Lyapunov exponent of the response system becomes negative, while, as explained above, identical chaotic systems synchronize by following the same chaotic trajectory, when their largest transverse Lyapunov exponent of the synchronized manifold decreases from positive to negative values.

This processus deserves to be investigated for our predator-prey food chain case. A more detailed analysis of such phenomena will be provided in the near future.

References

1. L'Enigme de l'Emergence (2005) *Sciences et Avenir*, 143, Juillet-Août.
2. Aziz-Alaoui M.A. (2006) Synchronization of Chaos *Encyclopedia of Mathematical Physics*. Elsevier, 5:213-226
3. Aziz-Alaoui M.A. (2002) Study of a LeslieGower-type tritrophic population model, *Chaos Solitons and Fractals*, 14:12751293.
4. Boccaletti S., Kurths J., Osipov G., Valladares D. and Zhou (2002) The synchronization of chaotic systems, *Physics Reports* 366: 1-101.
5. Chen G. and Dong X.(1998) *From chaos to order*, Singapore: World Scientific.
6. Daher Okiye M.. (2004) Etude et Analyse Asymptotique de certains Systèmes Dynamiques Non-Linéaires, *PhD Dissertation* University of Le Havre, France.
7. Derivière S. and Aziz-Alaoui M.A. (2003) Estimation of attractors and synchronization of generalized Lorenz systems, *Dynamics of Continuous, Discrete and Impulsive Systems series B: Applications and Algorithms* 10(6): 833-852.

8. Holmes P. (2005) Ninety plus thirty years of nonlinear dynamics: Less is more and more is different, *International Journal of Bifurcation and Chaos*, 15(9): 2703-2716.
9. Pecora L. and Carroll T. (1990) Synchronization in chaotic systems, *Physical Review letters* 64, pp. 821-824.
10. Pikovsky A., Rosenblum M. and Kurths J. (2001) *Synchronization, A Universal Concept in Nonlinear Science*, Cambridge University Press.
11. Rulkov N., Sushchik M., Tsimring L. and Abarbanel H. (1995) Generalized synchronization of chaos in directionally coupled chaotic systems, *Phys. Rev. E* 51(2): 980-994.

Robust H_∞ Filtering Based Synchronization for a Class of Uncertain Neutral Systems

Alif A.[1], Boutayeb M.[2], and Darouach M.[1]

[1] UHP-CRAN-CNRS-UMR 7039, IUT de Longwy
186 rue de Lorraine
54400 Cosnes et Romain, France.
Adil.Alif@iut-longwy.uhp-nancy.fr, darouach@iut-longwy.uhp-nancy.fr

[2] LSIIT-CNRS-UMR 7005, University of Louis Pasteur Strasbourg
Pole API, Bd.S.Brant
67400 ILLKIRCH, France. Mohamed.Boutayeb@ipst-ulp.u-strasbg.fr

Summary. In this work, a new delay dependent approach to deal with the robust H_∞ filtering based synchronization problem for a class of structured uncertain neutral systems with unknown delays is proposed. The new approach gives the maximal size of the interval of evolution of the delay within which the filtering based synchronization process can be guaranteed. The obtained criterion can guarantee the filtering based synchronization process within any interval in $I\!R^+$, representing the evolution of the delay, where the size is less than the maximal size value obtained while solving the obtained criterion. Sufficient conditions to reduce the effect of the disturbance input on the filtering based synchronization error output to a prescribed level are established and expressed in terms of linear matrix inequalities. A numerical example is provided to illustrate the validity of the obtained results.

Key words: neutral systems, robust H_∞-Filtering based synchronization, linear matrix inequality, descriptor model, unknown delay, new delay dependent approach, structured uncertainties.

1 Introduction

The synchronization problem of coupled systems has attracted a lot of attention during the two last decades. Indeed, this is due to its potential applications in several fields, see for instance [1] where the synchronization of coupled chaotic systems has been exploited for secure communications. In the last years, the lag synchronization problem has received an increasing interest also due to its several applications, see for instance [2] and [3]. However, in spite of the importance of all these works, others aspects of the synchronization problem remaind not much explored, and in some cases still unexplored, like the H_∞ filtering based synchronization problem for neutral systems (systems which contain delays in the system state and in the derivative of the state). For delayed systems (systems which contain delay only in the system state), several works have been reported in the literature dealing with the filtering based synchronization problem of these systems, see for instance [4] and [5] where the delay-dependent approach (the obtained criteria depend on the maximal value of the delay) has been adopted, and [6] where the delay independent approach (more conservative since the obtained criteria is independent on the delay information) has been followed. See also [11], [12] and [10] where in addition the delay is supposed to be unknown which bring additional difficulties to the filtering problem of such systems. A significance progress has been occurred concerning the first approach, and this is owing to the introduction of new and less conservative model transformations and boundedness tools. See for instance [7], [8] and [9] where the descriptor model transformation has been introduced, and less conservative results, compared to the others methods, have been obtained. In this paper, using the descriptor model transformation, a new delay dependent approach to deal with the robust H_∞ filtering based synchronization problem for a class of linear neutral systems subjected to structured uncertainties is tackled. The delays are supposed to be unknown, that in the derivative of the system state is supposed to be completely unknown, whereas that in the system state is assumed to be unknown but evolves in a pre-specified interval. Unlike the classical delay dependent approach where the goal is to seek the large value of the delay under which the filtering based synchronization process can be guaranteed, and the delay independent approach, where the filtering based synchronization process can be guaranteed for any value of the delay in $\mathbb{R}^+$, the proposed approach gives the maximal size of the interval of evolution of the delay within which the filtering based synchronization process can be guaranteed. This means that the filtering based synchronization process can be guaranteed within any given interval in $\mathbb{R}^+$, representing the interval of evolution of the delay, where the size is less than the maximal size value obtained while solving the obtained criterion. A new filter structure is introduced to ensure the robust H_∞ filtering based synchronization process. Also, several technics are used to transform the hard nonlinear obtained conditions into a convex problem given in terms of some linear matrix inequalities. The validity of the obtained results is illustrated via a numerical example.

2 Problem Statement

In this study, we consider a neutral system with a general form as follows:

$$\dot{x}(t) - E\dot{x}(t-\tau_d) = Ax(t) + A_d x(t-\tau) + B_p p(t) + B_w w(t) \quad (1)$$

$$x(t) = \phi(t), \quad t \in [-\hat{\tau}, 0] \quad (2)$$

$$y(t) = Cx(t) + C_d x(t-\tau) + C_p p(t) + C_w w(t) \quad (3)$$

$$z(t) = Lx(t) + L_w w(t) \quad (4)$$
$$q(t) = D_1 x(t) + D_d x(t-\tau) + D_p p(t) + D_w w(t) \quad (5)$$
$$p(t) = \Delta(t) q(t) \quad (6)$$

where $x(t) \in \mathbb{R}^n$ is the system state, $w(t) \in \mathcal{L}_2^q[0,\infty)$ is the external disturbances, $y(t) \in \mathbb{R}^m$ is the measurable output, and $z(t) \in \mathbb{R}^s$ is the regular output to be estimated where $s \leq n$. $p(t) \in \mathbb{R}^n$ represents the structured uncertainties and by adding the output $q(t) \in \mathbb{R}^n$ we describe the influence of these uncertainties on the system. $\Delta(t)$ is the uncertainty between $p(t)$ and $q(t)$ such that $\Delta(t)\Delta(t)^T \leq I$. A, A_d, E, B_p, B_w, C, C_d, C_p, C_w, D_1, D_d, D_p, D_w, L and L_w are known constant matrices with appropriate dimensions. τ and τ_d are two positive constant scalars. τ_d is supposed to be completely unknown, whereas τ is unknown but assumed to evolve in the interval $[\tau_1,\ \tau_2]$, where τ_1 and τ_2 are positive known scalars ($\tau_1 < \tau_2$). $\widehat{\tau} = max(\tau_d, \tau_2)$, and $\phi(t)$ is a continuous vector function.
Here we suppose that $\Delta(t)$ has the following structural property:

$$\Delta(t) \in \mathcal{D} := \{\Delta(t) = diag(\Delta_1(t), \cdots, \Delta_k(t), \eta_1 I, \cdots, \eta_l I) : k, l \in \mathbb{N}\} \quad (7)$$

where $\eta_j \in \mathbb{R}$ such that $|\eta_j| \leq 1$ and $\Delta_j(t)$ is full block such that $\Delta_j(t)\Delta_j^T(t) \leq I$.
We define the set of symmetric scaling matrix corresponding to arbitrary block diagonal structure $\mathcal{D}$ as follows:

$$\mathcal{S}_\Delta := \{S : S = S^T,\ \ S\Delta(t) = \Delta(t)S, for\ \ \Delta(t) \in \mathcal{D}\}. \quad (8)$$

also, the set of antisymmetric scaling matrix corresponding to arbitrary block diagonal structure $\mathcal{D}$ can be described by:

$$\mathcal{T}_\Delta := \{T : T^T = -T,\ \ T\Delta(t) = \Delta(t)^T T, for\ \ \Delta(t) \in \mathcal{D}\}. \quad (9)$$

In general, the symmetric scaling matrix denoted by $\mathcal{S}$ corresponding to the block diagonal uncertainty can be rewritten as:

$$S := diag(\delta_1 I, \cdots, \delta_k I, S_1, \cdots, S_l), k, l \in \mathbb{N}, \delta_j \in \mathbb{R} \text{ and } S_i = S_i^T \quad (10)$$

Similarly, the antisymmetric scaling matrix can be described as

$$T := diag(\underbrace{0, \cdots, 0}_{k}, T_1, \cdots, T_l), k, l \in \mathbb{N}, \delta_j \in \mathbb{R} \text{ and } T_i^T = -T \quad (11)$$

Assumption: To guarantee robustness of the results with respect to small changes of the delay τ_d, we assume that the difference operator $D : \mathcal{C}[-\tau_d, 0] \rightarrow \mathbb{R}^n$, given by $D(x_t) = x(t) - Ex(t-\tau_d)$ is asymptotically stable for all values of τ_d or, equivalently that

$$\rho(E) < 1.$$

This means that all the eigenvalues of E are inside the unite circle.

In this work, we consider a filter of order n with state-space realization of the following form:

$$\dot{\hat{x}}(t) = (N - A_d)\hat{x}(t) + A_d\widehat{x}(t-\tau_2) + Dy(t) \tag{12}$$
$$\hat{x}(t) = \widehat{\phi}(t), \quad t \in [-\widehat{\tau}, 0] \tag{13}$$
$$\widehat{z}(t) = K\hat{x}(t) \tag{14}$$

where $\widehat{x}(t) \in I\!R^n$ is the filter state, and $\widehat{z}(t) \in I\!R^s$ is the estimate of $z(t)$. N, D and K are the gain matrices with appropriate dimensions to be determined. $\widehat{\phi}(t)$ is a continuous vector function. The filter structure has not been taking arbitrarily, indeed, this structure plays an important role, as we will see, in the obtention of a special kind of augmented systems, for which our new delay dependent approach can be applied. To show that, let us first introduce the well known equality $x(t-\tau) = x(t) - \int_{t-\tau}^{t} \dot{x}(s)\, ds$ in (1). Thus, we deduce that

$$\begin{aligned} \dot{x}(t) - E\dot{x}(t-\tau_d) =& (A + A_d)x(t) - A_d \int_{t-\tau_2}^{t} \dot{x}(s) + A_d \int_{t-\tau_2}^{t-\tau} \dot{x}(s)\, ds \\ &+ B_p p(t) + B_w w(t) \end{aligned} \tag{15}$$

and remark that (12) can be equivalently rewritten as follows

$$\dot{\widehat{x}}(t) = N\widehat{x}(t) - A_d \int_{t-\tau_2}^{t} \dot{\widehat{x}}(s)\, ds + Dy(t) \tag{16}$$

Now let $e(t) = x(t) - \widehat{x}(t)$. Hence, we deduce from equations (15), (16) and (3) that

$$\begin{aligned} \dot{e}(t) - E\dot{x}(t-\tau_d) = & (A + A_d - N - DC)x(t) + Ne(t) - DC_d x(t-\tau) \\ & -A_d \int_{t-\tau_2}^{t} \dot{e}(s)\, ds + A_d \int_{t-\tau_2}^{t-\tau} \dot{x}(s)\, ds \\ & +(B_p - DC_p)p(t) + (B_w - DC_w)w(t) \end{aligned} \tag{17}$$

which can be rewritten as follows

$$\begin{aligned} \dot{e}(t) - E\dot{x}(t-\tau_d) = & (A + A_d - N - DC)x(t) + (N - A_d)e(t) \\ & -DC_d x(t-\tau) + A_d e(t-\tau) \\ & +A_d \int_{t-\tau_2}^{t-\tau} \dot{x}(s)\, ds - A_d \int_{t-\tau_2}^{t-\tau} \dot{e}(s)\, ds \\ & +(B_p - DC_p)p(t) + (B_w - DC_w)w(t) \end{aligned} \tag{18}$$

Thus, by introducing the filtering based synchronization error $\overline{z}(t) = z(t) - \widehat{z}(t)$, and by means of equations (1), (4) and (14), we get the following augmented system

$$\dot{x}_a(t) - \widehat{E}\dot{x}_a(t-\tau_d) = A_1 x_a(t) + A_2 x_a(t-\tau) - A_3 \int_{t-\tau_2}^{t-\tau} \dot{x}_a(s)\,ds + \widehat{B}_p p(t) + \widehat{B}_w w(t) \quad (19)$$

$$\overline{z}(t) = \overline{L}x_a(t) + L_w w(t) \quad (20)$$

$$q(t) = \overline{D}x_a(t) + \overline{D}_d x_a(t-\tau) + D_p p(t) + D_w w(t) \quad (21)$$

$$p(t) = \Delta(t) q(t) \quad (22)$$

where

$$x_a(t) = \begin{bmatrix} x(t) \\ e(t) \end{bmatrix}, \quad (23)$$

$$\widehat{E} = \begin{bmatrix} E & 0 \\ E & 0 \end{bmatrix}, \quad (24)$$

$$A_1 = \begin{bmatrix} A & 0 \\ A + A_d - N - DC & N - A_d \end{bmatrix}, \quad (25)$$

$$A_2 = \begin{bmatrix} A_d & 0 \\ -DC_d & A_d \end{bmatrix}, \quad (26)$$

$$A_3 = \begin{bmatrix} 0 & 0 \\ -A_d & A_d \end{bmatrix}, \quad (27)$$

$$\widehat{B}_p = \begin{bmatrix} B_p \\ B_p - DC_p \end{bmatrix}, \quad (28)$$

$$\widehat{B}_w = \begin{bmatrix} B_w \\ B_w - DC_w \end{bmatrix}, \quad (29)$$

$$\overline{D} = \begin{bmatrix} D_1 & 0 \end{bmatrix}, \quad (30)$$

$$\overline{D}_d = \begin{bmatrix} D_d & 0 \end{bmatrix}, \quad (31)$$

$$\overline{L} = \begin{bmatrix} L - K & K \end{bmatrix}. \quad (32)$$

Then the H_∞ filtering based synchronization problem becomes one of finding the parameters of the H_∞ filter (12)-(14) that guarantee the $\mathcal{L}_2$-norm γ attenuation level of the system (19)-(22), for any prescribed scalar γ. More explicitly, given a scalar $\gamma > 0$, our purpose will be to provide the conditions under which the performance index

$$\int_0^\infty \left(\overline{z}^T(t)\overline{z}(t) - \gamma^2 w^T(t) w(t) \right) dt < 0 \quad (33)$$

is achieved.

3 The Main Result

Sufficient conditions, given in terms of linear matrix inequalities, to solve the filtering based synchronization problem as has been defined above are stated in the following theorem.

Theorem 1. *Given $\gamma > 0$, the filtering based synchronization problem as has been described above is achieved for all $w(t) \in \mathcal{L}_2^q[0,\infty]$, and for all positive delays: $\tau_d > 0$ and $\tau \in [\tau_1, \tau_2]$, if there exist positive definite matrices Q_{11}, Q_{22}, R_{11}, R_2, V_1, V_2 and $R_3 \in I\!R^{n\times n}$, and symmetric positive definite matrix $S \in \mathcal{S}_\Delta$, and antisymmetric matrix $Y_a \in \mathcal{T}_{\Delta^T}$, and symmetric matrices $\overline{X}_1 \cdots \overline{X}_6$, $\overline{Z}_1 \cdots \overline{Z}_6$ and matrices Q_2, Q_3, X, Y, $\overline{Y}_1 \cdots \overline{Y}_6$ with appropriate dimensions that satisfy the following linear matrix inequalities:*

$$\begin{bmatrix} \overline{X}_1 & \overline{Y}_1^T \\ \overline{Y}_1 & \overline{Z}_1 \end{bmatrix} > 0, \begin{bmatrix} \overline{X}_2 & \overline{Y}_2^T \\ \overline{Y}_2 & \overline{Z}_2 \end{bmatrix} > 0, \begin{bmatrix} \overline{X}_3 & \overline{Y}_3^T \\ \overline{Y}_3 & \overline{Z}_3 \end{bmatrix} > 0$$
$$\begin{bmatrix} \overline{X}_4 & \overline{Y}_4^T \\ \overline{Y}_4 & \overline{Z}_4 \end{bmatrix} > 0, \begin{bmatrix} \overline{X}_5 & \overline{Y}_5^T \\ \overline{Y}_5 & \overline{Z}_5 \end{bmatrix} > 0, \begin{bmatrix} \overline{X}_6 & \overline{Y}_6^T \\ \overline{Y}_6 & \overline{Z}_6 \end{bmatrix} > 0 \tag{34}$$

$$\begin{bmatrix}
\Phi_{11} & * & * & * & * & * & * & * & * & * & * & * \\
\Phi_{21} & \Psi_1 & * & * & * & * & * & * & * & * & * & * \\
\Phi_{31} & 0 & \Psi_2 & * & * & * & * & * & * & * & * & * \\
\Phi_{41} & 0 & 0 & \Psi_3 & * & * & * & * & * & * & * & * \\
\Phi_{51} & 0 & \Phi_{53} & 0 & \Psi_4 & * & * & * & * & * & * & * \\
\Phi_{61} & 0 & 0 & 0 & 0 & \Psi_5 & * & * & * & * & * & * \\
\Phi_{71} & 0 & 0 & 0 & 0 & 0 & \Psi_6 & * & * & * & * & * \\
\Phi_{81} & 0 & \Phi_{83} & 0 & 0 & 0 & 0 & \Psi_7 & * & * & * & * \\
\Phi_{91} & 0 & 0 & 0 & 0 & 0 & 0 & 0 & \Psi_8 & * & * & * \\
\Phi_{101} & 0 & 0 & 0 & 0 & 0 & 0 & 0 & 0 & \Psi_9 & * & * \\
\Phi_{111} & 0 & 0 & 0 & 0 & 0 & 0 & 0 & 0 & 0 & \Psi_{10} & * \\
\Phi_{121} & 0 & \Phi_{123} & 0 & 0 & 0 & 0 & 0 & 0 & 0 & 0 & \Psi_{11}
\end{bmatrix} < 0 \tag{35}$$

where

$$\widehat{X} = \begin{bmatrix} A^T Q_{11} & (A+A_d)^T Q_{22} - X^T - C^T Y^T \\ 0 & X^T - A_d^T Q_{22} \end{bmatrix},$$
$$\Phi_{11} = \begin{bmatrix} Q_2 + Q_2^T & * \\ Q_3^T - Q_2 + \widehat{X} & -Q_3 - Q_3^T \end{bmatrix} + \sum_{k=1}^{5} \overline{X}_k + (\tau_2 - \tau_1)\overline{X}_6, \tag{36}$$

$$\Phi_{21} = \overline{Y}_1 + diag(V_1^{-1}, V_2^{-1}) \begin{bmatrix} 0 & \widehat{E} \end{bmatrix}, \tag{37}$$

$$\Phi_{31} = \overline{Y}_2 + S^{-1} \begin{bmatrix} 0 & 0 & D_1 & 0 \end{bmatrix}, \tag{38}$$

$$\Phi_{41} = \overline{Y}_3 + S^{-1} \begin{bmatrix} 0 & 0 & D_d & 0 \end{bmatrix}, \tag{39}$$

$$\Phi_{51} = \overline{Y}_4 + \begin{bmatrix} L_w B_w^T Q_{11} & L_w B_w^T Q_{22} - L_w C_w^T Y^T & L - K & K \end{bmatrix}, \tag{40}$$

$$\Phi_{53} = L_w D_w^T S^{-1}, \tag{41}$$

$$\Phi_{61} = \overline{Y}_5 + \begin{bmatrix} 0 & 0 & R_{11}A_d & 0 \\ 0 & 0 & -YC_d & Q_{22}A_d \end{bmatrix}, \tag{42}$$

$$\Phi_{71} = \sqrt{\tau_2 - \tau_1}\Big(\overline{Y}_6 - R_2^{-1}A_3\left[0\ I\right]\Big), \tag{43}$$

$$\Phi_{81} = \left[B_p^T Q_{11}\ B_p^T Q_{22} - C_p^T Y^T\ 0\ 0\right], \tag{44}$$

$$\Phi_{83} = D_P^T S^{-1} + Y_a^T, \tag{45}$$

$$\Phi_{91} = \begin{bmatrix} Q_{11} & 0 & 0\,0 \\ 0 & Q_{22} & 0\,0 \end{bmatrix}, \tag{46}$$

$$\Phi_{101} = \left[Q_2\ Q_3\right], \tag{47}$$

$$\Phi_{111} = \sqrt{\tau_2 - \tau_1}\left[Q_2\ Q_3\right], \tag{48}$$

$$\Phi_{121} = \left[B_w^T Q_{11}\ B_w^T Q_{22} - C_w^T Y^T\ 0\ 0\right], \tag{49}$$

$$\Phi_{123} = D_w^T S^{-1}, \tag{50}$$

$$\Psi_1 = \overline{Z}_1 - diag(V_1^{-1}, V_2^{-1}), \tag{51}$$

$$\Psi_2 = \overline{Z}_2 + R_3 - S^{-1},\ \ \Psi_3 = \overline{Z}_3 - R_3, \tag{52}$$

$$\Psi_4 = \overline{Z}_4 + L_w L_w^T - \gamma^2 I,\ \ \Psi_5 = \overline{Z}_5 - diag(R_{11}, Q_{22}), \tag{53}$$

$$\Psi_6 = \overline{Z}_6 - R_2^{-1},\ \ \Psi_7 = -S^{-1}, \tag{54}$$

$$\Psi_8 = -diag(R_{11}, Q_{22}),\ \ \Psi_9 = -diag(V_1^{-1}, V_2^{-1}), \tag{55}$$

$$\Psi_{10} = -R_2^{-1},\ \ \Psi_{11} = -I. \tag{56}$$

If these linear matrix inequalities hold, then the γ attenuation level performance is satisfied, namely $\|H_{\overline{z}w}(s)\|_\infty \le \gamma$, where $H_{\overline{z}w}(s)$ is the transfer matrix of system (19)-(22). The gain matrices N, D, K of the filter (12)-(14) are given by:

$$N = Q_{22}^{-1}X, \qquad D = Q_{22}^{-1}Y \tag{57}$$

and the gain K is a free parameter satisfying the linear matrix inequalities (34)-(35).

Proof. Firstly, without loss of generality, we consider that $\phi(t) = \widehat{\phi}(t) = 0$, $\forall t \in [-\widehat{z}\ 0]$. Now, remark that the H_∞-norm of the system described in (19)-(22), and the following system:

$$\begin{aligned} \dot{\xi}(t) - \widehat{E}^T\dot{\xi}(t-\tau_d) &= A_1^T\xi(t) + A_2^T\xi(t-\tau) - A_3^T\int_{t-\tau_2}^{t-\tau}\dot{\xi}(s)\,ds \\ &\quad + \overline{D}^T\widehat{q}(t) + \overline{D_d}^T\widehat{q}(t-\tau) + \overline{L}^T\widetilde{z}(t) \end{aligned} \tag{58}$$

$$\widehat{w}(t) = \widehat{B}_w^T\xi(t) + D_w^T\widehat{q}(t) + L_w^T\widetilde{z}(t) \tag{59}$$

$$\widehat{p}(t) = \widehat{B}_p^T\xi(t) + D_p^T\widehat{q}(t) \tag{60}$$

$$\widehat{q}(t) = \Delta^T(t)\widehat{p}(t) \tag{61}$$

are equal, where $\xi(t) \in I\!R^{2n}$, $\widehat{w}(t) \in I\!R^{q}$, $\widetilde{z}(t) \in I\!R^{s}$, $\widehat{p}(t) \in I\!R^{n}$. Indeed, $\overline{\sigma}(H_{\widehat{w}\widetilde{z}}(jw))=\overline{\sigma}(H^T_{\widehat{w}\widetilde{z}}(-jw)) = \overline{\sigma}(H_{\widetilde{z}w}(jw))$, where $\overline{\sigma}(\#)$ denotes the largest singular value of #. In fact (58)-(61) represents the backward adjoint of (19)-(22). An equivalent descriptor form representation of (58) is given by

$$\dot{\xi}(t) = \eta(t) \tag{62}$$

$$\begin{aligned} 0 &= -\eta(t) + \widehat{E}^T\eta(t-\tau_d) + A_1^T\xi(t) + A_2^T\xi(t-\tau)\,ds \\ &\quad - A_3^T\int_{t-\tau_2}^{t-\tau}\eta(s)\,ds + \overline{D}^T\widehat{q}(t) + \overline{D_d}^T\widehat{q}(t-\tau) + \overline{L}^T\widetilde{z}(t). \end{aligned} \tag{63}$$

In association with (62)-(63), we consider the following lyapunov-krasovskii functional candidate

$$V(t) = V_1(t) + V_2(t) + V_3(t) + V_4(t) \tag{64}$$

where

$$V_1(t) = \left[\xi^T(t)\eta^T(t)\right] FP \begin{bmatrix} \xi(t) \\ \eta(t) \end{bmatrix} \tag{65}$$

$$V_2(t) = \int_{t-\tau}^{t} \xi^T(s)R_1\xi(s)\,ds \tag{66}$$

$$V_3(t) = \int_{-\tau_2}^{-\tau}\int_{t+s}^{t} \eta^T(\theta)R_2\eta(\theta)\,\mathrm{d}\theta\mathrm{d}s \tag{67}$$

$$V_4(t) = \int_{t-\tau_d}^{t} \eta^T(s)V\eta(s)\,ds \tag{68}$$

where $F = \begin{bmatrix} I_{2n} & 0 \\ 0 & 0 \end{bmatrix}$, $P = \begin{bmatrix} P_1 & 0 \\ P_2 & P_3 \end{bmatrix}$ and $P_1 = P_1^T > 0$, $R_1 = R_1^T > 0$, $R_2 = R_2^T > 0$, and $V = V^T > 0$ are the weighting matrices to be determined. It should be pointed out that by the proposed Lyapunov function we aim in addition to the obtention of a stability criterion of $\xi(t)$, after performing the derivative of this function with respect to t, to avoid any possible emergence of terms which have τ or τ_d as coefficient, and to keep only those which have $\tau_2 - \tau$ as coefficient. To show that, let us first performing the derivative of $V_1(t)$ with respect to t along the trajectories of (62)-(63), we get

$$
\begin{aligned}
\dot{V}_1(t) &= 2\psi^T(t)P^T \begin{bmatrix} 0 & I \\ A_1^T & -I \end{bmatrix} \psi(t) \\
&+ 2\psi^T(t)P^T \begin{bmatrix} 0 \\ \widehat{E}^T \end{bmatrix} \eta(t-\tau_d) \\
&+ 2\psi^T(t)P^T \begin{bmatrix} 0 \\ \overline{D}^T \end{bmatrix} \widehat{q}(t) \\
&+ 2\psi^T(t)P^T \begin{bmatrix} 0 \\ \overline{D}_d^T \end{bmatrix} \widehat{q}(t-\tau) \\
&+ 2\psi^T(t)P^T \begin{bmatrix} 0 \\ \overline{L}^T \end{bmatrix} \widetilde{z}(t) \\
&+ 2\psi^T(t)P^T \begin{bmatrix} 0 \\ A_2^T \end{bmatrix} \xi(t-\tau)\, ds \\
&- 2\int_{t-\tau_2}^{t-\tau} \psi^T(t)P^T \begin{bmatrix} 0 \\ A_3^T \end{bmatrix} \eta(s)\, ds
\end{aligned} \tag{69}
$$

where

$$
\psi(t) = \begin{bmatrix} \xi(t) \\ \eta(t) \end{bmatrix} \tag{70}
$$

Now, for real matrices $X_i = X_i^T$, Y_i $i = 1, \cdots, 6$ and $Z_i = Z_i^T$ let

$$
\begin{bmatrix} X_i & Y_i^T \\ Y_i & Z_i \end{bmatrix} > 0 \tag{71}
$$

Then, it is readily checked that

$$
0 \le \left[\, \psi^T(t)\ \#^T \,\right] \begin{bmatrix} X_i & Y_i^T \\ Y_i & Z_i \end{bmatrix} \begin{bmatrix} \psi(t) \\ \# \end{bmatrix} = \psi^T(t)X_i\psi(t) + 2\psi^T Y_i^T \# + \#^T Z_i \# \tag{72}
$$

where # represents a given matrix. Thus, after completing the derivative of $V(t)$ with respect to t along the trajectories of (62)-(63), and adding the six inequalities (72), where # takes values in the following set:

$$
\{\, \eta(t-\tau_d),\ \widehat{q}(t),\ \widehat{q}(t-\tau),\ \widetilde{z}(t),\ \xi(t-\tau),\ \eta(s) \,\}
$$

and thereafter arranging terms, we obtain the following condition:

$$
\dot{V}(t) \le \int_{t-\tau_2}^{t-\tau} \Omega^T(t,s)\nabla_0(\tau)\Omega(t,s)\, ds \tag{73}
$$

where

$$\nabla_0(\tau) = \begin{bmatrix} \Sigma_0 & * & * \\ (\tau_2-\tau)^{-1}\left(Y_1 + \begin{bmatrix} 0 \ \widehat{E} \end{bmatrix} P\right) & (\tau_2-\tau)^{-1}(-V+Z_1) & * \\ (\tau_2-\tau)^{-1}\left(Y_2 + \begin{bmatrix} 0 \ \overline{D} \end{bmatrix} P\right) & 0 & (\tau_2-\tau)^{-1}Z_2 \\ (\tau_2-\tau)^{-1}\left(Y_3 + \begin{bmatrix} 0 \ \overline{D}_d \end{bmatrix} P\right) & 0 & 0 \\ (\tau_2-\tau)^{-1}\left(Y_4 + \begin{bmatrix} 0 \ \overline{L} \end{bmatrix} P\right) & 0 & 0 \\ (\tau_2-\tau)^{-1}\left(Y_5 + A_2 \begin{bmatrix} 0 \ I \end{bmatrix} P\right) & 0 & 0 \\ \left(Y_6 - A_3 \begin{bmatrix} 0 \ I \end{bmatrix} P\right) & 0 & 0 \end{bmatrix}$$

$$\begin{bmatrix} * & * & * & * \\ * & * & * & * \\ * & * & * & * \\ (\tau_2-\tau)^{-1}Z_3 & * & * & * \\ 0 & (\tau_2-\tau)^{-1}Z_4 & * & * \\ 0 & 0 & (\tau_2-\tau)^{-1}(-R_1+Z_5) & * \\ 0 & 0 & 0 & -R_2+Z_6 \end{bmatrix} \tag{74}$$

$$\Omega(t,s) = \begin{bmatrix} \psi(t)^T \ \eta^T(t-\tau_d) \ \widehat{q}^T(t) \ \widehat{q}^T(t-\tau) \ \widetilde{z}^T(t) \ \xi^T(t-\tau) \ \eta^T(s) \end{bmatrix}^T \tag{75}$$

$$\begin{aligned} \Sigma_0 = (\tau_2-\tau)^{-1}\Bigg(P^T \begin{bmatrix} 0 & I \\ A_1^T & -I \end{bmatrix} + \begin{bmatrix} 0 & A_1 \\ I & -I \end{bmatrix} P + X_1 + X_2 + X_3 + X_4 \\ + X_5 + \begin{bmatrix} I \\ 0 \end{bmatrix} R_1 \begin{bmatrix} I \ 0 \end{bmatrix} + \begin{bmatrix} 0 \\ I \end{bmatrix} V \begin{bmatrix} 0 \ I \end{bmatrix}\Bigg) + X_6 + \begin{bmatrix} 0 \\ I \end{bmatrix} R_2 \begin{bmatrix} 0 \ I \end{bmatrix} \end{aligned} \tag{76}$$

Now we are ready to deal with the prescribed γ attenuation level problem. The problem is to seek a criterion under which we have $\|H_{\overline{z}w}(s)\|_\infty \leq \gamma$, or equivalently $\|H_{\widehat{w}\widetilde{z}}(s)\|_\infty \leq \gamma$. This problem can be presented otherwise: seeking criterion that will ensure that the performance index

$$\int_0^\infty \left(\widehat{w}^T(t)\widehat{w}(t) - \gamma^2\widetilde{z}^T(t)\widetilde{z}(t)\right) dt < 0 \tag{77}$$

is satisfied. To this end, we shall introduce at the outset some useful results which will help us in our analysis. Firstly, let $S \in \mathcal{S}_\Delta$ and $T \in \mathcal{T}_{\Delta^T}$. It is easy to see that $S \in \mathcal{S}_\Delta$ is equivalent to $S \in \mathcal{S}_{\Delta^T}$. On the other hand, since $S > 0$, we may deduce also that $S^{\frac{1}{2}} \in \mathcal{S}_\Delta$ and $S^{\frac{1}{2}} \in \mathcal{S}_{\Delta^T}$. Thus, based on these results, it is readily checked that for $\widehat{q} = \Delta^T(t)\widehat{p}$, we have $\widehat{q}^T S\widehat{q} \leq \widehat{p}^T S\widehat{p}$ and $\widehat{q}^T T\widehat{p} + \widehat{p}^T T^T\widehat{q} = 0$. Indeed: $\widehat{q}^T S\widehat{q} = \widehat{p}^T \Delta(t)S\Delta^T(t)\widehat{p} = \widehat{p}^T \Delta(t)S^{\frac{1}{2}}S^{\frac{1}{2}}\Delta^T(t)\widehat{p} = \widehat{p}^T S^{\frac{1}{2}}\Delta(t)\Delta^T(t)S^{\frac{1}{2}}\widehat{p} \leq \widehat{p}^T S\widehat{p}$ since $\Delta(t)\Delta^T(t) \leq I$, and $\widehat{q}^T T\widehat{p} + \widehat{p}^T T^T\widehat{q} = \widehat{p}^T \Delta(t)T\widehat{p} + \widehat{p}^T T^T\Delta^T(t)\widehat{p} = \widehat{p}^T \Delta(t)T\widehat{p} - \widehat{p}^T T\Delta^T(t)\widehat{p} = 0$ since $T \in \mathcal{T}_{\Delta^T}$. Therefore, we may deduce that the two following functional terms:

$$\widehat{V}(t) = \int_0^t \begin{bmatrix} \widehat{p}(s) \\ \widehat{q}(s) \end{bmatrix}^T \begin{bmatrix} S & T^T \\ T & -S \end{bmatrix} \begin{bmatrix} \widehat{p}(s) \\ \widehat{q}(s) \end{bmatrix} ds \tag{78}$$

$$\overline{V}(t) = \int_{t-\tau}^{t} \widehat{q}^T(s) S R_3 S \widehat{q}(s)\, ds \tag{79}$$

are positives, where $S \in \mathcal{S}_\Delta$, $T \in \mathcal{T}_{\Delta^T}$, $\widehat{q} = \Delta^T(t)\widehat{p}$, and $R_3 > 0$. Also we have $V(0) = \widehat{V}(0) = \overline{V}(0) = 0$ ($\overline{V}(0) = 0$ since $\xi(t) = 0\ \forall t \in [-\widehat{\tau},\ 0]$ implies that $\widehat{q}(t) = 0\ \forall t \in [-\widehat{\tau},\ 0]$) , and this enables us to conclude that:

$$\begin{aligned} \int_0^\infty \left(\widehat{w}^T(t)\widehat{w}(t) - \gamma^2 \widetilde{z}^T(t)\widetilde{z}(t)\right) dt &\leq \int_0^\infty \Big(\widehat{w}^T(t)\widehat{w}(t) \\ &\quad -\gamma^2 \widetilde{z}^T(t)\widetilde{z}(t) + \dot{V}(t) + \dot{\widehat{V}}(t) + \dot{\overline{V}}(t)\Big)\, dt \\ = \int_0^\infty \Big[\int_{t-\tau_2}^{t-\tau} (\tau_2 - \tau)^{-1} [\widehat{w}^T(t)\widehat{w}(t) &- \gamma^2 \widetilde{z}^T(t)\widetilde{z}(t) \\ &+ \dot{\widehat{V}}(t) + \dot{\overline{V}}(t)]\, ds + \dot{V}(t)\Big]\, dt. \end{aligned} \tag{80}$$

Consequently, after performing the derivative of $V(t)$, $\widehat{V}(t)$ and $\overline{V}(t)$ with respect to t along the trajectories of (58)-(61), and arranging the terms, we obtain the following result:

$$\begin{aligned} &\int_0^\infty \left(\widehat{w}^T(t)\widehat{w}(t) - \gamma^2 \widetilde{z}^T(t)\widetilde{z}(t)\right) dt \\ &\qquad \leq \int_0^\infty \int_{t-\tau_2}^{t-\tau} \Omega^T(t,s) \nabla_1(\tau) \Omega(t,s)\, \mathrm{d}s\, \mathrm{d}t \end{aligned} \tag{81}$$

where

$$\nabla_1(\tau) = \begin{bmatrix} (1.1) & * & * & * & * & * & * \\ (2.1) & (\tau_2-\tau)^{-1}(-V+Z_1) & * & * & * & * & * \\ (3.1) & 0 & (3.3) & * & * & * & * \\ (4.1) & 0 & 0 & * & * & * & * \\ (5.1) & 0 & (\tau_2-\tau)^{-1}L_w D_w & (4.4) & * & * & * \\ (6.1) & 0 & 0 & 0 & (5.5) & * & * \\ Y_6 - A_3 \left[0\ I\right] P & 0 & 0 & 0 & 0 & (\tau_2-\tau)^{-1}(-R_1+Z_5) & * \\ & & & 0 & 0 & 0 & -R_2+Z_6 \end{bmatrix} \tag{82}$$

$$\Sigma_1 = \Sigma_0 + (\tau_2 - \tau)^{-1} \begin{bmatrix} I \\ 0 \end{bmatrix} \widehat{B}_w \widehat{B}_w^T \left[I\ 0\right] \tag{83}$$

$$(1.1) = \Sigma_1 + (\tau_2 - \tau)^{-1} \begin{bmatrix} I \\ 0 \end{bmatrix} \widehat{B}_p S \widehat{B}_p^T \left[I\ 0\right] \tag{84}$$

$$(2.1) = (\tau_2 - \tau)^{-1}\left(Y_1 + \left[0\ \widehat{E}\right] P\right) \tag{85}$$

$$(3.1) = (\tau_2 - \tau)^{-1} \begin{pmatrix} Y_2 + (D_w \widehat{B}_w^T + D_p S \widehat{B}_p^T + T\widehat{B}_p^T) \begin{bmatrix} I \; 0 \end{bmatrix} \\ + \begin{bmatrix} D_1 \; 0 \end{bmatrix} \begin{bmatrix} 0 \; I \end{bmatrix} P \end{pmatrix} \tag{86}$$

$$(3.3) = (\tau_2 - \tau)^{-1} \begin{pmatrix} Z_2 + D_p S D_p^T + D_p T^T \\ +T D_p^T + D_w D_w^T + S R_3 S - S \end{pmatrix} \tag{87}$$

$$(4.1) = (\tau_2 - \tau)^{-1} \Big(Y_3 + \begin{bmatrix} D_d \; 0 \end{bmatrix} \begin{bmatrix} 0 \; I \end{bmatrix} P \Big) \tag{88}$$

$$(4.4) = (\tau_2 - \tau)^{-1} (Z_3 - S R_3 S) \tag{89}$$

$$(5.1) = (\tau_2 - \tau)^{-1} \Big(Y_4 + \begin{bmatrix} L - K \; K \end{bmatrix} \begin{bmatrix} 0 \; I \end{bmatrix} P + L_w \widehat{B}_w^T \begin{bmatrix} I \; 0 \end{bmatrix} \Big) \tag{90}$$

$$(5.5) = (\tau_2 - \tau)^{-1} (Z_4 + L_w L_w^T - \gamma^2 I) \tag{91}$$

$$(6.1) = (\tau_2 - \tau)^{-1} \Big(Y_5 + A_2 \begin{bmatrix} 0 \; I \end{bmatrix} P \Big) \tag{92}$$

Therefore, the problem becomes one of satisfying the following condition

$$\nabla_1(\tau) < 0 \tag{93}$$

which becomes after pre and post multiplying by the following transformation

$$\Xi_1 = \mathrm{diag}\{\sqrt{\tau_2 - \tau} I, \sqrt{\tau_2 - \tau} I, \sqrt{\tau_2 - \tau} I, \sqrt{\tau_2 - \tau} I, \sqrt{\tau_2 - \tau} I, \sqrt{\tau_2 - \tau} I, I\} \tag{94}$$

and thereafter performing the Schur formula to the quadratic term on S in the (1,1) entry of $\nabla_1(\tau)$ to the following condition:

$$\left[\begin{array}{ccccccccc} (\tau_2 - \tau)\Sigma_1 & * & * & * & * & * & * & * \\ Y_1 + \begin{bmatrix} 0 \; \widehat{E} \end{bmatrix} P & -V + Z_1 & * & * & * & * & * & * \\ (3.1) & 0 & (3.3) & * & * & * & * & * \\ Y_3 + \begin{bmatrix} D_d \; 0 \end{bmatrix} \begin{bmatrix} 0 \; I \end{bmatrix} P & 0 & 0 & Z_3 - S R_3 S & * & * & * & * \\ \begin{pmatrix} Y_4 + \begin{bmatrix} L - K \; K \end{bmatrix} \begin{bmatrix} 0 \; I \end{bmatrix} P \\ + L_w \widehat{B}_w^T \begin{bmatrix} I \; 0 \end{bmatrix} \end{pmatrix} & 0 & L_w D_w^T & 0 & Z_4 + L_w L_w^T - \gamma^2 I & * & * & * \\ Y_5 + A_2 \begin{bmatrix} 0 \; I \end{bmatrix} P & 0 & 0 & 0 & 0 & -R_1 + Z_5 & * & * \\ \sqrt{\tau_2 - \tau} \Big(Y_6 - A_3 \begin{bmatrix} 0 \; I \end{bmatrix} P \Big) & 0 & 0 & 0 & 0 & 0 & -R_2 + Z_6 & * \\ \widehat{B}_p^T \begin{bmatrix} I \; 0 \end{bmatrix} & 0 & 0 & 0 & 0 & 0 & * & -S^{-1} \end{array} \right] < 0 \tag{95}$$

with

$$(3.1) = Y_2 + \Big(D_w \widehat{B}_w^T + D_P S \widehat{B}_p^T + T \widehat{B}_p^T \Big) \begin{bmatrix} I \; 0 \end{bmatrix} + \begin{bmatrix} D_1 \; 0 \end{bmatrix} \begin{bmatrix} 0 \; I \end{bmatrix} P \tag{96}$$

$$(3.3) = Z_2 + D_p S D_p^T + D_p T^T + T D_p^T + D_w D_w^T + S R_3 S - S \tag{97}$$

It is readily seen that this problem is computationally untractable since it is non convex. Thus next, we shall try to give rise to tractable problem. Firstly, observe from the (1.1) entry of (95) and from (83) and (76) that $-(P_3 + P_3^T)$ must be negative definite since Σ_1, and thus Σ_0, must be negative definite, which in conjunction with the requirement of $P_1 > 0$, implies that P is non-singular. Hence, let:

$$Q = P^{-1} = \begin{bmatrix} Q_1 & 0 \\ Q_2 & Q_3 \end{bmatrix} \tag{98}$$

$$\Xi_2 = \begin{bmatrix} Q^T & 0 & 0 & 0 & 0 & 0 & 0 & 0 \\ 0 & I & 0 & 0 & 0 & 0 & 0 & 0 \\ 0 & 0 & I & 0 & 0 & 0 & 0 & -D_p S - T \\ 0 & 0 & 0 & I & 0 & 0 & 0 & 0 \\ 0 & 0 & 0 & 0 & I & 0 & 0 & 0 \\ 0 & 0 & 0 & 0 & 0 & I & 0 & 0 \\ 0 & 0 & 0 & 0 & 0 & 0 & I & 0 \\ 0 & 0 & 0 & 0 & 0 & 0 & 0 & I \end{bmatrix} \tag{99}$$

Now, first pre and post multiplying (95) by Ξ_2 and Ξ_2^T, respectively, which leads to the following condition:

$$\begin{bmatrix} (1.1) & * & * & * & * & * & * & * \\ (2.1) & -V + Z_1 & * & * & * & * & * & * \\ (3.1) & 0 & (3.3) & * & * & * & * & * \\ (4.1) & 0 & 0 & (4.4) & * & * & * & * \\ (5.1) & 0 & L_w D_w^T & 0 & (5.5) & * & * & * \\ (6.1) & 0 & 0 & 0 & 0 & -R_1 + Z_5 & * & * \\ (7.1) & 0 & 0 & 0 & 0 & 0 & -R_2 + Z_6 & * \\ (8.1) & 0 & (8.3) & 0 & 0 & 0 & 0 & -S^{-1} \end{bmatrix} < 0 \tag{100}$$

where

$$(1.1) = (\tau_2 - \tau) Q^T \Sigma_1 Q \tag{101}$$

$$(2.1) = Y_1 Q + \begin{bmatrix} 0 & \widehat{E} \end{bmatrix} \tag{102}$$

$$(3.1) = Y_2 Q + D_w \widehat{B}_w^T \begin{bmatrix} I & 0 \end{bmatrix} Q + \begin{bmatrix} D_1 & 0 \end{bmatrix} \begin{bmatrix} 0 & I \end{bmatrix} \tag{103}$$

$$(3.3) = Z_2 + S R_3 S - S - T S^{-1} T^T + D_w D_w^T \tag{104}$$

$$(4.1) = Y_3 Q + \begin{bmatrix} D_d & 0 \end{bmatrix} \begin{bmatrix} 0 & I \end{bmatrix} \tag{105}$$

$$(4.4) = Z_3 - S R_3 S \tag{106}$$

$$(5.1) = Y_4 Q + \begin{bmatrix} L - K & K \end{bmatrix} \begin{bmatrix} 0 & I \end{bmatrix} + L_w \widehat{B}_w^T \begin{bmatrix} I & 0 \end{bmatrix} Q \tag{107}$$

$$(5.5) = Z_4 + L_w L_w^T - \gamma^2 I \tag{108}$$

$$(6.1) = Y_5 Q + A_2 \begin{bmatrix} 0 & I \end{bmatrix} \tag{109}$$

$$(7.1) = \sqrt{\tau_2 - \tau}\Big(Y_6 Q - A_3 \begin{bmatrix} 0 & I \end{bmatrix}\Big) \tag{110}$$

$$(8.1) = \widehat{B}_p^T \begin{bmatrix} I & 0 \end{bmatrix} Q \tag{111}$$

$$(8.3) = D_p^T + S^{-1} T^T \tag{112}$$

On the other hand, remark that the nonlinear term $(\tau_2 - \tau) Q^T \Sigma_1 Q$ can be rewritten as follows:

$$(\tau_2 - \tau) Q^T \Sigma_1 Q = \Sigma_2 + \Lambda^T \Omega \Lambda \tag{113}$$

where

$$\Sigma_2 = \begin{bmatrix} 0 & I \\ A_1^T & -I \end{bmatrix} Q + Q^T \begin{bmatrix} 0 & I \\ A_1^T & -I \end{bmatrix}^T + Q^T (X_1 + X_2 + X_3 + X_4 + X_5 + (\tau_2 - \tau) X_6) Q \tag{114}$$

$$\Lambda = \begin{bmatrix} Q^T \begin{bmatrix} I \\ 0 \end{bmatrix} & Q^T \begin{bmatrix} 0 \\ I \end{bmatrix} & \sqrt{\tau_2 - \tau} Q^T \begin{bmatrix} 0 \\ I \end{bmatrix} & Q^T \begin{bmatrix} I \\ 0 \end{bmatrix} \widehat{B}_w \end{bmatrix}^T \tag{115}$$

$$\Omega = diag\{R_1, V, R_2, I\}. \tag{116}$$

Thus, first applying the Schur formula to (1,1) entry of (100), and thereafter pre and post multiplying the resulting condition by Ξ_4 and Ξ_4^T, respectively, where

$$\Xi_4 = \begin{bmatrix} I & 0 & 0 & 0 & 0 & 0 & 0 & 0 & 0 & 0 & 0 & 0 \\ 0 & V^{-1} & 0 & 0 & 0 & 0 & 0 & 0 & 0 & 0 & 0 & 0 \\ 0 & 0 & S^{-1} & 0 & 0 & 0 & 0 & 0 & 0 & 0 & 0 & -S^{-1} D_w \\ 0 & 0 & 0 & S^{-1} & 0 & 0 & 0 & 0 & 0 & 0 & 0 & 0 \\ 0 & 0 & 0 & 0 & I & 0 & 0 & 0 & 0 & 0 & 0 & 0 \\ 0 & 0 & 0 & 0 & 0 & R_1^{-1} & 0 & 0 & 0 & 0 & 0 & 0 \\ 0 & 0 & 0 & 0 & 0 & 0 & R_2^{-1} & 0 & 0 & 0 & 0 & 0 \\ 0 & 0 & 0 & 0 & 0 & 0 & 0 & I & 0 & 0 & 0 & 0 \\ 0 & 0 & 0 & 0 & 0 & 0 & 0 & 0 & I & 0 & 0 & 0 \\ 0 & 0 & 0 & 0 & 0 & 0 & 0 & 0 & 0 & I & 0 & 0 \\ 0 & 0 & 0 & 0 & 0 & 0 & 0 & 0 & 0 & 0 & I & 0 \\ 0 & 0 & 0 & 0 & 0 & 0 & 0 & 0 & 0 & 0 & 0 & I \end{bmatrix} \tag{117}$$

and finally to conclude, pre and post multiplying the conditions given in (71) $i = 1 \cdots 6$ by

$$\text{diag}\{Q^T, V^{-1}\},\ \text{diag}\{Q^T, S^{-1}\},\ \text{diag}\{Q^T, S^{-1}\},$$
$$\text{diag}\{Q^T, I\},\ \text{diag}\{Q^T, R_1^{-1}\},\ \text{diag}\{Q^T, R_2^{-1}\}$$

and their transposes, respectively. Thus, we deduce that the H_∞ performance is guaranteed if the following matrix inequalities hold:

$$\nabla_3(\tau) < \Theta_1^T S \Theta_1 \tag{118}$$

$$\begin{aligned} &\begin{bmatrix} \overline{X}_1 & \overline{Y}_1^T \\ \overline{Y}_1 & \overline{Z}_1 \end{bmatrix} > 0, \begin{bmatrix} \overline{X}_2 & \overline{Y}_2^T \\ \overline{Y}_2 & \overline{Z}_2 \end{bmatrix} > 0, \begin{bmatrix} \overline{X}_3 & \overline{Y}_3^T \\ \overline{Y}_3 & \overline{Z}_3 \end{bmatrix} > 0 \\ &\begin{bmatrix} \overline{X}_4 & \overline{Y}_4^T \\ \overline{Y}_4 & \overline{Z}_4 \end{bmatrix} > 0, \begin{bmatrix} \overline{X}_5 & \overline{Y}_5^T \\ \overline{Y}_5 & \overline{Z}_5 \end{bmatrix} > 0, \begin{bmatrix} \overline{X}_6 & \overline{Y}_6^T \\ \overline{Y}_6 & \overline{Z}_6 \end{bmatrix} > 0 \end{aligned} \tag{119}$$

where

$$\nabla_3(\tau) = \begin{bmatrix} \Sigma_2 & * & * & * & * & * & * & * & * & * & * & * \\ (2.1) & -V^{-1}+\overline{Z}_1 & * & * & * & * & * & * & * & * & * & * \\ (3.1) & 0 & \overline{Z}_2 + R_3 - S^{-1} & * & * & * & * & * & * & * & * & * \\ (4.1) & 0 & 0 & \overline{Z}_3 - R_3 & * & * & * & * & * & * & * & * \\ (5.1) & 0 & L_w D_w^T S^{-1} & 0 & Z_4 + L_w L_w^T - \gamma^2 I & * & * & * & * & * & * & * \\ (6.1) & 0 & 0 & 0 & 0 & -R_1^{-1} + \overline{Z}_5 & * & * & * & * & * & * \\ (7.1) & 0 & 0 & 0 & 0 & 0 & -R_2^{-1} + \overline{Z}_6 & * & * & * & * & * \\ (8.1) & 0 & D_p^T S^{-1} + Y_a^T & 0 & 0 & 0 & 0 & -S^{-1} & * & * & * & * \\ \begin{bmatrix} I & 0 \end{bmatrix} Q & 0 & 0 & 0 & 0 & 0 & 0 & 0 & -R_1^{-1} & * & * & * \\ \begin{bmatrix} 0 & I \end{bmatrix} Q & 0 & 0 & 0 & 0 & 0 & 0 & 0 & 0 & -V^{-1} & * & * \\ (11.1) & 0 & 0 & 0 & 0 & 0 & 0 & 0 & 0 & 0 & -R_2^{-1} & * \\ (12.1) & 0 & D_w^T S^{-1} & 0 & 0 & 0 & 0 & 0 & 0 & 0 & 0 & -I \end{bmatrix} \tag{120}$$

$$\begin{aligned} \Sigma_2 = &\begin{bmatrix} 0 & I \\ A_1^T & -I \end{bmatrix} Q + Q^T \begin{bmatrix} 0 & I \\ A_1^T & -I \end{bmatrix}^T + \overline{X}_1 \\ &+ \overline{X}_2 + \overline{X}_3 + \overline{X}_4 + \overline{X}_5 + (\tau_2 - \tau)\overline{X}_6 \end{aligned} \tag{121}$$

$$(2.1) = \overline{Y}_1 + V^{-1} \begin{bmatrix} 0 & \widehat{E} \end{bmatrix} \tag{122}$$

$$(3.1) = \overline{Y}_2 + S^{-1} \begin{bmatrix} D_1 & 0 \end{bmatrix} \begin{bmatrix} 0 & I \end{bmatrix} \tag{123}$$

$$(4.1) = \overline{Y}_3 + S^{-1} \begin{bmatrix} D_d & 0 \end{bmatrix} \begin{bmatrix} 0 & I \end{bmatrix} \tag{124}$$

$$\begin{aligned} (5.1) = &\overline{Y}_4 + \begin{bmatrix} L - K & K \end{bmatrix} \begin{bmatrix} 0 & I \end{bmatrix} \\ &+ L_w \widehat{B}_w^T \begin{bmatrix} I & 0 \end{bmatrix} Q \end{aligned} \tag{125}$$

$$(6.1) = \overline{Y}_5 + R_1^{-1} A_2 \begin{bmatrix} 0 & I \end{bmatrix} \tag{126}$$

$$(7.1) = \sqrt{\tau_2 - \tau}\left(\overline{Y}_6 - R_2^{-1} A_3 \begin{bmatrix} 0 & I \end{bmatrix}\right) \tag{127}$$

$$(8.1) = \widehat{B}_p^T \begin{bmatrix} I & 0 \end{bmatrix} Q \tag{128}$$

$$(11.1) = \sqrt{\tau_2 - \tau} \begin{bmatrix} 0 & I \end{bmatrix} Q \tag{129}$$

$$(12.1) = \widehat{B}_w^T \begin{bmatrix} I & 0 \end{bmatrix} Q \tag{130}$$

$$\Theta_1 = \begin{bmatrix} 0 & 0 & Y_a^T & 0 & 0 & 0 & 0 & 0 & 0 & 0 \end{bmatrix} \tag{131}$$

where

$$\overline{X}_k = Q^T X_k Q, \quad k = 1 \ldots 6 \tag{132}$$

$$\overline{Y}_1 = V^{-1} Y_1 Q, \quad \overline{Y}_2 = S^{-1} Y_2 Q, \quad \overline{Y}_3 = S^{-1} Y_3 Q \tag{133}$$

$$\overline{Y}_4 = Y_4 Q, \quad \overline{Y}_5 = R_1^{-1} Y_5 Q, \quad \overline{Y}_6 = R_2^{-1} Y_6 Q \tag{134}$$

$$\overline{Z}_1 = V^{-1} Z_1 V^{-1}, \quad \overline{Z}_2 = S_1^{-1} Z_2 S_1^{-1}, \quad \overline{Z}_3 = S_1^{-1} Z_3 S_1^{-1} \tag{135}$$

$$\overline{Z}_5 = R_1^{-1} Z_5 R_1^{-1}, \quad \overline{Z}_6 = R_2^{-1} Z_6 R_2^{-1}, \quad Y_a = S^{-1} T S^{-1} \tag{136}$$

Remark that $Y_a \in \mathcal{T}_{\Delta^T}$. Indeed: lets $S \in \mathcal{S}_\Delta$, a simple pre and post multiplication of both sides of $S\Delta(t) = \Delta(t)S$ by S^{-1}, (S^{-1} exist since $S > 0$), allows us to deduce that $S^{-1} \in \mathcal{S}_\Delta$ and equivalently that $S^{-1} \in \mathcal{S}_{\Delta^T}$. Therefore, for $T \in \mathcal{T}_{\Delta^T}$, we have:

$$\begin{aligned} Y_a \Delta^T(t) &= S^{-1} T S^{-1} \Delta^T(t) = S^{-1} T \Delta^T(t) S^{-1} \\ &= S^{-1} \Delta(t) T S^{-1} = \Delta(t) S^{-1} T S^{-1} = \Delta(t) Y_a \end{aligned}$$

and it is readily checked that $Y_a^T = -Y_a$.
Now, to complet the linearization purpose, and thus render the problem a linear matrix inequalities feasible problem, we have just to take

$$Q_1 = \begin{bmatrix} Q_{11} & 0 \\ 0 & Q_{22} \end{bmatrix} > 0, \quad R_1 = \begin{bmatrix} R_{11} & 0 \\ 0 & Q_{22} \end{bmatrix}^{-1}, \quad V = \begin{bmatrix} V_1 & 0 \\ 0 & V_2 \end{bmatrix} > 0, \tag{137}$$

and expanding the terms of $\nabla_3(\tau)$. Until now, we have proved that the H_∞ performance is guaranteed for a given delay τ. To cope with all delays evolving in the interval $[\tau_1, \ \tau_2]$, we shall see that this is guaranteed if conditions (118), (119) hold for $\tau = \tau_1$. To see that, first let $h(\tau) = \sqrt{\tau_2 - \tau}$. It is readily checked that $0 \leq h(\tau) \leq h(\tau_1)$ and thus if conditions (118), (119) are guaranteed for $\tau = \tau_1$, then by convexity of the set $\{h(\tau) : (h(\tau))^2 A + h(\tau) B + C < 0\}$ if $A \geq 0$, and in our case

$$A = diag\{\overline{X}_6, 0, 0, 0, 0, 0, 0, 0, 0, 0\} \geq 0$$

conditions (118), (119) remain true for all $h = \sqrt{\tau_2 - \tau}$ evolving in the interval $[0, \ \sqrt{\tau_2 - \tau_1}]$, which is equivalent to τ evolving in the interval $[\tau_1, \ \tau_2]$. Therefore, to summarize the problem becomes one of satisfying conditions (118), (119) for $\tau = \tau_1$. It is worth noting that by the proposed method, we can ensure that the performance index (77) holds just inside the interval of evolution $[\tau_1, \ \tau_2]$, and outside this interval one can guarantee anything. To see that, let us take for instance an interval of evolution where the lower bound τ_1 is not nul and its size $\tau_2 - \tau_1$ is equal to the maximal allowable size obtained by conditions (118), (119) with $\tau = \tau_1$. Thereafter, suppose that, for instance, these conditions remain true for τ evolving in $[0, \ \tau_1[$, this imply that condition (118) holds for $\sqrt{\tau_2 - \tau_1} \leq h(\tau) \leq \sqrt{\tau_2}$ which is in contradiction with the fact that $\tau_2 - \tau_1$ is the maximal allowable size of the interval of evolution of the delay. On the other hand, one can see that $\nabla_3(\tau_1)$ is linear matrix which is required to be less than a positive term given by $\Theta_1^T S \Theta_1$. Hence, sufficient condition to solve this condition is to seek a criterion that will ensure that the linear matrix inequality $\nabla_3(\tau_1) < 0$ holds, which leads after developing terms to the LMI (35). This completes the proof of theorem 1. □

Remark 1. It is readily seen that the linear matrix inequalities in Theorem 1 depend on the size of the interval of evolution of the delay $(\tau_2 - \tau_1)$ and not on the maximal allowable delay τ_2 as it is used in the classical delay dependent approach. We can recover this case by taking $\tau_1 = 0$. Thus, the goal becomes one of seeking the largest interval $[\tau_1, \ \tau_2]$ under which the linear matrix inequalities (34) and (35) hold. Furthermore, since the problem did not depend on the delay, but on the size of its interval of evolution. We can take any interval in $I\!R^+$ where the size is less than the maximal allowable size value obtained while solving the linear matrix inequalities in Theorem 1, then the filtering based synchronization process can be guaranteed for any delay evolving in this interval since the linear matrix inequalities remain true.

Example 1. Consider a system with the same structure as in (1)-(6), where

$$
\begin{aligned}
E &= \begin{bmatrix} 0.3 & 0 \\ 0.2 & 0.2 \end{bmatrix}, A = \begin{bmatrix} -3 & 0 \\ -1 & -2 \end{bmatrix} \\
A_d &= \begin{bmatrix} 1 & 0 \\ -1 & 1 \end{bmatrix}, B_w = \begin{bmatrix} 0.2 \\ 0 \end{bmatrix}, C = \begin{bmatrix} 1 & 0 \end{bmatrix} \\
C_d &= \begin{bmatrix} 0 & 1 \end{bmatrix}, C_p = \begin{bmatrix} 0.3 & 0 \end{bmatrix}, \\
D_1 &= \begin{bmatrix} 0.1 & 0 \\ 0 & 0.5 \end{bmatrix}, D_d = \begin{bmatrix} 0.1 & 0 \\ 0 & 0.2 \end{bmatrix} \\
D_p &= \begin{bmatrix} 0.5 & 0 \\ 0 & 0.1 \end{bmatrix}, L = \begin{bmatrix} 1 & 0 \end{bmatrix}, L_w = 0.1, \\
B_p &= \begin{bmatrix} 0.5 & 0 \\ 0 & 0.1 \end{bmatrix}, C_w = 0.2, D_w = \begin{bmatrix} 0.3 & 0 \end{bmatrix}
\end{aligned}
$$

after solving linear matrix inequalities (34) and (35) of theorem 1, we obtain the following results

$$
Q_{11} = \begin{bmatrix} 5.2782 & 0.0005 \\ 0.0005 & 0.0003 \end{bmatrix}, R_{11} = \begin{bmatrix} 5.0368 & 0.0010 \\ 0.0010 & 0.0004 \end{bmatrix},
$$

$$Q_{22} = 1.0e{-}005 \begin{bmatrix} 0.1136\ 0.0025 \\ 0.0025\ 0.0001 \end{bmatrix},$$
$$Y_a = 1.0e-005 \begin{bmatrix} 0 & 0.1063 \\ -0.1063 & 0 \end{bmatrix},$$
$$V_1 = \begin{bmatrix} 0.0718 & -0.0538 \\ -0.0538 & 206.2 \end{bmatrix},$$
$$V_2 = \begin{bmatrix} 136.0869 & -92.3852 \\ -92.3852 & 756.1726 \end{bmatrix}.$$

For the sake of brevity, we will not present all the obtained results. The filter (12)-(14) parameters are given by:

$$N = 1.0e+003 \begin{bmatrix} -0.6288 & 0.0073 \\ 9.8577 & -0.6648 \end{bmatrix},$$
$$K = 1.0e-004 \begin{bmatrix} 0.5424\ 0.0092 \end{bmatrix},$$
$$D = \begin{bmatrix} -0.2156 \\ -15.8949 \end{bmatrix}.$$

The minimal attenuation level obtained is given by $\gamma_{min} = 0.3979$. The maximal allowable size of the interval of evolution of the delay obtained is $(\tau_2 - \tau_1)_{max} = 0.98$Sec. Which means that the obtained filter can guarantee the filtering based synchronization process of the considered system for any interval in $I\!R^+$ representing the interval of evolution of the delay where the size is less than 0.98Sec.

Discussion

The new delay dependent filtering based synchronization approach proposed in this work can be classified between the two classical delay independent and dependent filtering based synchronization approaches from conservative degree point of view. Indeed, it is less conservative than the delay independent approach since the information of the delay, given by its interval of evolution is taking into account, and more conservative than the delay dependent approach, since it guarantees the filtering based synchronization process for any interval of evolution of the delay in $I\!R^+$ with the size less than the maximal allowable size value of the interval of the evolution of the delay obtained. This is in contrast to the classical delay dependent approach, where the lower bound of the interval of evolution of the delay is always frozen to zeros. More explicitly, the proposed approach can especially applied to the cases of delays for which, the classical delay dependent filtering based synchronization approach can not be applied. Namely, if d is the maximal allowable delay under which the filtering based synchronization process can be guaranteed by the classical delay dependent approach, for $\tau > d$, our approach can be applied and this by just chosen an interval of evolution which contains the corresponding delay and where the size is less than the maximal allowable size value of the interval of evolution of the delay obtained while solving our problem. Hence, the comparaison to the others existing methods is not utile.

4 Conclusion

In this work, simple and useful technics to deal with the robust H_∞ filtering based synchronization problem for linear neutral systems subjected to structured uncertainties with unknown delays have been presented. Owing to the descriptor model transformation and a new delay dependent approach, sweet conditions have been provided in terms of linear matrix inequalities to reduce the effect of the external disturbances on the filtering based synchronization error to a certain prescribed level.

References

1. Kocarev L, Parlitz U (1995) General approach for chaotic synchronization with applications to communication. Phys. Res. Lett 5028.
2. Shahverdiev E.M et al (2002) Lag synchronization in time-delayed systems. Phys. Lett. A 292:320-324.
3. Peng J, Liao X.F (2003) Synchronization of coupled time-delay chaotic system and its application to secure communications. J. Comput. Res. Dev 40:263-268.
4. de Souza C.E, Palhares R.M, Peres P.L.D (2001) Robust H_∞ Filter Design for Uncertainty Linear Systems with Multiple Time-Varying State Delays. IEEE Transaction on Signal Processing 49:569-576 March.
5. Wang Z.D, Yang F.W (2002) Robust Filtering for Uncertain Linear Systems With Delayed States and Outputs. IEEE Transactions on Circuits And Systems–I: Fundamental Theory and Applications 49:125-130 January.
6. Ma W.B, Adachi N, Amemiya T (1995) Delay-independent stabilization of uncertain linear systems of neutral type. Journal of Optimization Theory and Application 84:393-405.
7. Fridman E, Shaked U, Xie L (2003) Robust H_∞ Filtering of Linear Systems with time-varying delay. IEEE Transaction on Automatic Control 48:159-165 January.
8. Fridman E, Shaked U (2002) A Descriptor System Approach to H_∞ Control of Linear Time-Delay Systems. IEEE Transaction on Automatic Control 47:253-270.
9. Fridman E, Shaked U (2004) An improved delay-dependent H-infinity filtering of linear neutral systems. IEEE Transactions on Signal Processing 52:668 – 673 March.
10. Alif A, Darouach M, Boutayeb M (2005) On the design of a reduced order H_∞ filter for systems with unknown delay. In Proc. *American Control Conference*, Portland, Oregon 2901 – 2906 January.

11. Alif A, Boutayeb M, Darouach M (2003) Observers-Based Control for Systems with Unknown Delays. In Proc. *4th IFAC Workshop on Time Delay Systems (TDS'03)*, Rocquencourt, France, September.
12. Alif A, Boutayeb M, Darouach M (2004) OOn the design of an observer-based controller for linear systems with unknown delay. In Proc. *2nd IFAC Symposium on System, Structure and Control*, Oaxaca, Mexico, 8 – 10 December.

Part IV

Decision Support System

Automata-Based Adaptive Behavior for Economic Modelling Using Game Theory

Rawan Ghnemat[1], Saleh Oqeili[1], Cyrille Bertelle[2], and Gérard H.E. Duchamp[3]

[1] Al-Balqa' Applied University,
Al-Salt, 19117
Jordan
rawan.ghnemat@gmail.com, saleh@bau.edu.jo

[2] LITIS – University of Le Havre,
25 rue Philippe Lebon
76620 Le Havre cedex, France
cyrille.bertelle@gmail.com

[3] LIPN – University of Paris XIII,
99 avenue Jean-Baptiste Clément
93430 Villetaneuse, France
gheduchamp@gmail.com

Summary. In this paper, we deal with some specific domains of applications to game theory. This is one of the major class of models in the new approaches of modelling in the economic domain. For that, we use genetic automata which allow to buid adaptive strategies for the players. We explain how the automata-based formalism proposed – matrix representation of automata with multiplicities – allows to define a semi-distance between the strategy behaviors. With that tools, we are able to generate an automatic processus to compute emergent systems of entities whose behaviors are represented by these genetic automata.

Key words: adaptive behavior, game theory, genetic automata, prisoner dilemma, emergent systems computing

1 Introduction: Adaptive Behaviour Modeling for Game Theory

Since the five last decades, game theory has become a major aspect in economic sciences modelling and in a great number of domains where strategical aspects has to be involved. Game theory is usually defined as a mathematical tool allowing to analyse strategical interactions between individuals.

Initially funded by mathematical researchers, J. von Neumann, E. Borel or E. Zermelo in 1920s, game theory increased in importance in the 1940s with a major work by J. von Neumann and O. Morgenstern and then with the works of John Nash in the 1950s [9]. John Nash has proposed an original equilibrium ruled by an adaptive criterium. In game theory, the Nash equilibrium is a kind of optimal strategy for games involving two or more players, whereby the players reach an outcome to mutual advantage. If there is a set of strategies for a game with the property that no player can benefit by changing his strategy while the other players keep their strategies unchanged, then this set of strategies and the corresponding payoffs constitute a Nash equilibrium.

We can understand easily that the modelization of a player behavior needs some adaptive properties . The computable model corresponding to genetic automata are in this way a good tool to modelize such adaptive strategy .

The plan of this paper is the following. In the next section, we present some efficient algebraic structures, the automata with multiplicities, which allow to implement powerful operators. We present in section 3, some topological considerations about the definition of distances between automata which induces a theorem of convergence on the automata behaviors. Genetic operators are proposed for these automata in section 4. For that purpose, we show that the relevant "calculus" is done by matrix representions unravelling then the powerful capabilities of such algebraic structures. In section 5, we focus our attention on the "iterated prisonner dilemma" and we buid an original evolutive probabilistic automaton for strategy modeling, showing that genetic automata are well-adapted to model adaptive strategies. Section 6 shows how we can use the genetic automata developed previously to represent agent evolving in complex systems description. An agent behavior semi-distance is then defined and allows to propose an automatic computation of emergent systems as a kind of self-organization detection.

2 Automata from Boolean to Multiplicies Theory (Automata with Scalars)

Automata are initially considered as theoretical tools. They are created in the 1950's following the works of A. Turing who previously deals with the definition of an abstract "machine". The aim of the Turing machines is to define the boundaries for what a computing machine could do and what it could not do.

The first class of automata, called finite state automata corresponds to simple kinds of machines [21]. They are studied by a great number of researchers as abstract concepts for computable building. In this aspect, we can recall the works of some linguist researchers, for example N. Chomsky who defined the study of formal grammars.

In many works, finite automata are associated to a recognizing operator which allows to describe a language [2,10]. In such works, the condition of a transition is simply a symbol taken from an alphabet. From a specific state S, the reading of a symbol a allows to make the transitions which are labeled by a

and *come fromS* (in case of a deterministic automaton – a DFA – there is only one transition – see below). A whole automaton is, in this way, associated to a language, the recognized language, which is a set of words. These recognized words are composed of the sequences of letters of the alphabet which allows to go from a specific state called initial state, to another specific state, called final state.

A first classification is based on the geometric aspect : DFA (Deterministic Finite Automata) and NFA (Nondeterministic Finite Automata).

- In Deterministic Finite Automata, for each state there is at most one transition for each possible input and only one initial state.
- In Nondeterministic Finite Automata, there can be none or more than one transition from a given state for a given possible input.

Besides the classical aspect of automata as machines allowing to recognize languages, another approach consists in associating to the automata a functional goal. In addition of accepted letter from an alphabet as the condition of a transition, we add for each transition an information which can be considered as an output data of the transition, the read letter is now called input data. We define in such a way an *automaton with outputs* or *weighted automaton.*

Such automata with outputs give a new classification of machines. *Transducers* are such a kind of machines, they generate outputs based on a given input and/or a state using actions. They are currently used for control applications. *Moore machines* are also such machines where output depends only on a state, i.e. the automaton uses only entry actions. The advantage of the Moore model is a simplification of the behaviour.

Finally, we focus our attention on a special kind of automata with outputs which are efficient in an operational way. This automata with output are called *automata with multiplicities.* An automaton with multiplicities is based on the fact that the output data of the automata with output belong to a specific algebraic structure, a semiring [13, 22]. In that way, we will be able to build effective operations on such automata, using the power of the algebraic structures of the output data and we are also able to describe this automaton by means of a matrix representation with all the power of the new (i.e. with semirings) linear algebra.

Definition 1. (Automaton with multiplicities)
An automaton with multiplicities over an alphabet A and a semiring K is the 5-uple (A, Q, I, T, F) where

- $Q = \{S_1, S_2 \cdots S_n\}$ *is the finite set of state;*
- $I : Q \mapsto K$ *is a function over the set of states, which associates to each initial state a value of K, called entry cost, and to non- initial state a zero value ;*
- $F : Q \mapsto K$ *is a function over the set states, which associates to each final state a value of K, called final cost, and to non-final state a zero value;*

- *T is the transition function, that is $T : Q \times A \times Q \mapsto K$ which to a state S_i, a letter a and a state S_j associates a value z of K (the cost of the transition) if it exist a transition labelled with a from the state S_i to the state S_j and and zero otherwise.*

Remark 1. Automata with multiplicities are a generalisation of finite automata. In fact, finite automata can be considered as automata with multiplicities in the semiring K, the boolan set $B = \{0, 1\}$ (endowed with the logical "or/and"). To each transition we affect 1 if it exists and 0 if not.

Remark 2. We have not yet, on purpose, defined what a semiring is. Roughly it is the least structure which allows the matrix "calculus" with unit (one can think of a ring without the "minus" operation). The previous automata with multiplicities can be, equivalently, expressed by a matrix representation which is a triplet

- $\lambda \in K^{1\times Q}$ which is a row-vector which coefficients are $\lambda_i = I(S_i)$,
- $\gamma \in K^{Q\times 1}$ is a column-vector which coefficients are $\gamma_i = F(S_i)$,
- $\mu : A^* \mapsto K^{Q\times Q}$ is a morphism of monoids (indeed $K^{Q\times Q}$ is endowed with the product of matrices) such that the coefficient on the q_ith row and q_jth column of $\mu(a)$ is $T(q_i, a, q_j)$

3 Topological Considerations

If K is a field, one sees that the space $\mathcal{A}_{(n)}$ of automata of dimension n (with multiplicities in K) is a K-vector space of dimension $k.n^2 + 2n$ (k is here the number of letters). So, in case the ground field is the field of real or complex numbers [3], one can take any vector norm (usually one takes one of the Hölder norms $||(x_i)_{i\in I}||_\alpha := \left(\sum_{i\in I} |x_i|^\alpha\right)^{\frac{1}{\alpha}}$ for $\alpha \geq 1$, but any norm will do) and the distance is derived, in the classical way, by

$$d(\mathcal{A}_1, \mathcal{A}_2) = norm(V(\mathcal{A}_1) - V(\mathcal{A}_2)) \tag{1}$$

where $V(\mathcal{A})$ stands for the vector of all coefficients of $\mathcal{A} = (\lambda, \mu, \gamma)$ arranged in some order one has then the result of Theorem 1. Assuming that K is the field of real or complex numbers, we endow the space of series/behaviours with the topology of pointwise convergence (Topology of F. Treves [23]).

Theorem 1. *Let $(\mathcal{A}_n)$ be a sequence of automata with limit $\mathcal{L}$ ($\mathcal{L}$ is an automaton), then one has*

$$Behaviour(\mathcal{L}) = \lim_{n\to\infty} Behaviour(\mathcal{A}_n) \tag{2}$$

where the limit is computed in the topology of Treves.

4 Genetic Automata as Efficient Operators

We define the chromosome for each automata with multiplicities as the sequence of all the matrices associated to each letter from the (linearly ordered) alphabet. The chromosomes are composed with alleles which are here the lines of the matrix [6].

In the following, genetic algorithms are going to generate new automata containing possibly new transitions from the ones included in the initial automata.

The genetic algorithm over the population of automata with multiplicities follows a reproduction iteration broken up in three steps [14, 17, 18]:

- *Duplication*: where each automaton generates a clone of itself;
- *Crossing-over*: concerns a couple of automata. Over this couple, we consider a sequence of lines of each matrix for all. For each of these matrices, a permutation on the lines of the chosen sequence is made between the analogue matrices of this couple of automata;
- *Mutation*: where a line of each matrix is randomly chosen and a sequence of new values is given for this line.

Finally the whole genetic algorithm scheduling for a full process of reproduction over all the population of automata is the evolutionary algorithm:

1. For all couple of automata, two children are created by duplication, crossover and mutation mechanisms;
2. The fitness for each automaton is computed;
3. For all 4-uple composed of parents and children, the performless automata, in term of fitness computed in previous step, are suppressed. The two automata, still living, result from the evolution of the two initial parents.

Remark 3. The fitness is not defined at this level of abstract formulation, but it is defined corresponding to the context for which the automaton is a model, as we will do in the next section.

5 Applications to Competition-Cooperation Modeling Using Prisoner Dilemma

We develop in this section how we can modelize competition-cooperation processes in a same automata-based representation. The genetic computation allows to make automatic transition from competition to cooperation or from coopeartion to competition. The basic problem used for this purpose is the well-known prisoner dilemma [1].

5.1 From Adaptive Strategies to Probabilistic Automata

The prisoner dilemma is a two-players game where each player has two possible actions: cooperate (C) with its adversary or betray him ($\overline{C}$). So, four outputs are possible for the global actions of the two players. A relative payoff is defined relatively to these possible outputs, as described in the following table where the rows correspond to one player behaviour and the columns to the other player one.

Table 1. Prisoner dilemma payoff

	C	$\overline{C}$
C	(3,3)	(0,5)
$\overline{C}$	(5,0)	(1,1)

In the iterative version of the prisoner's dilemma, successive steps can be defined. Each player do not know the action of its adversary during the current step but he knows it for the preceding step. So, different strategies can be defined for a player behaviour, the goal of each one is to obtain maximal payoff for himself.

In Figures 1 and 2, we describe two strategies with transducers. Each transition is labeled by the input corresponding to the player perception which is the precedent adversary action and the output corresponding to the present player action. The only inital state is the state 1, recognizable by the incoming arrow labeled only by the output. The final states are the states 1 and 2, recognizable with the double circles.

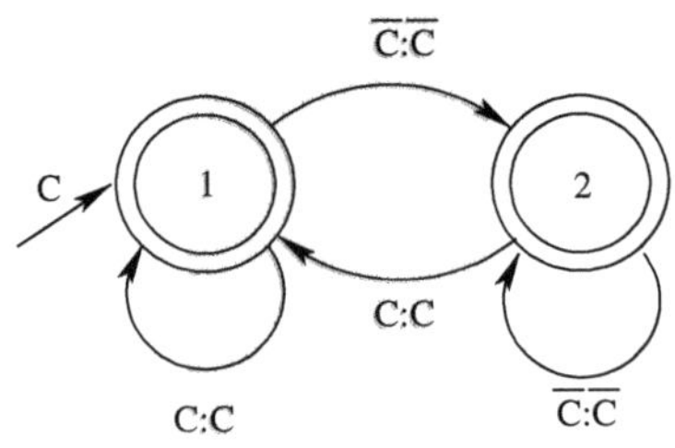

Fig. 1. Tit-for-tat strategy automaton

In the strategy of Figure 1, the player has systematically the same behaviour as its adversary at the previous step. In the strategy of Figure 2, the player chooses definitively to betray as soon as his adversary does it. The previous automaton represents static strategies and so they are not well adapted for the modelization of evolutive strategies. For this purpose, we propose a model based on a probabilistic automaton described by Figure 3 [5].

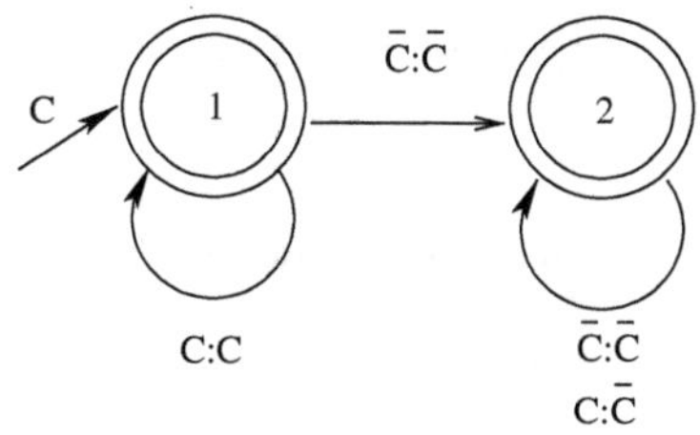

Fig. 2. Vindictive strategy automaton

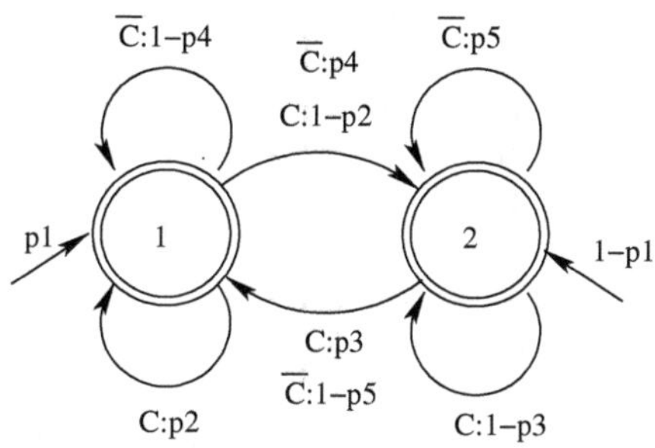

Fig. 3. Probabilistic multi-strategies two-states automaton

This automaton represents all the two-states strategies for cooperation and competitive behaviour of one agent against another in prisoner's dilemma.

The transitions are labeled in output by the probabilities p_i of their realization. The first state is the state reached after cooperation action and the second state is reached after betrayal.

For this automaton, the associated matrix representation, as described previously, is:

$$I = \begin{pmatrix} p_1 & 1 - p_1 \end{pmatrix} ; \tag{3}$$

$$F = \begin{pmatrix} p_6 \\ 1 - p_6 \end{pmatrix} ; \tag{4}$$

$$T(C) = \begin{pmatrix} p_2 & 1 - p_2 \\ p_3 & 1 - p_3 \end{pmatrix} ; \tag{5}$$

$$T(\overline{C}) = \begin{pmatrix} 1 - p_4 & p_4 \\ 1 - p_5 & p_5 \end{pmatrix} \tag{6}$$

5.2 From Probabilistic Automata to Genetic Automata

With the matrix representation of the automata, we can compute genetic automata as described in previous sections. Here the chromosomes are the sequences of all the matrices associated to each letter. We have to define the fitness in the context of the use of these automata. The fitness here is the value of the payoff.

5.3 General Genetic Algorithm Process for Genetic Automata

A population of automata is initially generated. These automata are playing against a predefined strategy, named S_0.

Each automaton makes a set of plays. At each play, we run the probabilistic automaton which gives one of the two outputs: (C) or $(\overline{C})$. With this output and the S_0's output, we compute the payoff of the automaton, according with the payoff table.

At the end of the set of plays, the automaton payoff is the sum of all the payoffs of each play. This sum is the fitness of the automaton. At the end of this set of plays, each automaton has its own fitness and so the selection process can select the best automata. At the end of these selection process, we obtain a new generation of automata.

This new generation of automata is the basis of a new computation of the 3 genetics operators.

This processus allows to make evolve the player's behavior which is modelized by the probabilistic multi-strategies two-states automaton from cooperation to competition or from competition to cooperation. The evolution of the strategy is the expression of an adaptive computation. This leads us to use this formalism to implement some self-organisation processes which occurs in complex systems.

6 Extension to Emergent Systems Modeling

In this section, we study how evolutive automata-based modeling can be used to compute automatic emergent systems. The emergent systems have to be understood in the meaning of complex system paradigm that we recall in the next section. We have previously defined some way to compute the distance between automata and we use these principles to define distance between agents behaviours that are modeled with automata. Finally, we defined a specific fitness that allows to use genetic algorithms as a kind of reinforcement method which leads to emergent system computation [15].

6.1 Complex System Description Using Automata-Based Agent Model

According to General System Theory [4,19], a complex system is composed of entities in mutual interaction and interacting with the outside environment. A system has some characteristic properties which confer its structural aspects, as schematically described in part (a) of Figure 4:

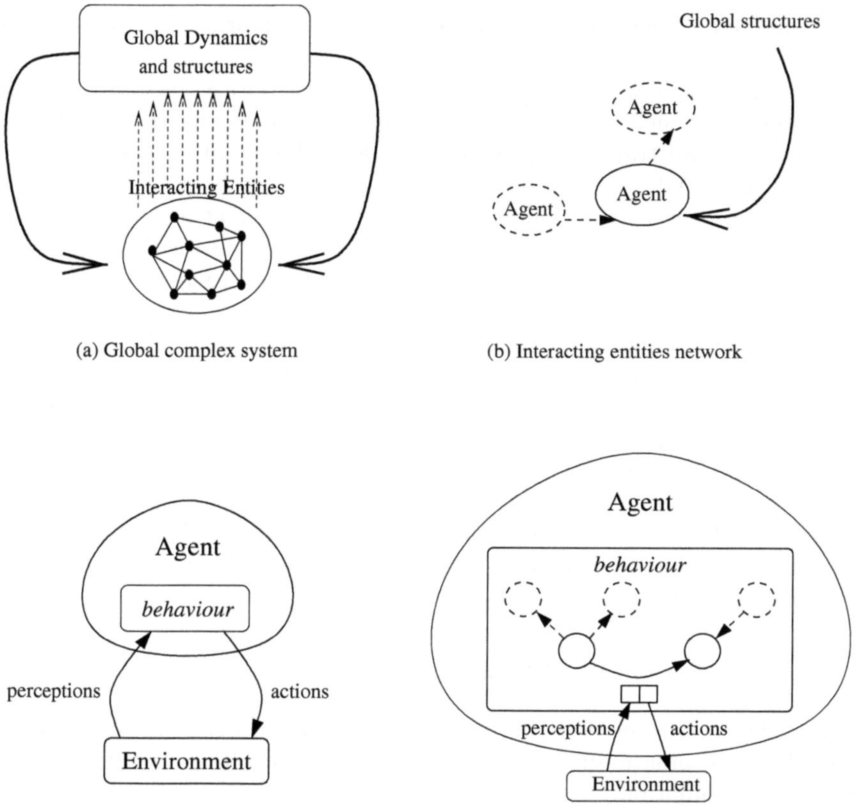

(a) Global complex system (b) Interacting entities network

(c) Agent–based model for entity (d) automata–based model for agent behaviour

Fig. 4. Multi-scale complex system description: from global to individual models

- The set elements or entities are in interactive dependance. The alteration of only one entity or one interaction reverberates on the whole system.
- A global organization emerges from interacting constitutive elements. This organization can be identified and carries its own autonomous behavior while it is in relation and dependance with its environment. The emergent organization possesses new properties that its own constitutive entities don't have. "The whole is more than the sum of its parts".
- The global organization retro-acts over its constitutive components. "The whole is less than the sum of its parts" after E. Morin.

The interacting entities network as described in part (b) of Figure 4 leads each entity to perceive informations or actions from other entities or from the whole system and to act itself.

A well-adapted modeling consists of using an agent-based representation which is composed of the entity called agent as an entity which perceives and acts on an environment, using an autonomous behaviour as described in part (c) of Figure 4.

To compute a simulation composed of such entities, we need to describe the behaviour of each agent. This one can be schematically described using internal states and transition processes between these states, as described in part (d) of Figure 4.

There are several definitions of "agents" or "intelligent agents" according to their behaviour specificities [11,24]. Their autonomy means that the agents try to satisfy a goal and execute actions, optimizing a satisfaction function to reach it.

For agents with high level autonomy, specific actions are realized even when no perception are detected from the environment. To represent the process of this deliberation, different formalisms can be used and a behaviour decomposed in internal states is an effective approach. Finally, when many agents operate, the social aspects must also be taken into account. These aspects are expressed as communications through agent organisation with message passing processes. Sending a message is an agent action and receiving a message is an agent perception. The previous description based on the couple: perception and action, is well adapted to this.

6.2 Agent Behavior Semi-Distance

We describe in this section the bases of the genetic algorithm used on the probabilistic automata allowing to manage emergent self-organizations in the multi-agent simulation.

For each agent, we define e an evaluation function of its own behaviour returning the matrix M of values such that $M_{i,j}$ is the output series from all possible successive perceptions when starting from the initial state i and ending at the final state j, without cycle. It will clearly be 0 if either i is not an initial state or j is not a final one and the matrix $M_{i,j}$ is indeed a matrix of evaluations [2] of subseries of

$$M^* := (\sum_{a\in A} \mu(a)a)^* \tag{7}$$

Notice that the coefficients of this matrix, as defined, are computed whatever the value of the perception in the alphabet A on each transition on the successful path[4]. That means that the contribution of the agent behaviour for collective organization formation is only based, here, on probabilities to reach a final state from an initial one. This allows to preserve individual characteristics in each agent behaviour even if the agent belongs to an organization.

Let x and y two agents and $e(x)$ and $e(y)$ their respective evaluations as described above. We define $d(x, y)$ a semi-distance (or pseudometrics, see [3] ch IX) between the two agents x and y as $||e(x) - e(y)||$, a matrix norm of

[4] A *succesful path* is a path from an initial state to a final state

the difference of their evaluations. Let $\mathcal{V}_x$ a neighbourhood of the agent x, relatively to a specific criterium, for example a spatial distance or linkage network. We define $f(x)$ the agent fitness of the agent x as :

$$f(x) = \begin{cases} \dfrac{card(\mathcal{V}_x)}{\sum\limits_{y_i \in \mathcal{V}_x} d(x,y_i)^2} & \text{if } \sum\limits_{y_i \in \mathcal{V}_x} d(x,y_i)^2 \neq 0 \\ \infty & \text{otherwise} \end{cases}$$

6.3 Evolutive Automata for Automatic Emergence of Self-Organized Agent- Based Systems

In the previous computation, we defined a semi-distance between two agents. This semi-distance is computed using the matrix representation of the automaton with multiplicities associated to the agent behaviour. This semi-distance is based on successful paths computation which needs to define initial and final states on the behaviour automata. For specific purposes, we can choose to define in some specific way, the initial and final states. This means that we try to compute some specific action sequences which are chararacterized by the way of going from some specific states (defined here as initial ones) to some specific states (defined here as final ones).

Based on this specific purpose which leads to define some initial and final states, we compute a behaviour semi-distance and then the fitness function defined previously. This fitness function is an indicator which returns high value when the evaluated agent is near, in the sense of the behaviour semi-distance defined previously, to all the other agents belonging to a predefined neighbouring.

Genetic algorithms will compute in such a way to make evolve an agent population in a selective process. So during the computation, the genetic algorithm will make evolve the population towards a newer one with agents more and more adapted to the fitness. The new population will contain agents with better fitness, so the agents of a population will become nearer each others in order to improve their fitness. In that way, the genetic algorithm reinforces the creation of a system which aggregates agents with similar behaviors, in the specific way of the definition of initial and final states defined on the automata.

The genetic algorithm proposed here can be considered as a modelization of the feed-back of emergent systems which leads to gather agents of similar behaviour, but these formations are dynamical and we cannot predict what will be the set of these aggregations which depends of the reaction of agents during the simulation. Moreover the genetic process has the effect of generating a feed- back of the emergent systems on their own contitutive elements in the way that the fitness improvement lead to bring closer the agents which are picked up inside the emergent aggregations.

For specific problem solving, we can consider that the previous fitness function can be composed with another specific one which is able to measure the capability of the agent to solve one problem. This composition of fitness functions leads to create emergent systems only for the ones of interest, that is, these systems are able to be developed only if the aggregated agents are able to satisfy some problem solving evaluation.

7 Conclusion

The aim of this study is to develop a powerful algebraic structure to represent behaviors concerning cooperation-competition processes and on which we can add genetic operators. We have explained how we can use these structures for modeling adaptive behaviors needed in game theory. More than for this application, we have described how we can use such adaptive computations to automatically detect emergent systems inside interacting networks of entities represented by agents in a simulation.

References

1. R. Axelrod (1997) *The complexity of cooperation*, Princeton University Press
2. J. Berstel and G. Reutenauer (1988) *Rational series and their language*, EATCS
3. N. Bourbaki (1998) *Elements of Mathematics: General Topology*, Chapters 5-10, Springer-Verlag Telos
4. L. von Bertalanffy (1968) *General System Theory*, Georges Braziller Ed.
5. C. Bertelle, M. Flouret, V. Jay, D. Olivier, and J.-L. Ponty (2002) "Adaptive behaviour for prisoner dilemma strategies based on automata with multiplicities." In *ESS 2002 Conf., Dresden*, Germany
6. C. Bertelle, M. Flouret, V. Jay, D. Olivier, and J.-L. Ponty (2001) "Genetic algorithms on automata with multiplicities for adaptive agent behaviour in emergent organizations" In *SCI'2001*, Orlando, Florida, USA
7. G.H.E. Duchamp, H. Hadj-Kacem and E. Laugerotte (2005) "Algebraic elimination of ϵ-transitions", *DMTCS*, 7(1):51-70
8. G. Duchamp and J-M Champarnaud (2004) *Derivatives of rational expressions and related theorems*, Theoretical Computer Science **313**
9. N. Eber (2004) *Théorie des jeux*, Dunod
10. S. Eilenberg (1976) *Automata, languages and machines*, Vol. A and B, Academic press
11. J. Ferber (1999) *Multi-agent system*, Addison-Wesley
12. L.J. Fogel, A.J. Owens, M.J. Welsh (1966) *Artificial intelligence through simulated evolution*, John Wiley
13. J.S. Golan (1999) *Power algebras over semirings*, Kluwer Academic Publishers
14. D.E. Goldberg (1989) *Genetic Algorithms*, Addison-Wesley
15. J. H. Holland (1995) *Hidden Order – How adaptation builds complexity*, Persus books ed.
16. J.E. Hopcroft, R. Motwani, J.D. Ullman (2001) *Introduction to automata theory, Languages and Computation*, Addison-Wesley
17. J. Koza (1997) *Genetic programming*, Encyclopedia of Computer Sciences and Technology

18. M. Mitchell (1996) *An introduction to Genetic Algorithms*, The MIT Press
19. J.-L. Le Moigne (1999) *La modélisation des systèmes complexes*, Dunod
20. I. Rechenberg (1973) *Evolution strategies*, Fromman-Holzboog
21. M.P. Schutzenberger (1961) "On the definition of a family of automata", Information and Control, 4:245-270
22. R.P. Stanley (1999) *Enumerative combinatorics*, Cambridge University Press
23. F. Treves (1967) *Topological Vector Spaces, Distributions and Kernels*, Acad. Press
24. G. Weiss, ed. (1999) *Multiagent Systems*, MIT Press

A Novel Diagnosis System Specialized in Difficult Medical Diagnosis Problems Solving

Barna Laszlo Iantovics

Petru Maior University of Targu Mures
Targu Mures, Romania
ibarna@upm.ro

Summary. The purpose of the study consists in the development of a medical diagnosis system, capable of solving difficult medical diagnosis problems. In this paper we propose a novel medical diagnosis system. The medical diagnosis system is a heterogeneous system with human and artificial agents members specialized in medical diagnosis and assistant agents. The proposed diagnosis system can solve difficult medical diagnosis problems that cannot be solved by doctors or artificial systems specialized in medical diagnosis that operate in isolation. The problems solving by the diagnosis system is partially based on the blackboard-based problems solving.

Key words: complex system, medical diagnosis system, cooperative problem solving, heterogeneous multiagent system, blackboard-based problem solving, expert system agent, assistant agent

1 Introduction

In medical domains, many medical diagnosis systems that operate in isolation or cooperate are proposed and used [1, 2, 4, 7, 9–13, 15]. The paper [13] describes the state of the art medical information systems and technologies at the beginning of the 21st century; the complexity of construction of full-scaled clinical diagnoses is also analyzed. In the following, we enumerate some medical diagnosis systems.

Various diagnosis technologies are studied and used. It is necessary to develop automatic diagnosing and processing systems in many medical domains. The paper [10] presents a cardiac disease analyzing method using neural networks and fuzzy inferences.

In the paper [7] a large-scale heterogeneous medical diagnosis system is proposed. The diagnosis system is composed from doctors, expert system agents and mobile agents with medical specializations. The agents' members of the diagnosis system solve cooperatively the overtaken diagnosis problems. The proposed diagnosis system may have a large number of members. In the diagnosis system, new members can be introduced, the inefficient members can be eliminated.

The paper [4] presents a cooperative heterogeneous multiagent system specialized in medical diagnoses. The diagnosis system is composed from doctors and artificial agents. The knowledge necessary to an overtaken diagnosis problem solving is not specified in advance. The members of the diagnosis system must discover cooperatively the problem solving.

The paper [1] analyzes different aspects of the multiagent systems specialized in medical diagnosis. The understanding of such systems needs a high-level visual view of the systems' operation as a whole to achieve some application related purpose. The paper analyzes a method of visualizing, understanding, and defining the behaviour of a medical multiagent system.

The paper [11] presents a holonic medical diagnosis system that combines the advantages of holonic systems and multiagent systems. The presented system is an Internet-based diagnosis system for diseases. The proposed holonic medical diagnosis system consists of a tree-like structured alliance of agents specialized in medical diagnosis that collaborate in order to provide viable medical diagnoses.

The paper [9] proposes a methodology, based on computer algebra and implemented in CoCoA language, for constructing rule-based expert systems that can be applied to the diagnosis of some illnesses.

The paper [2] describes intelligent medical diagnosis systems with built-in functions for knowledge discovery and data mining. The implementation of machine learning technology in the medical diagnosis systems seems to be well suited for medical diagnoses in specialized medical domains.

The paper [12] presents a self-organizing medical diagnosis system, mirroring swarm intelligence to structure knowledge in holonic patterns. The system sets up on an alliance of agents specialized in medical diagnosis that self-organize in order to provide viable medical diagnoses.

The paper [15] presents a system called FELINE composed of five autonomous intelligent agents. These agents cooperate to identify the causes of anemia at cats. The paper presents a tentative development methodology for cooperating expert systems.

2 Assistant Agents, Expert System Agents

Systems that operate in isolation cannot solve many difficult problems [3, 5, 14]. The solution of these problems is based on the cooperation of more systems with different [6] *capabilities* and *capacities*. The capability of a system consists in the *specializations* detained by the system. A specialization is a problem solving method [3]. The capacity of a system consists in the amount of problems that can be solved in deadline by the system using the detained resources.

The *agents* represent systems with properties like [6]: increased autonomy in operation, capability of communication and cooperation with other systems. In the following, we call agents the humans and the artificial systems with agents' properties. The systems composed from more agents are called *multiagent systems.*

Expert systems can be endowed with medical diagnosis capability. We propose the endowment of the expert systems specialized in medical diagnosis with agents' capabilities. We call these agents, *expert system agents* [4–7].

As examples of advantages of the expert system agents as opposed to the expert systems, we mention [4–7]:

- the expert system agents can perceive and interact with the environment. They can learn and execute different actions in the environment autonomously;
- the expert system agents can communicate with other agents and humans, which allow the cooperative problems solving;
- the expert system agents can solve more flexibly problems than the traditional expert systems. If an expert system agent cannot solve a problem (doesn't have the necessary capability and/or capacity), then he can transmit the problem for solving to another agent.

The *knowledge-based agents* can be endowed with capabilities to assist the agents (human and artificial) specialized in medical diagnosis in the problems solving processes. We call the agents mentioned before *assistant agents* [5–7]. The expert system agents can be endowed with capability to help the doctors in different problems solving.

As examples of assistance that can be offered by an assistant agent to a medical specialist (artificial or human), we mention [5–7]:

- the assistant agent can analyze details that are not observed by a doctor. For example, we mention the suggestion of a doctor for a patient to use a medicine without analyzing important contraindications of that medicine;
- the assistant agent can verify the correctness of a problem's solution obtained by the medical specialist. The assistant agent knows the problem that is solved by the specialist. This way, the specialist and the assistant agent can solve simultaneously the same problem using different problem solving methods. For example, a doctor specialized in cardiology and an assistant expert system agent specialized in cardiology can try to identify simultaneously a cardiology related illness. The same solution obtained by the assistant agent and the medical specialist increases the certitude in the correctitude of the obtained solution. The accuracy in detecting the same illness by different agents may be different;
- the specialist can require the assistant agent's help in solving subproblems of an overtaken problem. This cooperating manner allows the problems solving faster.

3 The Proposed Medical Diagnosis System

A *medical diagnosis problem* consists in the description of one or more illnesses. A person may have more illnesses, each of them with specific symptoms. The *solution of the problem* represents the identified illness or illnesses. A medical diagnosis elaboration may have many difficulties. The symptoms of more illnesses may have some similarities, which make their identification difficult. The symptoms of the same illness may be different at different persons. In the case of some illnesses, the causes of the illnesses are not known. A medicine to an illness may have different effects at different persons that suffer from the illness. A person may have allergy to a medicine.

In this paper we propose a cooperative heterogeneous medical diagnosis complex system for difficult medical diagnosis problems solving. A doctor or an expert system agent that operates in isolation cannot solve some difficult medical diagnosis problems [5]. In many medical diagnosis problems solving knowledge from more medical domains have to be used. The problems solving by the proposed diagnosis system is partially based on the *blackboard-based problem solving* [8, 14]. Figure 1 illustrates the proposed medical diagnosis system.

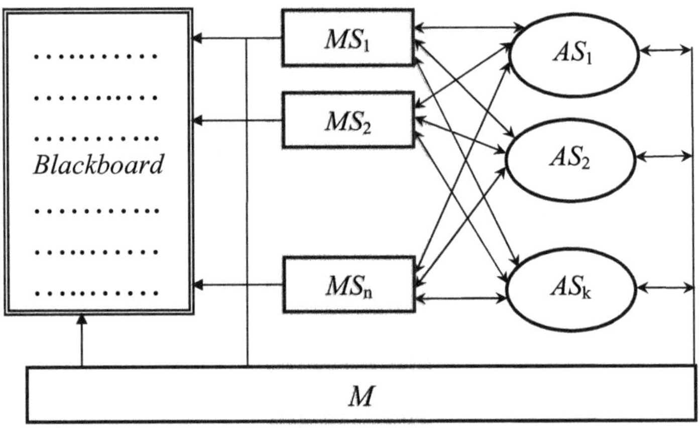

Fig. 1. The proposed medical diagnosis system

The proposed medical diagnosis system *MDS* is composed from a set of agents.

$$MDS = MS \cup \{M\} \cup AS.$$

$MS = \{MS_1, MS_2, \ldots, MS_n\}$ represent agents (human and artificial) specialized in medical diagnosis called *knowledge sources*. As examples of knowledge sources, we mention: the doctors and the expert system agents. M represents a doctor called *moderator*. $AS = \{AS_1, AS_2, \ldots, AS_k\}$ represent the assistant agents (human and artificial). The artificial assistant agents are knowledge-based agents. As examples of artificial assistant agents, we mention the Internet agents, the interface agents and the assistant robots. An Internet agent may collect knowledge from distributed knowledge bases. As an example of knowledge that can be collected by an Internet agent, we mention the description of the symptoms of an illness. Assistant robots can realize different medical analyzes. An assistant interface agent can help a doctor in the communication with the artificial agents. As examples of assistance offered by an interface agent to a doctor, we mention: the translation of the knowledge communicated by artificial agents specialized in medical diagnosis into an understandable form to the doctor, the indication of the assistant agents that has a medical specialization, the indication of the human and artificial agents. As examples of human assistants, we mention the medical assistants. As an example of assistance, an expert system agent can require from a human medical assistant the realization of some medical analyses necessary in increasing the accuracy of an illness identification.

The agents' members of the diagnosis system have different capabilities. Each agent is endowed with a specialization set $SP = \{S_1, S_2, \ldots, S_m\}$. An agent can be endowed with a limited number of specializations [3, 6]. Each knowledge source is specialized in some aspects of the problems solving that can be solved by the diagnosis system. The knowledge sources have different specializations sets in medical domains. The moderator is specialized in the coordination of the problems solving. The agents from the set AS have different specializations sets that allow the assistance of the agents from the set $MS \cup \{M\}$. The specializations set of an agent can be different from the specializations set of the other agent. If an agent AG_i is endowed with the specializations set SP_i, there may exist an agent AG_j with the specializations set SP_j, where $SP_i \neq SP_j$.

The algorithm *Problem Solving* describes a medical diagnosis problem solving cycle by the proposed medical diagnosis system. A problem solving cycle consists in overtaking and solving a problem. The problem is solved onto a *blackboard*. The blackboard represents a memory shared by all the members of the diagnosis system. A problem solving begins when the problem is written onto the blackboard. The knowledge sources watch the blackboard, looking for an opportunity to apply their expertise to develop the solution. When a knowledge source finds sufficient information to contribute, he records the contribution onto the blackboard. This additional information may enable other knowledge sources to apply their expertise. This process of adding contributions onto the blackboard continues until the problem is solved or the problem cannot be solved. The moderator supervises the problem solving process. The assistant agents may help the knowledge sources and the moderator during their operation. All the members of the medical diagnosis system solve the diagnosis problem cooperatively.

The artificial software agents can access the content of the blackboard (memory). The humans can view the content of the blackboard via output devices. Only the knowledge sources can write onto the blackboard. The writing onto the blackboard consists in adding, changing or retracting knowledge from the blackboard. A knowledge source can write onto the blackboard if it has writing right. The right for the knowledge sources to write onto the blackboard is allowed and retracted by the moderator. In the selection of the best-fitted knowledge source to process a problem result, the moderator uses the informations contained in the announcements parameters.

Algorithm – Problem Solving

The problem P that has to be solved is written onto the blackboard.

While ((the problem is not solved) *and* (the solving can continue)) *do*
{

- Each knowledge source watches the blackboard and sends a message to the moderator that contains the specification of its capability and capacity to process the knowledge from the blackboard.
- The moderator, using the informations contained in the received messages from the knowledge sources, chooses the best-fitted knowledge source MS_i capable to process the knowledge from the blackboard.
- The moderator allows the writing right to the selected agent MS_i.
- The knowledge source MS_i processes the knowledge from the blackboard. The obtained result is written onto the blackboard.
- After the knowledge source MS_i finishes the knowledge processing, the moderator retracts the writing right from the knowledge source MS_i.

}
End.

The announcement, in which a knowledge source MS_i specifies how it can contribute to a problem solving, has the following parameters:

$$< Capability, Capacity, Relevance > .$$

Capability represents the capability of the agent MS_i (the specialization that can use MS_i in the problem processing). *Capacity* represents the processing capacity of MS_i (the problem processing time). *Relevance* specifies the estimated importance of the processing that can be realized by MS_i (the measure in which the processing approaches the problem solution).

As examples of knowledge that can be added onto the blackboard by a knowledge source at a problem solving cycle we mention:

- the results of some medical analyses;
- a new supposed illness. The knowledge source supposes that the patient has an illness;
- new questions that must be answered by the patient;

- new questions that must be answered by other knowledge sources. The knowledge source is limited in knowledge, but is specialist in a certain medical domain.

As examples of knowledge that can be retracted from the blackboard by a knowledge source at a problem solving cycle, we mention:

- useless informations. Some information written onto the blackboard are not relevant in the diagnosis process;
- a supposed illness. The knowledge source demonstrates that the patient does not have the supposed illness written onto the blackboard.

As examples of knowledge that can be modified onto the blackboard by a knowledge source at a problem solving cycle, we mention:

- the knowledge that is changed in time. Some medical analysis results are changing in time (a diagnosis process may have a longer duration). Some patients do not describe correctly the symptoms of their illness.

The knowledge from the blackboard must be understandable to all the members (human and artificial) of the diagnosis system. For the representation of the knowledge written onto the blackboard, the agents must use the same *knowledge representation language* and must share the same *ontology* (dictionary of the used terms). The notions knowledge representation language and ontology are defined in the papers [3, 14]. If a doctor cannot understand the knowledge written onto the blackboard, then he can require the help of an assistant agent in the translation of the knowledge into an understandable form for it. A doctor can requires an artificial assistant agent's help in the writing of some knowledge onto the blackboard. If it is necessary, the assistant agent can translate the knowledge transmitted by the doctor into an understandable form to the other agents' members of the diagnosis system.

4 An Example of a Problem Solving

In the following, we present a scenario that illustrates how a proposed medical diagnosis system *MDS* solves an overtaken medical diagnosis problem P (the patient suffers from two illnesses, a cardiology and an urology related illness).

$$P = <\textit{description of a cardiology related illness, description of an urology related illness}>.$$

$$MDS = \{AGg, AGc, AGu\} \cup \{M\} \cup \{ASi\}.$$

AGg, AGc and AGu are the knowledge sources. AGg is an expert system agent specialized in general medicine. AGc is a doctor specialized in cardiology. AGu is a doctor specialized in urology. M is a moderator doctor specialized in the coordination of the problem solving. ASi is an assistant agent specialized in informations search about the patients in distributed databases.

The solution *SOL* of the problem that must be obtained represent the identified two illnesses of the patient.

SOL = <*the identified cardiology related illness,*
the identified urology related illness>.

In the following, we describe step by step the scenario of the overtaken problem *P* solving by the medical diagnosis system *MDS* using the cooperative problem solving described in the previous section.

Problem solving

Step 1

- The problem *P* is written onto the blackboard.

Step 2

- Each knowledge source watches the blackboard.
- *AGg* requires the assistance of *ASi* in obtaining the description of the patient's prior illnesses.
- *ASi* transmits the required information to *AGg*.

Step 3

- Each knowledge source announces its capability and capacity to contribute to the problem *P* processing.
- *M* based on the announcements parameters values chooses the best-fitted knowledge source. Let *AGg* be the selected knowledge source (the contribution of *AGg* is the most relevant, *AGg* can write onto the blackboard the patient's prior illnesses and different observations related to the patient's current illnesses).
- *M* allows the writing right to *AGg*.
- *AGg* processes the problem *P* obtaining the result P_1 (P_1 represents a new problem) that is written onto the blackboard.
- *M* retracts the writing right from *AGg*.

Step 4

- Each knowledge source announces its capability and capacity to contribute to the problem P_1 processing.
- *M* chooses the best-fitted knowledge source capable to contribute to the problem P_1 solving. Let *AGc* be the selected knowledge source (*AGc* can establishes the patient's cardiology related illness).
- *M* allows the writing right to *AGc*.
- *AGc* processes the problem P_1 obtaining the result P_2 that is written onto the blackboard.
- *M* retracts the writing right from *AGc*.

Step 5

- Each knowledge source announces its capability and capacity to contribute to the problem P_2 processing.
- M chooses the best-fitted knowledge source AGu capable to contribute to the problem solving (AGu can establishes the patient's urology related illness).
- M allows the writing right to AGu.
- AGu processes the problem P_2 obtaining the solution SOL that is written onto the blackboard.
- M retracts the writing right from AGu.

End.

The problem P solving process can be described as follows:

$$AGg(P) \Rightarrow AGc(P_1) \Rightarrow AGu(P_2) \Rightarrow SOL.$$

The result P_1 represents the patient illnesses symptoms, the descriptions of the patient's prior illnesses and different general observations related to the patient's current illnesses elaborated by AGg. The result P_2 represents the cardiology related illness of the patient identified by AGc, the symptoms of the patient's illnesses, the description of the patient's prior illnesses and different general observations related to the patient's current illnesses elaborated by AGg. The result SOL represents the identified two illnesses of the patient. The urology related illness is identified by AGu.

5 Advantages of the Proposed Diagnosis System

The elaboration of a medical diagnosis by a doctor or an expert system that operates in isolation may have many difficulties [5]. Many difficult medical diagnosis problems solving require cooperation in their solving [4,5,7,11,15]. The main advantage of the proposed medical diagnosis problems solving consists in solving difficult diagnosis problems, which requires knowledge from different medical domains. The problems solving specializations in the diagnosis system are distributed between the human and artificial agents' members of the system. The knowledge necessary to the diagnosis problems solving are not specified in advance, the diagnosis system members must discover cooperatively the problems solving.

The problems solving difficulty is distributed between the agents members of the diagnosis system. Each knowledge source can use its specializations in certain circumstances, when it finds knowledge on the blackboard that can process. The moderator is responsible to decide which knowledge source will overtake a problem result processing at a moment of time. The knowledge sources and the moderator can require the help of assistant agents.

The artificial agents' members of the medical diagnosis system can be endowed with new specializations. The inefficient specializations can be eliminated or improved. The adaptation of a cooperative multiagent system in the efficient solving of a problem many times is easier than the adaptation of an agent that solves the same problem [3, 6, 14].

New agents can be introduced in the diagnosis system. It is not necessary for a newly introduced agent (human or artificial) in the system to understand the knowledge representation language used in the system. Assistant agents may help the new agent in its interaction with the agents from the system by translating the knowledge used in the system into an understandable form to the agent.

6 Conclusions

We propose the endowment of the expert systems specialized in medical diagnosis with agents' capabilities. We call these agents expert system agents. Assistant human and artificial agents can be endowed with the capability to help the doctors and expert system agents in the problems solving processes. In this paper, we have proposed a novel cooperative heterogeneous medical diagnosis system composed from agents (human and artificial) specialized in medical diagnosis and assistant agents (human and artificial). The cooperative problem solving by the proposed diagnosis system combines the human and artificial agents advantages in the problems solving. The humans can elaborate decisions using their knowledge and intuition. The intuition allows the elaboration of the decisions without the use of all the necessary knowledge, some problems for which does not exist elaborated solving methods can be solved this way. The artificial thinking allows the problems solving based on existent problems solving methods sometimes verifying many conditions. This way, the artificial agents can solve problems precisely, verifying conditions that can be ignored by the humans.

References

1. Abdelaziz T, Elammari M, Unland R (2004) Visualizing a Multiagent-Based Medical Diagnosis System Using a Methodology Based on Use Case Maps. MATES 2004, In: Lindemann G (ed) LNAI 3187, Springer-Verlag, Berlin Heidelberg, 198–212
2. Alves V, Neves J, Maia M, Nelas L (2001) A Computational Environment for Medical Diagnosis Support Systems. ISMDA 2001, In: Crespo J, Maojo V, Martin F (eds) LNCS 2199, Springer-Verlag, Berlin Heidelberg, 42–47
3. Ferber J (1999) Multi-Agent Systems: An introduction to Distributed Artificial Intelligence. Addison Wesley, London
4. Iantovics BL (2005) A Novel Medical Diagnosis System. Proceedings of the European Conference on Complex Systems, ECCS 2005, In: Bourgine P, Kepes F, Schoenauer M (eds) Paris, France, 165–165

5. Iantovics BL (2005) Cooperative Medical Diagnosis Systems. Proceedings of the International Conference Interdisciplinarity in Engineering, Targu Mures, Romania, 669–674
6. Iantovics BL (2004) Intelligent Agents. PhD Dissertation, Babes Bolyai University of Cluj Napoca, Cluj Napoca, Romania
7. Iantovics BL (2005) Medical Diagnosis Systems. Proceedings of the International Conference Several Aspects of Biology, Chemistry, Informatics, Mathematics and Physics, Oradea, Romania
8. Jagannathan V, Dodhiawala R, Baum LS (eds) (1989) Blackboard Architectures and Applications. Academic Press, San Diego
9. Laita LM, Gonzlez-Paez G, Roanes-Lozano E, Maojo V, de Ledesma L, Laita L (2001) A Methodology for Constructing Expert Systems for Medical Diagnosis. ISMDA 2001, In: Crespo J (ed) LNCS 2199, Springer-Verlag, Berlin Heidelberg, 146–152
10. Mitsukura Y, Mitsukura K, Fukumi M, Akamatsu N, Witold-Pedrycz W (2004) Medical Diagnosis System Using the Intelligent Fuzzy Systems. KES 2004, In: Negoita MGh (ed) LNAI 3213, Springer-Verlag, Berlin Heidelberg, 807–826
11. Unland R (2003) A Holonic Multi-agent System for Robust, Flexible, and Reliable Medical Diagnosis. OTM Workshops 2003, In: Meersman R, Tari Z (eds) LNCS 2889, Springer-Verlag, Berlin Heidelberg, 1017–1030
12. Unland R, Ulieru M (2005) Swarm Intelligence and the Holonic Paradigm: A Promising Symbiosis for Medical Diagnostic Systems Design. Proceedings of the 9th International Conference on Knowledge-Based and Intelligent Information and Engineering Systems, Melbourne, Australia
13. Vesnenko AI, Popov AA, Pronenko MI (2002) Topo-typology of the structure of full-scaled clinical diagnoses in modern medical information systems and technologies. Cybernetics and Systems Analysis 38(6), Plenum Publishing Corporation: 911–920
14. Weiss G (ed) (2000) Multiagent Systems: A Modern Approach to Distributed Artificial Intelligence. MIT Press Cambridge Massachusetts, London, England
15. Wooldridge M, O'Hare GMP, Elks R (1991) FELINE – A case study in the design and implementation of a co-operating expert system. Proceedings of the International Conference on Expert Systems and their Applications, Avignon, France

Constraint Programming and Multi-Agent System Mixing Approach for Agricultural Decision Support System

Sami Al-Maqtari[1], Habib Abdulrab[1], and Ali Nosary[2]

[1] LITIS – INSA of ROUEN
Place Émile Blondel, BP 8
76131 Mont-Saint-Aignan Cedex, France
http://www.insa-rouen.fr/psi/
sami.almaqtari@insa-rouen.fr, abdulrab@insa-rouen.fr

[2] Engineering Faculty – Sana'a University
Yemen
ali.nosary@univ-rouen.fr

Summary. The objective of this paper is to present the grand schemes of a model to be used in an agricultural Decision support System. We start by explaining and justifying the need for a hybrid system that uses both Multi-Agent System and Constraint Programming paradigms. Then we show our approach for Constraint Programming and Multi-Agent System mixing based on the concept of *Controller Agent*. Also, we present some concrete constraints and agents to be used in an application based on our proposed approach for modeling the problem of water use for agricultural purposes.

Key words: multi-agent system (MAS), constraint programming (CP), decision support system (DSS), controller agent, water management

1 Introduction

Water is the most vital resource in human life and a critical economic factor in the developing countries. And Yemen is considered as one of the most water-scarce countries in the world. According to worldwide norms, domestic uses and food self-sufficiency require up to 1100 m^3/capita/year. However, in Yemen the available water quantity amounts to little more than 130 m^3/capita/year [12]. Moreover, resources are unevenly distributed, 90% of the population has access to less than 90 m^3/capita/year. Table 1 shows a comparison of the annual quota per capita between some countries in the region and the global average.

The decrement in annual water quota per capita in Yemen is due to (among other causes) the high population growth rate which has been about 3.7% (1980-1997 average) and is expected to be about 2.6% (1997-2015) [12].

Most water use goes for irrigation purposes [10]. The average share of agriculture in total water use is about 92% because of the rapid progress of irrigated agriculture in Yemen at a pace faster than any comparator country (see Table 2).

Table 1. Renewable Water Resources Per Capita

Country name	Renewable Water Resources Per Capita (m^3/capita/year) 1980	1997	2015
Egypt	1,424	966	735
Jordan	396	198	128
Morocco	1,531	1,088	830
Saudi Arabia	257	120	69
Yemen	246	130	82
World	10,951	8,336	6,831

Sa'dah basin is one of the four principle basins in Yemen and one of the most touched regions by water crises. Well inventory [11] shows a total of 4589 waterpoints in Sa'dah Region, of which 99.78% wells and the rest represents springs and other water-point (tanks, dams). These water-points are either owned by one owner (46.61%) or shared between two or more owners (53.39%). The abstraction from agricultural wells in Sa'dah region represent over 90% of the annual water abstraction of the basin.

Table 2. Water use share

Country name	Water use Agriculture	Industry	Domestic
Egypt	86%	8%	6%
Jordan	75%	3%	22%
Morocco	92%	3%	5%
Saudi Arabia	90%	1%	9%
Yemen	92%	1%	7%
World	69%	22%	9%

Well inventory shows also that for 82% of the wells is used for irrigation, while 1% is used for domestic needs and 0.5% for feeding water supply networks. About 16.5% of the inventoried wells are not in use. In which consider irrigation status we can find that only 2.64% of the sampled farmland is rainfed land while the rest (97.36%) is irrigated land.

The complexity of such situation requires reflect the need for a good decision support system. Such system has to take into account all necessary constraints such as the respect of shared water-points using agreement. It has also to model and to simulate the interaction between the different actors in the whole process such as the negotiations between consumers and water suppliers, and to model decision taking process, like the criteria and strategy of water allocation that are used by water suppliers. By making an analogy, constraint programming, therefore, looks a good approach in order to help finding a solution that satisfies the constraints of the problem, while multi-agent system approach can help in describing the interaction between the various actors in the model.

In the next sections we give a short introduction to the Constraint Programming and Multi-Agent System, after that we describe our approach for mixing both paradigms in order to model the problem of water using for irrigation purposes.

2 Constraint Programming

2.1 Introduction

Constraint programming is an emergent software technology for declarative description and effective solving of large, particularly combinatorial, problems. It is a programming paradigm in which a set of constraints that a solution must meet are specified rather than set of steps to obtain such a solution. A constraint is simply a logical relation among several unknowns (or variables), each taking a value in a given domain. A constraint thus restricts the possible values that variables can take; it represents some partial information about the variables of interest. The idea of constraint programming is to solve problems by stating constraints (conditions, properties) which must be satisfied by the solution.

2.2 Constraint Satisfaction Problem (CSP)

A Constraint Satisfaction Problem (CSP) consists of:

1. a set of n variables $X = x_1, x_2, , x_n$,
2. for each variable x_i, a finite set D_i of possible values (its domain),
3. and a set of constraints restricting the values that a set of variables can take simultaneously.

A solution to a CSP is an assignment of a value to every variable from its domain, in such a way that every constraint is satisfied. We may want to find: (i) just one solution, with no preference as to which one, (ii) all solutions, or (iii) an optimal, or at least a good solution, given some objective function defined in terms of some or all of variables.

Thus, the CSP is a combinatorial problem which can be solved by search. Clearly, with a large number of variable simple algorithms of searching all possible combinations take a long time to run. So the researches in the constraint satisfaction area concentrate on finding adhoc algorithms which solve the problem more efficiently, especially by using techniques like *global constraints.*

2.3 Global Constraint

A global constraint encapsulates several simple constraints [2], [3], and we can achieve stronger pruning of domains by exploiting semantic information about this set of constraints. Filtering algorithms for global constraints are based on methods of graph theory, discrete mathematics, or operation research so they make the bridge between these mathematical areas and searchbased constraint programming with origins in artificial intelligence. [2] has proposed a dynamic view of global constraints. Such a dynamic global constraint allows adding a new variable(s) during the course of problem solving and removing this variable(s) upon backtracking. Thus, a dynamic global constraint can be posted before all the constrained variables are known which brings the advantage of earlier domain pruning mainly for a system where not all information is necessarily known a priori.

2.4 Over-Constrained Problems and Constraint Hierarchies

In many cases, a solution of CSP does not exist, and we can not make a valuation of variables that satisfies all the constraints. Constraint hierarchies [9] were introduced for describing such over-constrained systems by specifying constraints with hierarchical strengths or preferences. It allows us to specify declaratively not only the constraints that are required to hold, but also weaker constraints at an arbitrary but finite number of strengths. Weakening the strength of constraints helps to find a solution of previously over-constrained system of constraints. Intuitively, the hierarchy does not permit to the weakest constraints to influence the result. Constraint hierarchies define the so called comparators aimed to select solutions (the best assignment of values to particular variables) via minimizing errors of violated constraints.

2.5 Constraint Programming and Agricultural Water Management

In his work, Jaziri [7] has applied a methodology for constraints and optimization modeling in order to minimize the risk of water streaming. We can note that the constraints in the classical definition of CSP are relations between sample variables. However, in a complex system like agricultural water uses, the constraints represent relations between different instances of the system (parcels, resources...) acting on the variables that characterize these instances (parcel crop, demanded water quality...).

According to Jaziri [7] we can distinguish three layers: constraints, instances and variables. These layers are divided according two different points of view: user level and system level. At user level, a constraint represents a condition that links some system instances. At system level, the constraint is defined as a restriction over the values of a set of simple variables that characterize the instances linked by this constraint.

In the case of agricultural water use, the user expresses the system constraints such as water provision which is a relation between the supplier and the consumer. This relation is translated at system level as constraints relating various variables such as consumer required water, supplier available water quantity, transport mean capacity, etc...

3 Multi-Agent System

3.1 Introduction

The Agent-Oriented (AO) approach gives the ability to construct flexible systems with complex and sophisticated behavior by combining highly modular components. These components represent agents having autonomy and interaction characteristics.

What is an agent? The term agent has many definitions. According to Wooldridge [13] an agent is a software system that is (i) situated in some environment, (ii) capable of autonomous actions in order to meet its objectives and (iii) capable of communicating with other agents. From this definition we can say that an agent is an entity that can act and react in his environment and interact with other agents.

A multi-agent system is made up of a set of several agents (representing different tasks and/or entities in the system) that exist at the same time, share common resources and communicate with each other. For simplicity a multi-agent system can be viewed as a network of agents (problem solvers) coupled lightly, who work together to solve problems that are beyond their individual capacities [5].

The research on the agents is also a research on: (i) Decision – what are the mechanisms of the agent decision? What is relation between their perception, their representations and their actions? How the agents break down their goals and tasks? (ii) Control – what are the relations between agents? How are they coordinated? This coordination can be represented as a cooperation to fulfill a common task or as a negotiation between agents having different interests. (iii) Communication – what types of message do they exchange? What syntax these messages obey?

3.2 MAS and the Simulation of Resource Management

The simulation of the management of common resources poses the problem of correlation between groups of agents and dynamic resources. In the multi-agent system paradigm we look at the simulated system from a distributed and cooperative point of view.

In the domain of water use for agricultural purposes, we can find various actors (different farmers, resource managers...). Using multi-agent system paradigm allows us to simulate these actors decision mechanisms and their evolution, the interactions, the negotiations, and the cooperation between these actors in the model.

4 MAS and CP Mixing Approach

4.1 Introduction

We can notice that the model of the simulated system using Constraints Programming is built as a set of variables that represent the simulated system variables, and different constraints relating between these variables. All these constraints will be inserted into a solver to be manipulated and treated together as a whole unit in order to find and assign values to the system variables. In the other side, in the multi-agent system, agents are mainly characterised by the autonomy, i.e. each agent tries independently to achieve its own goal. The agent could interact, cooperate and/or negotiate with other agents either directly or via its effects on the environment. Combining the both paradigm defines the *Distributed CSP*.

4.2 Distributed CSP

Distributed constraint satisfaction problems (DCSP) are an appropriate abstraction for multi-agent cooperative problem solving [6]. They are characterized by multiple reasoning agents making local independent decisions about a global common constraint satisfaction problem (CSP).

In a DCSP, the variables are distributed among agents (see Figure 1). Yokoo [14], [15] has proposed solving distributed CSP by using an asynchronous backtracking algorithm. This can be done by allowing agents to run concurrently and asynchronously. Each agent proposes values for its own variables and communicates these values with relevant agents.

We can note from Figure 1 that the constraints are represented by oriented arcs between the agents. Agents propose their variables values according to the oriented arcs. In this example A_3 receives values propositions from both agent A_1 and A_2 and test them according to their respective constraints with its own variable possible value.

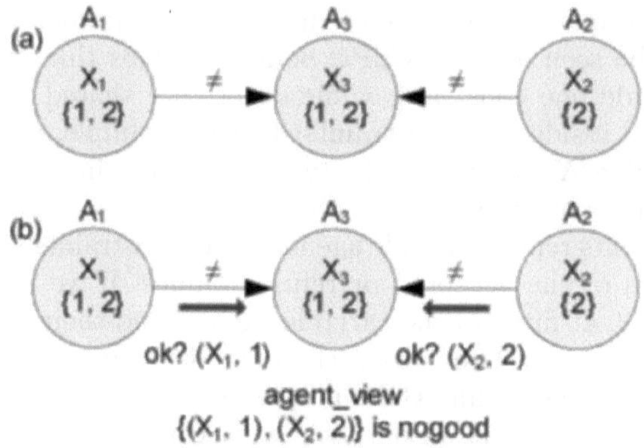

Fig. 1. an example of DCSP

4.3 Another Approach of MAS and CSP mixing

We propose here another approach for DCSP resolution. It consists of adapting the methodology like that proposed by [7] for constraints modelling with multi-agent system paradigm. Our approach is based on a special kind of agents called controller agent that takes in charge the responsibility of verifying the constraint satisfaction. We will show this approach through the following sections:

Our model

Our approach [1] of mixing MAS and CSP is based on the concept of a controller agent which has the responsibility to verify the constraint satisfaction. By comparing between Figure 1 and Figure 2 we can note that the agents in the system (A_1, A_2, and A_3) do not communicate directly the proposed values of their variables. Instead, they send these values to controller agents (C_1 and C_2) who verify these values according to the constraint which they hold.

Using a controller agent for verifying a constraint satisfaction allows separating constraints verifying process from the functionality of other agents. In This case we designate the agents which represent the different entities in the system separately from the controller agents which will be designated for verifying the constraint satisfaction only. Modifing the manner of constraint satisfaction testing involves contoller agents and does not affect the tasks of the other agents.

Another advantage of using the controller agent is treating the whole CSP as multiple *quasi-independent* local CSPs. The notion of quasi-independent CSP can be informally defind as a set of local CSPs, each one has its own *local* variables and shares some *shared* variables with the other local CSPs. The fact of giving values for these shared variables separates these local CSPs completely. If we consider the example in Figure 2, the constraints inside the controllers can be seen as local CSPs. Some variables (like X_1 for example) are involved inside one constraint while others are shared variables like X_3 which is involved inside two constraints. In this example, giving a value for the shared variable X_3 separates the two constraints inside C_1 and C_2 into two independent constraints.

Controller agents report variable agents of the constraints satisfaction state and inform them of the accepted or refused values. If we take the example shown in Figure 2, at first (a) the variable agents send their value proposition to the controller agents. Then (b) each controller agent verifies the validity of its own constraint according these propositions. C_2 accepts the values sent by A_2 and A_3; therefore it sends back an acceptance message. While the values sent by A_1 and A_3 do not satisfy the controller C_1, therefore it sends a message of refuse back to them and wait for other propositions. In (c) A_3 has already satisfied a constraint and prefers to wait a little before proposing another value that may affect both constraints, while A_1 can freely propose another value from its variable domain. This new value is verified by C_1 (d) which sends back an acceptance message.

Application of this approach for agricultural water use management

Agricultural water use management implies various actors constrained by different constraints. The constraints in such domain are represented by relations between the different actors (or actors' attributes). For example, shared water-points user should have an agreement for water use (they are constrained by this agreement). Applying our approach for agricultural water use is described as follows:

Application needs and parameters

In Sa'dah basin we can find:

- Different modes of water consuming: agricultural (crop et livestock), domestic, industrial, urban and municipal...
- Different types of water resources (water-points), some of them are owned by farm owner and others are shared between several farmers.
- Several means of water transport.
- Several irrigation techniques.

We can note here that there are a lot of parameters need to be manipulated in order to achieve a stable and durable situation of water consuming. To simplify and minimize the problem parameters we can note that the agricultural water use represents about 92% of the total water use in Sa'dah basin, so we can consider (for the moment) the use of water for agricultural purposes only. Another note is coming from the fact that wells represent 99.7% of water-points used for irrigation purposes. So we can neglect other water-points. This

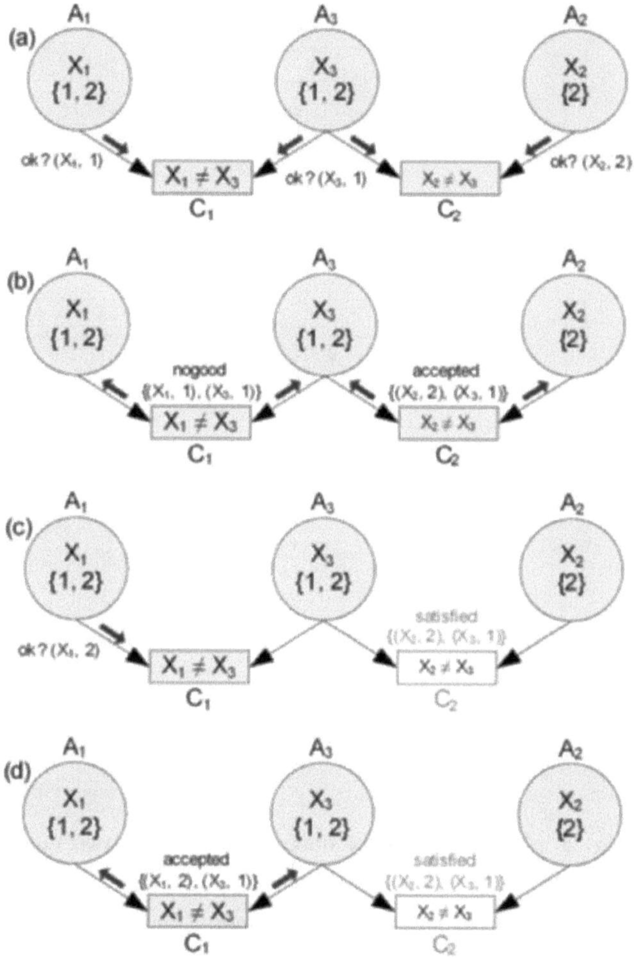

Fig. 2. an example of our mixing approach

can help to reduce first and second points of the above description into: Two ownership types of wells used for irrigation purposes.

Figure 3 shows a general UML class diagram of the proposed model. The two main parts are *Consumer* and *Supplier* and both are related by a *TransportMean*. The supplier supervises various types of water resources, so *WaterResource* is an interface allowing the user to add other types of resources. The consumers could be *Domestic* or *Framers* (or other types). The farmer owns one or more *Farm* containing one or more *Parcel*. The farmer chooses and plants a *Crop* in each parcel, chooses its *IrrigationMethod*, and calculates water needs according to these choices and other parameters (like *Weather*).

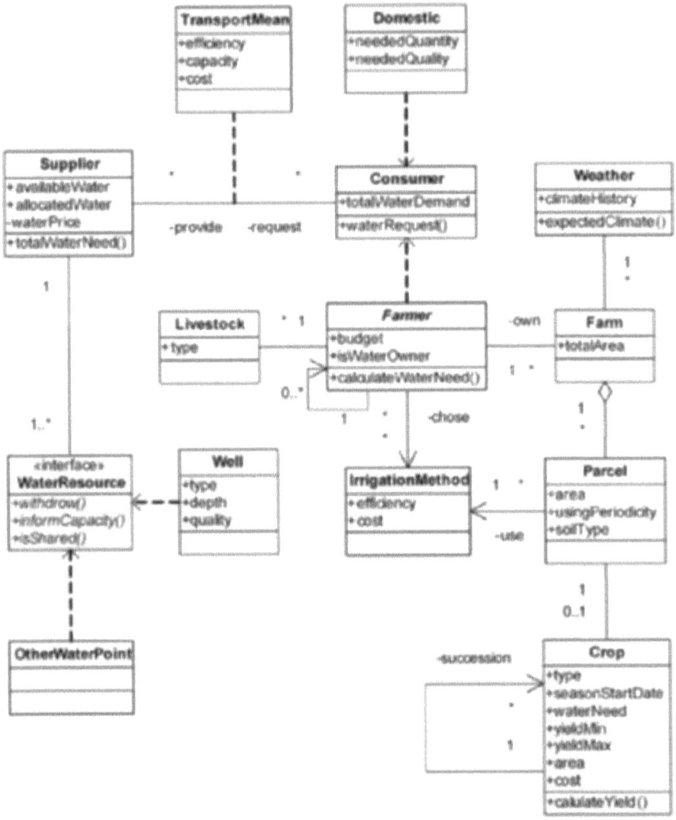

Fig. 3. general UML class diagram of the proposed model

A farmer may be related to several other farmers (farmers who share the same water resource for example).

In order to clarify the needs and the effictive attributes in the system we can describe a scenario of the situation as follows: Water authorities want to keep water use at its durable limits. Farmers who own their water resources want firstly to increase their profits and try not to overexploit their own water resources. While farmers who share common water-points want to have more convenient access to water with acceptable costs and keep their profitability. Other farmers need only to be satisfied by the proposed solution. Water and local authorities encourage the use of more efficient water transport means and irrigation techniques. Framers are willing to use water saving technique but they do not want to affect their profitability with extra costs.

from this scenario we can cite here the needs of the different deciders in the system:

1. Increasing the profitability (water owner farmers).
2. Limiting water overexploiting (water owner farmers).

3. More convenient water access (shared-water farmers).
4. Keeping the profitability (shared-water farmers).
5. Framers satisfaction (farmers).
6. Achieving durable water use (water authorities).
7. encouraging the use of water saving techniques (water and local authorities,...)

These needs can be translated into the following objectives:

1. Enhancing water exploiting (needs 1, 3, 6, and 7).
2. Keeping the profitability (needs 1 and 4).
3. Framers satisfaction (need 5).

The aspects of these objectives can be shown as follows:

1. Enhancing water exploiting: can be view from two different aspects.
   - Reducing water wastage at irrigation technique level (aspect 1).
   - Reducing water wastage at the transport means level (aspect 2).
   - Enhancing water sharing agreement (aspect 3).
2. Keeping the profitability: one aspect (aspect 4).
3. Framers satisfaction: satisfaction in profitability (same aspect 4).

These aspects have the following criterions:

1. Reducing water wastage at irrigation technique level: the waste water can be estimated by considering the water used for irrigation and the actual crop water requirements. It is also related to the efficiency of irrigation techniques.
2. Reducing water wastage at the transport means level: it is a function of the efficiency of transport means.
3. Enhancing water sharing agreement: it can be calculated by considering farmer budget and shared water cost.
4. Keeping the profitability: it is a function of crop yield and planted area in each farm and the cost of irrigation and transport means.

We can deduce finally the attributes of the system for evaluating these criterions:

1. Crop water needs: crop.waterNeed (criterion 1).
2. Parcel water use: totalWaterUse (criterion 1).
3. Irrigation method efficiency: irrigationMethod.efficiency (criterions 1 and 4).
4. Irrigation method cost: irrigationMethod.cost (criterion 4).
5. Transport means efficiency: transportMean.efficiency (criterions 2 and 4).
6. Transport means cost: transportMean.cost (criterion 4).
7. Farmer budget: farmer.budget (criterion 3).
8. Shared water price: supplier.waterPrice (criterion 3).
9. Crop yield: crop.yieldMax, crop.yieldMin (criterion 4).
10. Parcel planted area: parcel.area (criterion 4).

The constraints in the application

As we mentioned in section 2.5, the constraints in our case represent relations between different instances of the system. We can cite here some constraints in the model:

1. Crop water requirement: every crop has its own water requirement that must be satisfied. This requirement varies according to the start date and the current stage of crop evolution.
2. Compatible soil type: the soil type of a parcel determines what kind of crop can be planted on this parcel.
3. Parcel crop type preference: some farmer can accept only planting specific types of crop, but he may also accept a reasonable proposition out of these choices (constraint relaxation). This constraint can be seen as specifying a specific domain for each parcel.
4. Profitability of crop type: farmers privilege the crops with the highest profitability. This implies the water prices and crop planting cost (land preparation, seed cost, labor cost...).
5. Crop succession: some type of crops can not be planted in the same land without knowing the previous planted crop type in the same land.
6. Water sharing agreement: shared water points can not be used except in a certain quantity and in a certain slice of time.
7. Transport means capacity: even if we have sufficient quantity of water we can not transport a quantity superior to transport mean capacity.

The agents in the application

The objective of the model is to assign a crop to every parcel in a way that respects the constraints of the system and try to optimize water exploitation. We can distinguish here three main types of agents:

1. Farmer agent: this agent takes in charge the following tasks:
   - Choosing parcels' crops type and calculating their water requirements.
   - Choosing the irrigation method.
   - Negotiating with other farmer agents (who share the same water-points).
   - Demanding water supplying from supplier agents.
2. Supplier agent: it represents the water resources manager. It decides water quantity and quality to be provided to each consumer (farmer in this case).
3. Controller agent: they are the agents who control the validation of constraints of the system. In the following section we will explain controller agent functionality in more details.

Controller agent in our approach

Controller agent represents essential party of our approach in mixing Multi-Agent System and Constraints programming paradigms. We have seen in Figure 3 a general UML class diagram of proposed model. In our approach, each controller agent is attached to a constraint in the system. The controller agent has the responsibility to assure the satisfaction its constraint according to the constraint strength level (see section 2.4). It has the ability to decide if either (i) the constraint is satisfied, (ii) the constraint is not satisfied, or (iii) the constraint is not totally satisfied but it is accepted (for the constraints with a low level hierarchy or week strength). This gives the model a kind of flexibility because the system can achieve a stable state and have evaluation values for its variable even if some of its "week" constraints are not satisfied.

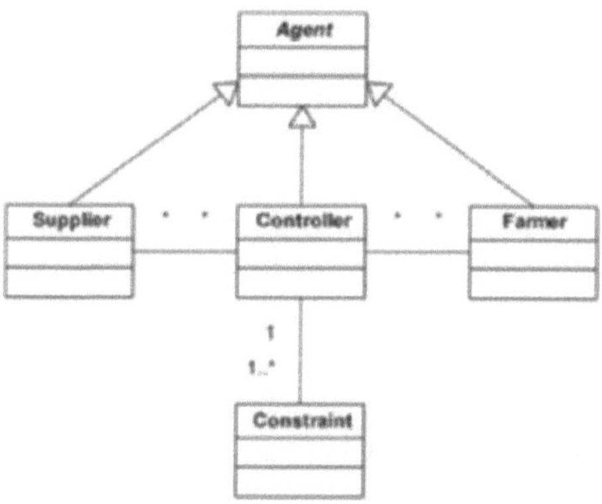

Fig. 4. model agents and their inter-connection

As shown in Figure 4, the system agents are interconnected between them. Each *Controller* tests and validates one *Constraint* (or more). These constraints rule the relation *Farmer-Farmer*, *Supplier-Supplier*, and *Supplier-Farmer*. Farmer and Supplier agents send their variables values to a controller agent according to their respective constraints. Each controller agent communicates then its constraint satisfaction state with both consumer and/or supplier agents.

Note that a constraint does not link only farmers and suppliers; it may link between different suppliers or different farmers only. In fact it joins only the agents who are related to (or constrained by) the controller constraint. For instance, crop succession is represented by a controller who is linked to the only one farmer but it constrains the farmer decision on what crop to be chosen for the next period of plantation.

Assigning a controller agent for a constraint makes it easy to manipulate system constraints. When a constraint is defined we actually define not only the variables and the constraint relation between them, but also the agents which are involved in this constraint. In other word, the agents which are intended to participate in a constraint have to provide the variable required by this constraint.

If we take the transport means capacity for example we can define this constraint as follows: the sum of water allocated by the water supplier (k) to all consumers connected by the same transport mean should not exceed this last one capacity. In this case, controller agent that checks this constraint would have a water supplier and some consumer (farmers in our case) as participants and every time a supplier try to provide his consumers with a quantity greater than transport mean capacity the controller send a message to it refusing this value.

The modelled constraints take many variables and have a sort of symmetry. This symmetry allows us to investigate the use of global constraint in order to make the resolution process more efficient. For example, although we have defined one constraint representing crop water requirement, this constraint will be instantiated for every simulated parcel. So the verification of this constraint satisfaction is in fact the verification of all instants of this constraint type. This implies the processing of very large number of constraints, and this is another aspect of the usefulness of global constraint techniques. Equally, as proposed by [2], [3], the use of dynamic global constraint allows adding new variables to the system dynamically. This looks very useful for a system where not all the information and variables values are entirely known.

5 Conclusion and Perspectives

Mixing the capabilities of Constraint Programming and Multi-Agent System represents an interesting approach for constructing a decision support system for the management of water use for agricultural purposes. The advantage of this approach is the fact of dealing with CSP is not done as a group of constraints and variables that have to be processed as a whole together. The solution of the system is emerged by the interactions between the different actors. Such approach allows endowing controller agents with the ability of controlling system constraints and as a result- satisfying the constraints locally. It facilitates adding not only new constraints in the system, but also other sort of consuming mode.

We have the intention to detail and extend the system components (the different constraints to be implemented, and the various agents to be added). Completing this model allows us not only to implements this model and to test it on the real situation of Sa'dah, but also to extend the model in order to be used with other types of consumer (domestic, industrial...) and their relative new constraints.

References

1. S. Al-Maqtari, H. Abdulrab and A. Nosary (2004) "A Hybrid System for Water Resources Management", GIS third international conference & exhibition
2. Roman Barták (2001) "Dynamic Global Constraints: A First View", Proceedings of CP-AI-OR 2001 Workshop, 39-49, Wye College
3. Roman Barták (2003) "Dynamic Global Constraints in Backtracking Based Environments", Annals of Operations Research, no.118, 101-119
4. Christian Bessiere, E. Hebrard, Brahim Hnich and Toby Walsh (2004) "The Complexity of Global Constraints", American Association for Artificial Intelligence
5. E. H. Durfee and T. A. Montgomery (1991) "MICE: A Flexible Testbed for Intelligent Coordination Experiments", Proceeding of 9th International AAAI Workshop on Distributed Artificial Intelligence, 25-40
6. William S. Havens (1997) "NoGood Caching for MultiAgent Backtrack Search", American Association for Artificial Intelligence (www.aaai.org)
7. Jaziri Wassim (2004) "Modélisation et gestion des contraintes pour un problème d'optimisation sur-contraint – Application à l'aide à la décision pour la gestion du risque de ruissellement", PhD Thesis, INSA-ROUEN – France
8. M. Le Bars, J.M. Attonaty and S. Pinson (2002) "An agent-Based simulation for water sharing between different users", X^{th} Congress of the European Association of Agricultural Economists (EAAE), Zaragoza, Spain
9. Hana Rudova (2001) "Constraint Satisfaction with Preferences", PhD Thesis, Faculty of Informatics, Masaryk University, Brno, Czech Republic
10. TECHNIPLAN, Unigeo & Technocenter (2004) "Studies For Regional Water Management Planning – Agricultural Survey in Sa'dah Water Region – Final Report", Rome and Sana'a
11. TECHNIPLAN, Unigeo & Technocenter (2004) "Studies For Regional Water Management Planning – Well Inventory in Sa'dah Region – Final Report", Rome and Sana'a
12. Christopher Ward, Satoru Ueda and Alexander McPhail (2000) "Water Resources Management in Yemen"
13. M. Wooldridge (1997) "Agent-based software engineering", IEEE Proceedings on Software Engineering, 144(1):26-37
14. Makoto Yokoo, Toru Ishida, Edmund H. Durfee and Kazuhiro Kuwabara (1992) "Distributed constraint satisfaction for formalizing distributed problem solving", 12th IEEE International Conference on Distributed Computing Systems, 614-621
15. Makoto Yokoo, Edmund H. Durfee, Toru Ishida and Kazuhiro Kuwabara (1998) "The Distributed Constraint Satisfaction Problem: Formalization and Algorithms", IEEE Transactions on Knowledge and DATA Engineering, 10(5):673-685

Part V

Spline Functions

Complex Systems Representation by C^k Spline Functions

Youssef Slamani[1], Marc Rouff[12], and Jean Marie Dequen[1*]

[1] Laboratoire des Sciences Appliquées de Cherbourg (LUSAC)
BP 78, rue Louis Aragon
50130 Cherbourg Octeville, France

[2] Laboratoire d'Ingénierie des Systèmes de Versailles (LISV)
Centre Universitaire de Technologie, 10/12 Avenue de l'Europe
78140 Velizy, France

Summary. This work presents the principal algebraic, arithmetic and geometrical properties of the C^k spline functions in the temporal space as well as in the frequencial space. Thanks to their good properties of regularity, of smoothness and compactness in both spaces, precise and powerful computations implying C^k spline functions can be considered. The main algebraic property of spline functions is to have for coefficients of their functional expansion of a considered function, the whole set of partial or total derivatives up to the order k of the considered function. In this way C^k spline function can be defined as the interpolating functions of the set of the all Taylor Mac Laurin expansion up to the degree k defined at each point of discretization of the considered studying function. This fundamental property allows a much easier representation of complex systems in the linear case as well as in the nonlinear case. Then traditional differential and integral calculus lead in the C^k spline functional spaces to new functional and invariant calculus.

Key words: C^k spline, temporal space, frequencial space, interpolation, C^k wavelet, complex systems

1 Introduction

Spline functions appear now as a powerful tool, not only for numerical computation, but also for the formal calculus. The basic algebraic property of these functions is to have, for coefficients of functional development of a considered function, the whole partial or total set of derivatives of this function up to the order k [7].

* Thanks to FEDER european agency for funding.

Thus, C^k spline functions expansion can be view as interpolating functions of the set of the local Taylor-Maclaurin expansions up to the order k defined at each point of discretization of the considered function whereas the known traditional splines (B splines....) are in general preset functions with no valuable adaptative properties. This property which is specific to C^k spline functions leads to a uniform convergence of the solution in function of k and the number of point of dicretization I. This uniform convergence is to our knowledge only seen by using C^k spline functions for computation [2].

Moreover, the Fourier transforms of these C^k spline functions are C^k wavelets which have the same remarkable algebraic property for their coefficients of the spectrum functional expansion of a considered function, i.e. these coefficients defined the set of the whole partial or total derivatives up to the degree k of the dual function. This fact opens the way, for example, to a new time-frequency signal analysis and especially to a new functional and invariant calculation replacing the differential and integral calculus by simple index manipulation. This allows to simplify representation of complex system as well as linear or nonlinear.

An other important advantage of these functions (splines or wavelets) is their remarkable property to obtain a functional representation with a very high accuracy with a small number of coefficients, and this in both spaces with the same precision. In this article, first we present the main properties of Spline functions in the direct space. In a second part, we gather the properties of Splines spectra or C^k wavelets [3]. In the third part, an example of wavelet expansion applied to sin(x) is shown and provides the uniform convergence with excellent rate and accuracy. Only one dimensional spline functions are considered but C^k spline functions and C^k wavelets can be easily defined to multivariate cases.

Finally in the fourth part, we show that the representation of the state equations of as well linear systems as nonlinear systems leads, in the space of splines, to a functional and invariant calculation at place of the classical differential and integral calculus [4].

2 C^k Spline Functions. Temporal Space

2.1 The Equally Spaced Nodes Representation

Let u(x) be an analytical or C^∞ or less monodimensional function defined on a monodimensional open set $\Omega = [0, X]$ which contains I+1 equally spaced nodes $x_{00}, x_{01}, \ldots, x_{0I}$ with $x_{00} = 0$ and $x_{0I} = X.\tilde{u}_{k,I}(x)$ the C^k approximation of this function on Ω is written as,

$$\tilde{u}_{k,I}(x) = \sum_{\nu=0}^{k} \left\{ \sum_{i=1}^{I-1} u^{(\nu)}[i] S_i^{k,\nu}(x) + u^{(\nu)}[0] P_{R0}^{k,\nu} + u^{(\nu)}[I] P_{LI}^{k,\nu} \right\} \tag{1}$$

Where $u^{(\nu)}[i]$, is the νth derivative of u(x) with respect to x at the node x_{0i}, $S_i^{k,\nu}(x)$ is the νth C^k spline function centered at the node x_{0i} and defined on the set $[x_{0(i-1)}, x_{0,(i+1)}]$, $P_{Ri}^{k,\nu}(x)$ and $P_{Li}^{k,\nu}(x)$ are respectivelly the right and left side of the C^k spline function $S_i^{k,\nu}(x)$ and are defined respectivelly on the sets $[x_{0,i}, x_{0,(i+1)}]$ and $[x_{0,(i-1)}, x_{0,i}]$, these definitions are done $\forall i \in [0, I]$, $\forall \nu \in [0, k]$.

2.2 The Unequally Spaced Nodes Representation

Under an arbitarilly nodes ditribution, i.e. an unequally spaced nodes assumption, $\tilde{u}_{k,I}(x)$ can be written as,

$$\tilde{u}_k(x) = \sum_{\nu=0}^{k}\sum_{i=1}^{I}\left(u^\nu[i-1]\, P_{R(i-1)}^{k,\nu}(x) + u^\nu[i]\, P_{Li}^{k,\nu}(x)\right) \tag{2}$$

Where $P_{R(i-1)}^{k,\nu}(x)$ and $P_{Li}^{k,\nu}(x)$ have the same support $[x_{0(i-1)}, x_{0i}]$ and then the same Δx_i, i.e. $\Delta x_i = x_{0i} - x_{0(i-1)}$.

$\tilde{\mathcal{U}}_{k,I}(f)$, the Fourier transform of $\tilde{u}_{k,I}(x)$ in these both cases, leads to the study of $\mathcal{S}_i^{k,\nu}(f)$, $\mathcal{P}_{Ri}^{k,\nu}(f)$ and $\mathcal{P}_{Li}^{k,\nu}(f)$, which are respectivelly the Fourier transform of $S_i^{k,\nu}(x)$, $P_{Ri}^{k,\nu}(x)$ and $P_{Li}^{k,\nu}(x)$ where f is the dual Fourier variable of x.

2.3 Algebraic Properties

Referring to [7], the C^k spline functions are defined by,

$$\left.\frac{d^l\, S_i^{k,\nu}(x)}{d\,x^l}\right|_{x=x_j} \equiv \delta[i-j]\,\delta[\nu-l]$$

which leads to

$$P_{Ri}^{k,\nu}(x) = \sum_{d=0}^{2k+1} a_{Rd}^{\nu}\left[\frac{x - x_{0i}}{\Delta x_{(i+1)}}\right]^d$$

on $[x_{0i}, x_{0(i+1)}]$,

$$P_{Li}^{k,\nu}(x) = \sum_{d=0}^{2k+1} a_{Ld}^{\nu}\left[\frac{x - x_{0i}}{\Delta x_i}\right]^d,$$

on $[x_{0(i-1)}, x_{0i}]$, and,

$$S_i^{k,\nu}(x) = P_{Ri}^{k,\nu}(x) + P_{Li}^{k,\nu}(x)$$

on $[x_{0(i-1)}, x_{0(i+1)}]$. Where a_{Rd}^{ν} and a_{Ld}^{ν} are the polynomial coefficients generating the C^k spline functions. The main properties of these coefficients are given in paragraph 4 of the reference [7], but let us remember five of them,

- $|a^{\nu}_{Ld}| = |a^{\nu}_{Rd}| \equiv a^{\nu}_{d},\ \forall\,\nu\,\in\,[0,k],\ \forall\,d\,\in\,[0,2k+1]$
- The a^{ν}_{Ld}, a^{ν}_{Rd}, and a^{ν}_{d} are stable in $\mathbb{Q}$
- $\forall\,\nu\,\in\,[0,k],\ \forall\,d\,\in\,[0,k],\ a^{\nu}_{Ld}\,\equiv\,a^{\nu}_{Rd}\,\equiv\,a^{\nu}_{d}\,\equiv\delta(d-\nu)\,\frac{\Delta x^{\nu}}{\nu!}$
- $a^{\nu}_{Rd} = (-1)^{d-k+\nu}\,|a^{\nu}_{Ld}|\,,\ \forall\,\nu\,\in\,[0,k],\ \forall\,d\,\in\,[k+1,2\,k+1]$
- $a^{\nu}_{Ld}|_{\Delta x\neq 1} = a^{\nu}_{Ld}|_{\Delta x=1}\,.\Delta x^{\nu}$ *and,* $a^{\nu}_{Rd}|_{\Delta x\neq 1} = a^{\nu}_{Rd}|_{\Delta x=1}\,.\Delta x^{\nu}$

For example, the following list gives a^{ν}_{Rd} and a^{ν}_{Ld} for $k=2$ and for all Δx

$a^{0}_{R0}=1,\ a^{0}_{R1}=0,\ a^{0}_{R2}=0,\ a^{0}_{R3}=-10,\ a^{0}_{R4}=15,\ a^{0}_{R5}=-6$

$a^{1}_{R0}=0,\ a^{1}_{R1}=\Delta x,\ a^{1}_{R2}=0,\ a^{1}_{R3}=-6\,\Delta x,\ a^{1}_{R4}=8\,\Delta x,\ a^{1}_{R5}-3\,\Delta x$

$a^{2}_{R0}=0,\ a^{2}_{R1}=0,\ a^{2}_{R2}=\frac{\Delta x^2}{2},\ a^{2}_{R3}=-\frac{3}{2}\,\Delta x^2,\ a^{2}_{R4}=\frac{3}{2}\,\Delta x^2,\ a^{2}_{R5}=-\frac{\Delta x^2}{2}$

$a^{0}_{L0}=1,\ a^{0}_{L1}=0,\ a^{0}_{L2}=0,\ a^{0}_{L3}=10,\ a^{0}_{L4}=15, a^{0}_{L5}=6$

$a^{1}_{L0}=0,\ a^{1}_{L1}=\Delta x,\ a^{1}_{L2}=0,\ a^{1}_{L3}=-6\,\Delta x,\ a^{1}_{L4}=-8\,\Delta x,\ a^{1}_{L5}=-3\,\Delta x$

$a^{2}_{L0}=0,\ a^{2}_{L1}=0,\ a^{2}_{L2}=\frac{\Delta x^2}{2},\ a^{2}_{L3}=\frac{3}{2}\,\Delta x^2,\ a^{2}_{L4}=\frac{3}{2}\,\Delta x^2,\ a^{2}_{L5}=\frac{\Delta x^2}{2}$

2.4 Representation of the C^k Spline Functions

Let us consider, using Figure 1, the six $S^{5,\nu}_{1}(x)$, $\nu\in[0,5]$, $x\in[0,2]$, centered at the point $x_1=1$, we can easily see that,

$$\left.\frac{d^l}{d\,x^{\ell}}S^{5,\nu}_{1}(x)\right|_{x=x_j} \equiv \delta[j-1]\,\delta[\nu-\ell]$$

i.e. each spline function assumes the representation of a fixed normalized derivative ν at the point $x=1$ and has a zero derivative for $\ell\neq\nu$ and for $x_j\neq 1$, in other words the $S^{k,\nu}_{i}$ are orthonormal functions for the Sobolev space $H_k(\Omega)$ with the scalar product,

$$< f,g >\equiv \sum_{i=0}^{I}\sum_{\nu=0}^{k}\left.\frac{d^{\nu}}{d\,x^{\nu}}f\right|_{x=x_i}\cdot\left.\frac{d^{\nu}}{d\,x^{\nu}}g^*\right|_{x=x_i}$$

On Figure 1, $P^{5,\nu}_{R,0}(x)$ and $P^{5,\nu}_{L,0}$ can be defined respectively as the right and left sides of the $S^{5,\nu}_{1}(x)$, i.e. as the polynomial functions defined respectively on $[0,1]$ and $[1,2]$ for each $S^{k,\nu}_{1}(x)$.

3 C^k Spline Spectra. The Frequencial Approach

Let $u(x)$ be a k time continuous and differentiable one-dimensional function defined on an appropriate set Ω, which contains $I+1$ nodes $x_0,\,x_1,\,\ldots,\,x_I$.

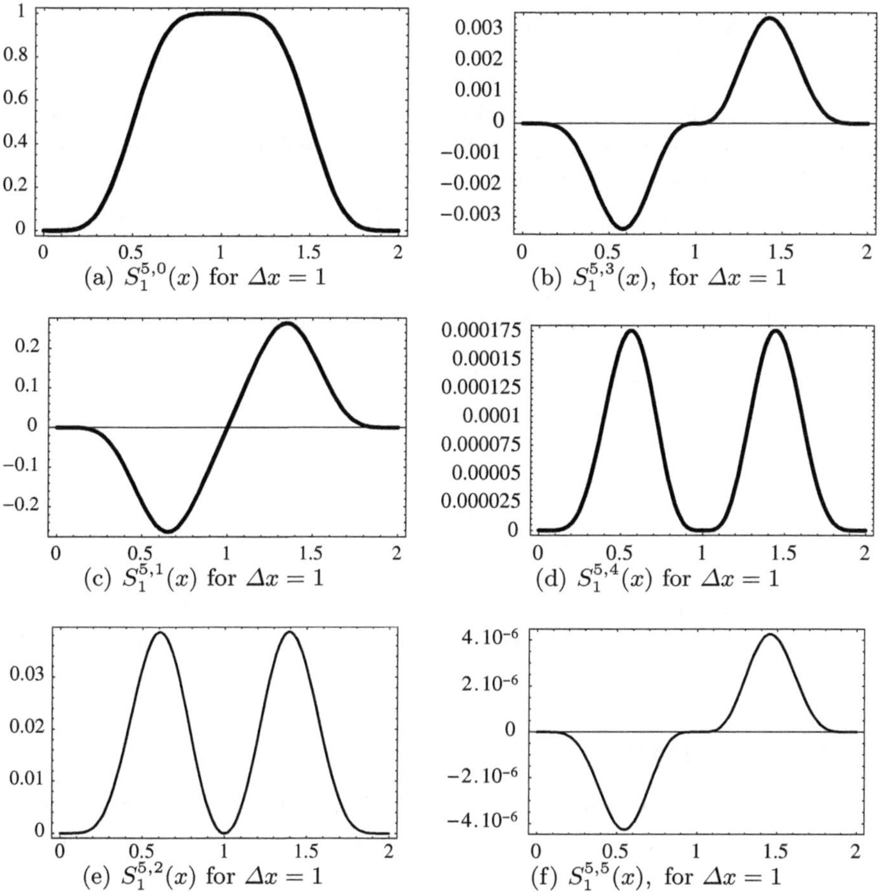

Fig. 1. C^k splines functions

Defining $\tilde{u}_{k,I}(\theta)$, the Fourier Transform of $u_{k,I}(x)$, where θ is the dual Fourier variable of x, we have,

$$\tilde{u}_{k,I}(\theta) \equiv \int_{\Omega} u_{k,I}(x)\, e^{-i\theta\, x}\, d\,x \tag{8}$$

3.1 The Equally Spaced Nodes Spectra Interpolation

When the $I+1$ nodes $x_0, x_1, \ldots, x_I$ of the set Ω are equally spaced, by using equation (1), we have,

$$\tilde{u}_{k,I}(\theta) = \sum_{\nu=0}^{k}\sum_{i=1}^{I-1} u^\nu[i]\, \mathcal{S}_0^{k,\nu}(\theta) e^{-\theta i} + u^\nu[0]\, \mathcal{P}_{R0}^{k,\nu}(\theta) + u^\nu[I]\, \mathcal{P}_{L0}^{k,\nu}(\theta)\, e^{-\theta I}$$

if $u(x)$ is a periodic function of period X then $\theta = i2\pi m \frac{\Delta x}{X}$ where m is the frequency index, if $u(x)$ is non periodic on $[0, X]$ then $\theta = 2\pi i f \Delta x$ where f is the frequency variable. $\mathcal{S}_0^{k,\nu}(\theta)$, $\mathcal{P}_{R0}^{k,\nu}(\theta)$, $\mathcal{P}_{L0}^{k,\nu}(\theta)$ are respectivelly the Fourier transforms of $S_0^{k,\nu}(x)$, $P_{R,0}^{k,\nu}(x)$ and $P_{L,0}^{k,\nu}$ all defined at the nodes x_0, with $\Delta x = x_i - x_{i-1}\ \forall i \in [1, I]$.

3.2 The Unequally Spaced Nodes Spectra Interpolation

Under an arbitrary nodes distribution, i.e. an unequally spaced nodes distribution $x_0, x_1, \ldots, x_I$, using equation (2), $\tilde{u}_{k,I}(\theta)$, the Fourier transform of $u_{k,I}(x)$ can be written as,

$$\tilde{u}_{k,I}(\theta) = \sum_{\nu=0}^{k}\sum_{i=1}^{I} \left\{ u_{i-1}^\nu\, \mathcal{P}_{R(i-1)}^{k,\nu}(\theta) + u_i^\nu\, \mathcal{P}_{R(i-1)}^{k,\nu}(\theta) \right\}$$

if $u(x)$ is periodic of period X then $\theta = i2\pi m \frac{\Delta x_i}{X}$ where m is the frequency index, if $u(x)$ is non periodic on $[0, X]$ then $\theta = 2\pi i f \Delta x_i$ where f is the frequency variable.

3.3 C^k Spline Spectra

As we saw in section 3.1 and 3.2, calculating $\tilde{u}_{k,I}(\theta)$ the Fourier Transform of $u_{k,I}(x)$, leads to study , $\mathcal{S}_0^{k,\nu}(\theta)$, $\mathcal{P}_{R(i-1)}^{k,\nu}(\theta)$ and $\mathcal{P}_{L(i-1)}^{k,\nu}(\theta)$ respectively the Fourier Transform of $S_0^{k,\nu}(x)$, $P_{R(i-1)}^{k,\nu}(x)$ and $P_{L(i-1)}^{k,\nu}(x)$.

3.3.1 C^k Spline $\tilde{S}_0^{k,\nu}(\theta)$ Relations

The following lemmas result from these preceeding properties and from classical algebraic computations.

LEMMA 3.3.1.1.
Let $\mathcal{S}_0^{k,\nu}(\theta)$ be the Fourier transform of $S_0^{k,\nu}(x)$, we have,

$$\mathcal{S}_0^{k,\nu}(\theta) = (-1)^{k+1} 2\, \Delta t^{\nu+1} \left\{ \sum_{\ell=1}^{2k+2} \frac{\cos(\theta + \ell\,\pi/2)}{\theta^\ell} \alpha[\ell,\nu] \right.$$

$$\left. - \sum_{j=E[\frac{k+1}{2}]}^{k} \frac{a_{2j+1}^{2\nu}\, (2j+1)!\, (-1)^j}{\theta^{2\ell+2}} \right\}$$

for ν even, and,

$$\mathcal{S}_0^{k,\nu}(\theta) = i(-1)^{k+1} 2\,\Delta t^{\nu+1} \left\{ \sum_{\ell=1}^{2k+2} \frac{\sin(\theta+\ell\,\pi/2)}{\theta^{\ell}} \alpha[\ell,\nu] - \sum_{j=E[\frac{k}{2}]+1}^{k} \frac{a_{2j}^{2\nu+1}\,(2j)!\,(-1)^j}{\theta^{2\ell+1}} \right\}$$

for ν odd, with,

$$\alpha[\ell,\nu] = \sum_{j=k+1+max[\ell-k-2,0]}^{2k+1} \frac{a_j^{\nu}\, j!\,(-1)^j}{(j+1-\ell)!} + \frac{(-1)^k\,\mathcal{U}[\nu+1-\ell]}{(\nu+1-\ell)!}$$

where $\mathcal{U}[.]$, max[.,.] and E[.] are respectively the Heaviside, the maximum and the Floor functions. i is defined as $i^2=-1$.

LEMMA 3.3.1.2.
The $\alpha[\ell,\nu]$ are stable in $\mathbb{Z}$ for all $\nu\in[0,k]$, and for all $\ell\in[1,2k+2]$.

LEMMA 3.3.1.3.
$\tilde{\mathcal{S}}_0^{k,\nu}$ defined by the lemma 2.1.1.1 is singular at $\theta=0$. Near this point we have for ν even the following Taylor developpment,

$$\tilde{\mathcal{S}}_0^{k,\nu} = (-1)^k 2\,\Delta t^{\nu+1} \sum_{\ell=0}^{+\infty} (-1)^{\ell}\,\theta^{2\ell}\,\beta_e[\ell,\nu] \qquad \text{with}$$

$$\beta_e[\ell,\nu] = \frac{(-1)^k a_{\nu}^{\nu}\,(2\ell+\nu)!}{(2\ell+\nu+1)!(2\ell)!} + \sum_{j=E[\frac{k}{2}]+1}^{k} \frac{a_{2j}^{\nu}(2j+2\ell)!}{(2j+2\ell+1)!(2\ell)!} - \sum_{j=E[\frac{k+1}{2}]}^{k} \frac{a_{2j+1}^{\nu}(2j+2\ell+1)!}{(2j+2\ell+2)!(2\ell)!}$$

for ν odd we have,

$$\tilde{\mathcal{S}}_0^{k,\nu} = i(-1)^{k+1} 2\,\Delta t^{\nu+1} \sum_{\ell=0}^{+\infty} (-1)^{\ell}\,\theta^{2\ell+1}\,\beta_o[\ell,\nu] \qquad \text{with}$$

$$\beta_o[\ell,\nu] = \frac{(-1)^k a_{\nu}^{\nu}\,(2\ell+\nu+1)!}{(2\ell+\nu+2)!(2\ell+1)!} + \sum_{j=E[\frac{k}{2}]+1}^{k} \frac{a_{2j}^{\nu}(2j+2\ell+1)!}{(2j+2\ell+2)!(2\ell+1)!} - \sum_{j=E[\frac{k+1}{2}]}^{k} \frac{a_{2j+1}^{\nu}(2j+2\ell+2)!}{(2j+2\ell+3)!(2\ell+1)!}$$

Clearly the $\beta_e[\ell,\nu]$ and the $\beta_o[\ell,\nu]$ are stable in $\mathbb{Q}$.

3.3.2 Representation of $\tilde{S}_0^{k,\nu}(\theta)$

The Fourier transform of these C^k spline functions are C^k waveletts, and are given for $k = 5$ by Figure 2. As show in Figure 1, it is not difficult to see that the $\mathcal{S}_0^{k,\nu}(\theta)$ are real functions for ν even and imaginary functions for ν odd. We can noticed their excellent compactness.

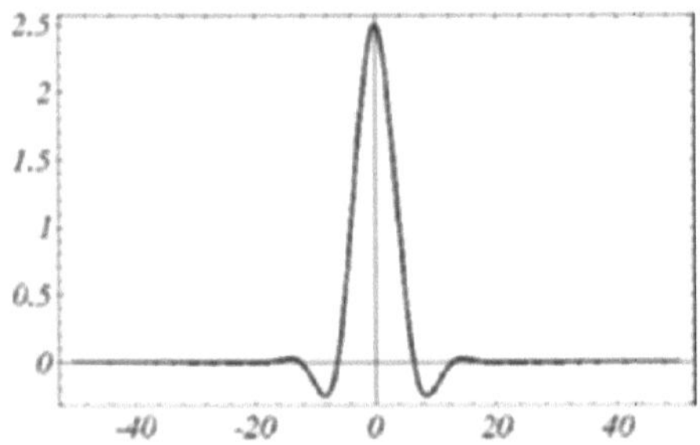

(a) Real part of $\mathcal{S}_1^{5,0}(\theta)$

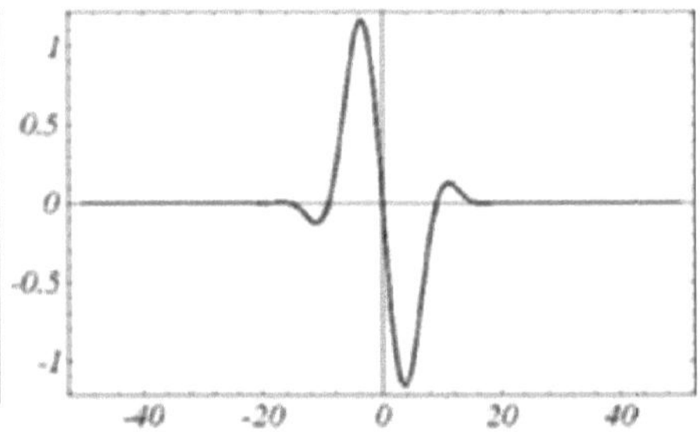

(b) Imaginary part of $\mathcal{S}_1^{5,1}(\theta)$

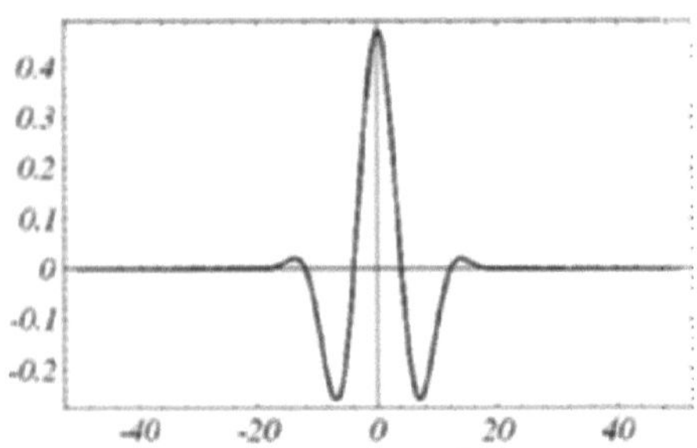

(c) Real part of $\mathcal{S}_1^{5,2}(\theta)$

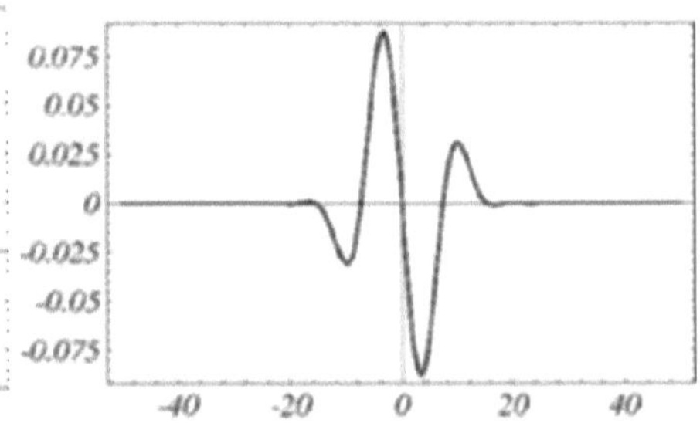

(d) Imaginary part of $\mathcal{S}_1^{5,3}(\theta)$

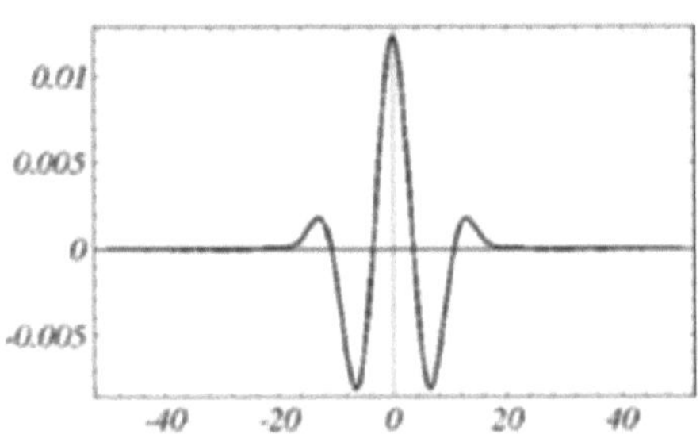

(e) Real part of $\mathcal{S}_1^{5,4}(\theta)$

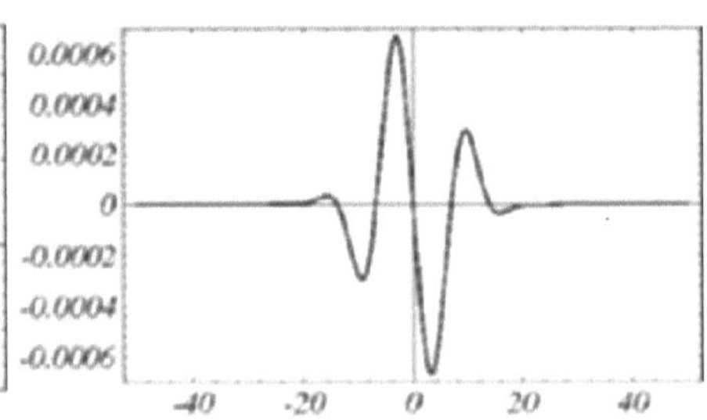

(f) Imaginary part of $\mathcal{S}_1^{5,5}(\theta)$

Fig. 2. Representation of $\mathcal{S}_1^{k,\nu}(\theta)$

3.3.3 $\mathcal{P}_{R,i}^{k,\nu}(\theta)$ and $\mathcal{P}_{L,i}^{k,\nu}(\theta)$ Algebraic Definitions

In this case we define in Ω a serie of intervals $\Delta x_i = x_{0i} - x_{0(i-1)}$. Same considerations as in section **3.3.1.** lead to the following lemmas.

LEMMA 3.3.3.1

Let $\mathcal{P}_{Ri}^{k,\nu}(\theta)$ be the Fourier transform of $P_{Ri}^{k,\nu}(x)$, for ν even we have,

$$\mathcal{P}_{Ri}^{k,\nu}(\theta) = (-1)^k {\Delta x_i}^{\nu+1} \left\{ i \left(\frac{(-1)^{k-\frac{\nu}{2}+1}}{\theta^{\nu+1}} - \sum_{d=E[\frac{k+1}{2}]}^{k} \frac{a_{2d}^{\nu}(2d)!(-1)^d}{\theta^{2d+1}} \right.\right.$$
$$\left. + \sum_{\ell=1}^{2k+2} \frac{sin(\theta+\ell\pi/2)}{\theta^\ell}\gamma[\ell] \right)$$
$$\left. + \left(\sum_{d=E[\frac{k}{2}]+1}^{k} \frac{a_{2d+1}^{\nu}(2d+1)!(-1)^d}{\theta^{2d+2}} - \sum_{\ell=1}^{2k+2} \frac{cos(\theta+\ell\pi/2)}{\theta^\ell}\gamma[\ell] \right) \right\} e^{-\theta i}$$

for ν odd we have,

$$\mathcal{P}_{Ri}^{k,\nu}(\theta) =$$
$$(-1)^{k+1} {\Delta x_i}^{\nu+1} \left\{ \left(\frac{(-1)^{k-(\frac{\nu+1}{2})}}{\theta^{\nu+1}} + \sum_{d=E[\frac{K+1}{2}]}^{k} \frac{(2d+1)!a_{(2d+1)}^{\nu}(-1)^{d+1}}{\theta^{2d+2}} \right.\right.$$
$$\left. + \sum_{\ell=1}^{2k+2} \frac{cos(\theta+\ell\pi/2)}{\theta^\ell}\gamma[\ell] \right)$$
$$\left. + i \left(\sum_{d=E[\frac{k+1}{2}]}^{k} \frac{a_{(2d)}^{\nu}(2d)!(-1)^d}{\theta^{2d+1}} - \sum_{\ell=1}^{2k+2} \frac{sin(\theta+\ell\pi/2)}{\theta^\ell}\gamma[\ell] \right) \right\} e^{-\theta i}$$

with,

$$\gamma[\ell] = \sum_{j=k+1+Max[\ell-(k+2),0]}^{2k+2} \frac{a_j^{\nu} j!(-1)^j}{(j+1-\ell)!} + \frac{(-1)^k \mathcal{U}[\nu+1-\ell]}{(\nu+1-\ell)}$$

The various coefficients of these functions are clearly stable in $\mathbb{Q}$.

LEMMA 3.3.3.2.

Let $\mathcal{P}_{Li}^{k,\nu}(\theta)$ be the Fourier transform of $P_{Li}^{k,\nu}(x)$, for ν even we have,

$$\Re e[\mathcal{P}_{Li}^{k,\nu}(\theta)e^{\theta i}] = \Re e[\mathcal{P}_{Ri}^{k,\nu}(\theta)e^{\theta i}]$$

and,

$$\Im m[\mathcal{P}_{Li}^{k,\nu}(\theta)e^{\theta i}] = -\Im m[\mathcal{P}_{Ri}^{k,\nu}(\theta)e^{\theta i}]$$

for ν odd we have,

$$\Re e[\mathcal{P}_{Li}^{k,\nu}(\theta)e^{\theta i}] = -\Re e[\mathcal{P}_{Ri}^{k,\nu}(\theta)e^{\theta i}]$$

and,

$$\Im m[\mathcal{P}_{Li}^{k,\nu}(\theta)e^{\theta i}] = \Im m[\mathcal{P}_{Ri}^{k,\nu}(\theta)e^{\theta i}]$$

Then as in the lemma 2.1.2.1., the various coefficients of these functions are also stable in $\mathbb{Q}$.

LEMMA 3.3.3.3.

$\mathcal{P}_{Ri}^{k,\nu}(\theta)$ is singular at $\theta = 0$. Near this point we have for all ν the following Taylor developpment,

$$\mathcal{P}_{Ri}^{k,\nu}(\theta) = \Delta x_i^{\nu+1}\left\{\left(\sum_{\ell=0}^{+\infty}\theta^{2\ell}\delta_r[\ell]\right) - i\left(\sum_{\ell=0}^{+\infty}\theta^{2\ell+1}\delta_i[\ell]\right)\right\}$$

with

$$\delta_r[\ell] = \frac{a_\nu^\nu(-1)^\ell(2\ell+\nu)!}{(2\ell+\nu+1)!(2\ell)!} + \sum_{d=k+1}^{2k+1}\frac{a_d^\nu(-1)^{d+k+\ell}(2\ell+d)!}{(2\ell+d+1)!(2\ell)!}$$

$$\delta_i[\ell] = \frac{a_\nu^\nu(-1)^\ell(2\ell+\nu+1)!}{(2\ell+\nu+2)!(2\ell+1)!} + \sum_{d=k+1}^{2k+1}\frac{a_d^\nu(-1)^{d+k+\ell}(2\ell+d+1)!}{(2\ell+d+2)!(2\ell+1)!}$$

The $\delta_r[\ell]$ and the $\delta_i[\ell]$ are stable in $\mathbb{Q}$.

LEMMA 3.3.3.4.

The Taylor developpement of $\mathcal{P}_{Li}^{k,\nu}(\theta)$ is given by the lemma 3.3.3.3. and the relation,

$$P_{Ri}^{k,\nu} = (-1)^\nu P_{Li}^{k,\nu*}$$

which is valid for all $\nu \in [0,k]$ and for all $i \in [0,I]$.

3.3.4 Representation of the $\mathcal{P}_{Ri}^{3,\nu}(\theta)$ and $\mathcal{P}_{Li}^{3,\nu}(\theta)$

The real part and the imaginary part of $\mathcal{P}_{R0}^{3,0}(\theta)$, $\mathcal{P}_{L0}^{3,0}(\theta)$, $\mathcal{P}_{R0}^{3,1}(\theta)$, and $\mathcal{P}_{L0}^{3,1}(\theta)$ are given in Figure 3 for $\Delta x = 1$. We can easily verify the properties of the lemma 3.3.3.2

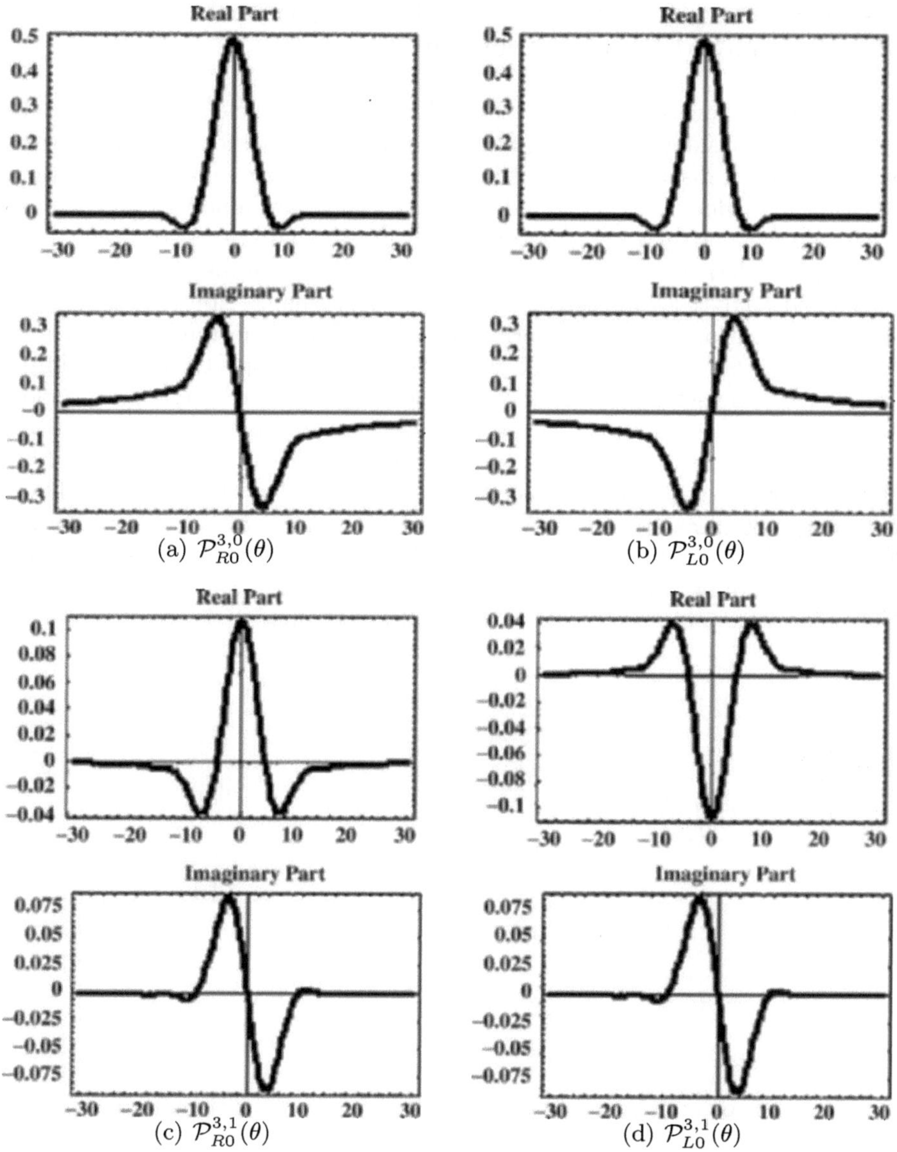

(a) $\mathcal{P}_{R0}^{3,0}(\theta)$ (b) $\mathcal{P}_{L0}^{3,0}(\theta)$

(c) $\mathcal{P}_{R0}^{3,1}(\theta)$ (d) $\mathcal{P}_{L0}^{3,1}(\theta)$

Fig. 3. The real and imaginary parts of $\mathcal{P}_{R0}^{3,0}(\theta)$, $\mathcal{P}_{L0}^{3,0}(\theta)$, $\mathcal{P}_{R0}^{3,1}(\theta)$, $\mathcal{P}_{L0}^{3,1}(\theta)$

4 Example : C^k Wavelets and the Representation of a Truncated Sinus Spectrum

This example shows one of the remarkable property of the C^k spline representation, if we consider an analytical function $f(x)$ and the set of its derivatives at each point of discretization until an arbitrarily large number of derivation. By using C^k spline functions and C^k wavelets we are able to build respectively the nth derivative of $f(x)$ on Ω and its associated spectra immediatly by using the derivatives from n to $n+k$ at each point of discretization of the considered analytical function $f(x)$ and by using the C^k spline functions for the representation of the nth derivatives of $f(x)$ with no limitation on n or by using the C^k wavelets which gives the associated spectrum by a simple switching of the set of functions of representation. The example given below shows that these computations can be done with an arbitrarily precision and with a finite small number of terms. Figure 4 gives left, the exact analytic reprentation of the Fourier transform of $sin(x)$ on $\Omega = [0, 10]$, and right its wavelets approximation for k=5, I=5, and with $\Delta x = 5/2$. Because Δx belongs to $\mathbb{Q}$, all computations are stable in $\mathbb{Q}$ and are exact numerical solutions. In fact if the draw $\ell og_{10}(\|e\|) = \varepsilon$, with

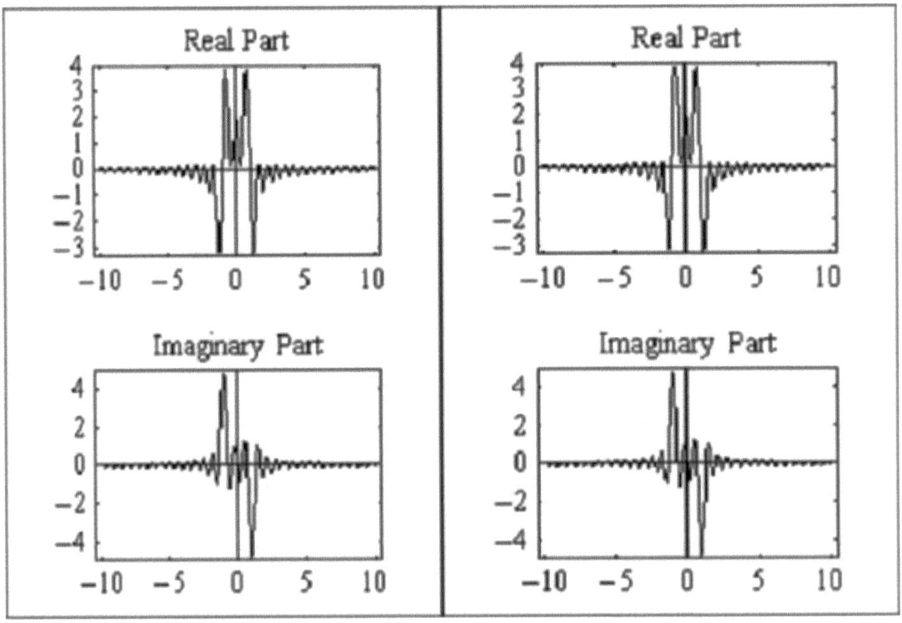

Fig. 4. Exact Fourier Transform of $sin(x)$, $x \in [0, 10]$ (left) and its C^k wavelets approximation (right)

$$\|e\| = \int_{-\infty}^{+\infty} \{TF\{Sin(x)\} - \tilde{u}_{k,I}(\theta)\}^2 \, d\theta$$

we see in Figure 5 that ε converge uniformly to $-\infty$ as we can expect from a representation which is only the interpolating expansion of the set of the Taylor-Mac Laurin expansions up to the degree k of the considered function $f(x)$ at each points of discretization. We can noticed from Figure 5 that, for $k = 6$ and $I = 9$, $\varepsilon = 10^{-32}$.

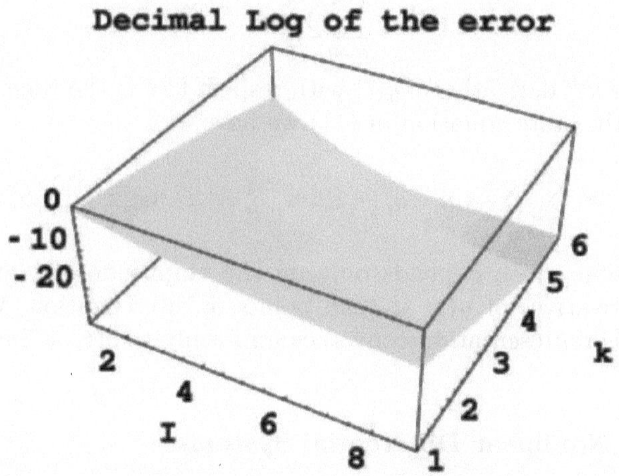

Fig. 5. Error ε versus k and I

5 Representation of Complex Systems

Another interesting property of C^k spline functions is their remarkable property of representating dynamical systems or dynamical constraints with the properties of precision and representation shown above. This allows to transform $H_k(\Omega)$ the natural space of representation of C^k spline function into dedicated manifold associated to a differentiel problem by introducing differential constraints of the problem including the dynamical system itself as simple algebraic relations in $H_k(\Omega)$, we will see later that these properties transpose the classical differential and integral calculus in a functional and invariant calculus.

5.1 Linear Systems

Let us consider a linear system represented by its state equation,

$$q(t) = A.q(t) + B.u(t)$$

$$y(t) = C.q(t)$$

where $q(t)$ is the state vector of dimension N, $u(t)$ is the control vector of dimension m and $y(t)$ the output vector of dimension r. A is the dynamical matrix of dimension $N \times N$, B the control matrix of dimension $N \times m$ and C the output matrix of dimension $r \times N$.

$q_{k,I}(t)$ the C^k spline approximation of the state vector $q(t)$ can be written symbolically by,

$$q_{k,I}(t) = \sum_{\nu=0}^{k} \sum_{i=0}^{I} q_i^{\nu} \, S_i^{k,\nu}(t) \tag{11}$$

where q_i^{ν} is the νth derivative of $q(t)$ with respect to t at the time t_i, $i \in [0, I]$. By including the state equation in (11) we have,

$$q_{k,I}(t) = \sum_{\nu=0}^{k} \sum_{i=0}^{I} \left\{ A^{\nu} \, . \, q_i^0 + Sum_{j=0}^{\nu-1} \, A^j.B.u_i^{(\nu-j-1)} \right\} \, S_i^{k,\nu}(t)$$

i.e. an expansion which depends only on the amplitude of $q(t)$ and on the $\nu - 1$ first derivatives of $u(t)$ at each points of discretization. We can show easily that this representation converges uniformly to $q(t)$ as in the example of section 4.

5.2 Explicit Nonlinear Differential Systems

Let us consider an explicit nonlinear differential system defined by its state equation,

$$\dot{q}(t) = F(q(t))$$

where $q(t)$ is the state vector of dimension N and F a C^k or more continous and derivable nonlinear vectorial function of the state $q(t)$ and of dimension N. We defined as $\mathcal{F}[\,]$ the vector field associated to $F(q(t))$, and we have,

$$\mathcal{F}[\,] \equiv \sum_{i=1}^{N} F^i(q(t)) \, \frac{\partial}{\partial x_i}$$

where $F^i(q(t))$ is the ith component of the vectorial function $F(q(t))$.

By using the state equation and the vector field properties, we have,

$$q_{k,I}(t) = \sum_{\nu=0}^{k} \sum_{i=0}^{I} \mathcal{F}[q(t)] \Big|_{t=t_i} \quad S_i^{k,\nu}(t)$$

which depends only on $q(t_i)$, i.e. the amplitude of the state vector at times t_i, $i \in [0, I]$. For example, if we consider the one dimensional nonlinear system $\dot{q}(t) = q(t)^2$, then its k, I approximation can be written as :

$$q_{k,I}(t) = \sum_{\nu=0}^{k} \sum_{i=0}^{I} \nu! \, \left(q_i^0\right)^{\nu-1} \, S_i^{k,\nu}(t)$$

5.3 Implicit Nonlinear Differential Systems

Let us consider an implicit nonlinear differential system defined by,

$$F\left(\dot{q}(t), q(t), u(t)\right) = 0$$

where $q(t)$ is the state vector of dimension N, $\dot{q}(t)$ the first derivative of the state vector, $u(t)$ the control vector of dimension m and F a C^k or more continuous and derivable nonlinear vectorial function of $q(t)$, $\dot{q}(t)$ and $u(t)$ are of dimension N. Same considerations as in section 5.1 and 5.3 lead to,

$$\sum_{i=0}^{I} \sum_{\nu=0}^{k} \left. \frac{d^\nu}{d\,t^\nu} F\left(\dot{q}(t), q(t), u(t)\right) \right|_{t=t_i} S_i^{k,\nu}(t) = 0$$

6 Conclusion

We have presented in this paper the main algebraic expressions and properties of the one-dimensional spline functions in temporal space as well as in frequency space. We have also shown that for as well linear as non linear systems, the classical differential and integral calculus can be replaced in splines space by functional and invariant calculus. The major advantages of c^k spline functions can be summarized as follows:

- The Coefficients of functional expansion by C^k spline functions are the same in time space as well as in frequency space. These coefficients are the set of the $k+1$ relevant derivatives of the considered function, at each point of discretization.
- C^k spines have excellent approximation properties. Interpolation by C^k spline functions converges uniformly with an excellent rate and accuracity. We have shown that for $k=5$ and $I=10$ the error estimate of sinus spectra approximation is around 10^{-30}. Interpolation by C^k spline functions is also available for non uniformly spaced nodes without any inconvenience.
- C^k spline functions and C^k wavelets have excellent localization properties. They open a new way for time-frequency signal analysis.
- The representation of linear as well as non linear complex systems in the space of C^k spline functions leads to a new functional and invariant calculation instead of the classical differential and integral calculus.
- The method is to our knowledge the only one which generates a functional space stable through every integro-differential operators and opens the way to a new numerical analysis.
- This allows to simplify and to rewrite a large number of classical problems, as for example nonlinear optimal control, where the transposition of this differential problem leads to a functional problem in the C^k spline functional spaces, permitting easily the complete elimination of the adjoint vector with the well known associated numerical or differential troubles.

- Dynamical equations, differential constrains or boundary conditions can be introduced in the C^k spline functional space as algebraic relations transforming the initial space in a dedicated manifolds.
- The Fourier transforms of C^k splines are C^k wavelets with the same good properties, opening the way to a new frequency analysis of nonlinear systems.

References

1. Marc Rouff (1992) "The computation of C^k spline functions",*Computers and Mathematics with Applications*, 23(1):103-110
2. Marc Rouff and Wenli Zhang (1994) "C^k spline functions and linear operators", *Computers and Mathematics with Applications*, 28(4):51-59
3. Marc Rouff and Youssef Slamani (2003) "C^k spline functions spectra : definitions and first properties", *International Journal of Pure and Applied Mathematics*, 8(3):307-333
4. Michael Unser (1999) "Splines : A perfect fit for signal and image processing", *IEEE Signal Processing Magazine*, 22-38

Computer Algebra and C^k Spline Functions: A Combined Tools to Solve Nonlinear Differential Problems

Zakaria Lakhdari[1], Philippe Makany[1], and Marc Rouff[21*]

[1] Laboratoire des Sciences Appliquées de Cherbourg (LUSAC)
BP 78, rue Louis Aragon
50130 Cherbourg Octeville, France

[2] Laboratoire d'Ingénierie des Systèmes de Versailles (LISV)
Centre Universitaire de Technologie, 10/12 Avenue de l'Europe
78140 Velizy, France

Summary. In this article, we describe a method to solve differential equations involved in dynamical nonlinear problems with C^k spline functions and their algebraic properties. C^k spline functions generate k time continuous and derivable approximations. These functions seem to be very efficient in several applied nonlinear differential problems. The computation of functions involved in nonlinear dynamical problem uses the algebraic methods known as Gröbner Bases, which are more robust, more accurate and often more efficient than the purely numerical methods. An example on solving nonlinear differential equations using these methods is presented.

Key words: C^k spline, differential equation, dynamical non linear, Gröbner bases, computer algebra

1 Introduction

The idea of computing an approximated solution of differential problem is not new. Since Fourier and his heat equations, various functions have been proposed for this task, trigonometric functions of course, exponential functions, various polynomials functions like Legendre, Chebitcheff, Laguerre or spline functions, and some discontinuous functions like Walsh functions.

In the first time, we present in this article the C^k spline functions which generate k time continuous and derivable approximations. These functions seem to be a very efficient tool for various applied nonlinear differential problems, and particularly in nonlinear control, functional differential equations problems, delayed differential equations, etc.

* Thanks to FEDER european agency for funding.

In the second time, in order to obtain an approximated solutions, we solve these spline functions wih computer algebra and Gröbner bases techniques. The plan of the paper is as follows. The main Section **2** is devoted to presenting the general form of the Differential Equations and their approximations by C^k spline functions, we review the necessary mathematical background. Section **3** contains an introduction to computer algebra and Gröbner bases techniques. In section **4** we report an evaluation of this approach by presenting an example. Finally, in section **5** we outline the main features of the combination of C^k spline functions and algebra computation in order to solve Ordinary Differential Equations (ODE) and open issues.

2 General Form of ODE's and Spline Functions

Let us consider the following nonlinear differential system of dimension N on $[0, T]$,

$$\dot{q}(t) = F(q(t)) \tag{1}$$

The representation of C^k spline functions of $q(t)$ in the base k is given by,

$$q_{k,I}(t) = \sum_{i=0}^{I} \sum_{\nu=0}^{k} q_i^\nu \, S_i^{k,\nu}(t)$$

where i is the discretization index on [0,T] and,

$$q_i^\nu \equiv \left. \frac{d^\nu}{d\,t^\nu} q(t) \right|_{t=t_i}$$

or given by,

$$q_{k,I}(t) = \sum_{i=0}^{I} \sum_{\nu=0}^{k} \left. \mathcal{F}^\nu [q(t)] \right|_{t=t_i} \quad S_i^{k,\nu}(t)$$

with $\mathcal{F}^\nu[\,]$ the νth iteration of the vector fields $\mathcal{F}[\,]$ defined by :

$$\mathcal{F}[\,] = \sum_{i=1}^{N} F^i(q(t)) \, \frac{\partial}{\partial\, q_i}$$

Then by derivation we have,

$$\dot{q}(t) = \sum_{i=0}^{I} \sum_{\nu=0}^{k} q_i^\nu \, \dot{S}_i^{k,\nu}(t)$$

and,

$$F(q(t)) = \sum_{i=0}^{I} \sum_{\nu=0}^{k} \left. \frac{d^\nu\, F(q(t))}{d\, t^\nu} \right|_{t=t_i} \quad S_i^{k,\nu}(t)$$

Then in this kind of problems the C^k spline functional expansion of $q(t)$ involves only the values of $q(t)$ at each point of discretization t_i, $i \in [0, I]$, and not their derivatives, and k can be understood as the induced topology in the functional space of the solution of (1).
The algorithm used in this paper is based on the minimization of the canonical functional distance represented by H the error of the method on $[0, T]$. This quadratic error H is defined as the following quadratic functional,

$$H(q_i^\nu) \equiv \int_0^T \; [\dot{q}(t) - F(q(t))]^2 \, dt$$

H can be written as,

$$H(q_i^\nu) = \sum_{i=0}^{I} \sum_{\nu=0}^{k} \sum_{i'=0}^{I} \sum_{\nu'=0}^{k} q_i^\nu \, q_{i'}^{\nu'} \; < \dot{S}_i^{k,\nu}(t), \dot{S}_{i'}^{k,\nu'}(t) >$$

$$-2 \sum_{i=0}^{I} \sum_{\nu=0}^{k} \sum_{i'=0}^{I} \sum_{\nu'=0}^{k} q_i^\nu \left. \frac{d^{\nu'} F(q(t))}{d t^{\nu'}} \right|_{t=t_{i'}} < \dot{S}_i^{k,\nu}(t), S_{i'}^{k,\nu'}(t) >$$

$$+ \sum_{i=0}^{I} \sum_{\nu=0}^{k} \sum_{i'=0}^{I} \sum_{\nu'=0}^{k} \left. \frac{d^{\nu} F(q(t))}{d t^{\nu}} \right|_{t=t_i} \left. \frac{d^{\nu'} F(q(t))}{d t^{\nu'}} \right|_{t=t_{i'}} < S_i^{k,\nu}(t), S_{i'}^{k,\nu'}(t) >$$

This quadratic error can be solved by two methods, a first one consists in using numerical methods as simulated annealing, a second one which uses algebraic methods.

In this paper, we use Gröbner bases methods (algebraic methods) to compute these functions, in this case we use the minimization of this quadratic error given by the following condition:

$$\frac{\partial \, H(q_{i'}^{\nu'})}{\partial \, q_i^\nu} = 0$$

This condition allows to obtain a polynomial system which can be solved by Gröbner bases, by assuming that $F(q(t))$ can be always tramsformed in a polynomial functions of $q(t)$. To our knowledge this asumption did not restrict the generality of the problem.

3 Computer Algebra and Gröbner Basis

In order to give an intuitive presentation of these notions, we frequently use analogies with well known linear algebra concepts. In the following, a polynomial is a finite sum of terms, and each term is the product of a coefficient by a monomial. Refer to [5,6,7,8] for a more detailed introduction.

3.1 Simplification of Polynomial System

Solving linear systems consists of studying vector spaces, and similarly, solving polynomial systems consists of studying ideals. More precisely, we define a system of polynomial equations $P_1 = 0, \ldots, P_R = 0$ as a list of multivariate polynomials with rational coefficients in the algebra $Q[X_1, \ldots, X_N]$. To such a system, we associate I, which is the ideal generated by $P_1, \ldots, P_R$; it is the smallest ideal containing these polynomials, as well as the set of $\sum_{k=1}^{R} P_k U_k$, where $U_k \in Q[X_1, \ldots, X_N]$. Since the P_k vanish exactly at points where all polynomials of I vanish, it is equivalent to studying the system of equations or the ideal I.

For a set of linear equations, one can compute an equivalent triangular system by "canceling" the leading term of each equation. A similar method can also be done for multivariate polynomials. Of course, we have to define the leading term of a polynomial or, in other words, order the monomials. Thus, we choose an ordering on monomials compatible with the multiplication. In this paper, we only use three kinds of ordering:

- "lexicographic" order: (Lex)

$$x^\alpha \equiv x^{(\alpha_1,\ldots,\alpha_N)} <_{Lex} x^\beta \equiv x^{(\beta_1,\ldots,\beta_N)} \Leftrightarrow \exists\, i_0\, \forall\, i = 1 \ldots i_0 - 1,$$

$$\alpha_i = \beta_i \text{ and, } \alpha_{i_0} < \beta_{i_0}$$

- "degree reverse lexicographic" order: (DRL)

$$x^{(\alpha_1,\ldots,\alpha_N)} <_{DRL} x^{(\beta_1,\ldots,\beta_N)} \Leftrightarrow x^{[(\sum_k \alpha_k),\beta_N,\ldots,\beta_1]} <_{Lex} x^{[(\sum_k \beta_k),\alpha_N,\ldots,\alpha_1]}$$

- "DRL by blocks" order: (DRL, DRL)

We split variables into two blocks $\alpha = (\alpha_1, \ldots, \alpha_N) = [(\alpha_1, \ldots, \alpha_{N'-1}), (\alpha_{N'}, \ldots, \alpha_N)] = (\alpha', \alpha")$ for some $N' < N$.

$$x^{(\alpha',\alpha")} <_{DRL,DRL} x^{(\beta',\beta")} \Leftrightarrow \left(x^{\alpha'} <_{DRL} x^{\beta'}\right) \text{ or } [(\alpha' = \beta') \text{ and}$$

$$x^{\alpha"} <_{DRL} x^{\beta"}]$$

Now, we can define the leading monomial (resp., term) of a polynomial as its monomial (resp., term) with highest degree.

3.2 Gröbner Bases

For solving systems of algebraic equations, we use Gröbner bases. Let us recall very briefly what it is, leaving details and precise definitions to [11, 8].

Given a set of polynomials, a Gröbner basis is another set of polynomials which has the same common roots in a very strong sense (the multiplicities are the same, the generated ideal is the same). A Gröbner basis may be viewed as a compiled form for a system of equations in the sense that no information is lost and, on the opposite, many properties of the solutions, such as their number or their values may easily be deduced from the Gröbner basis.

A Gröbner basis is a canonical form for a system of equations which depends only on the input equations and on a total ordering on the monomials (power products). Two orderings are especially important, the degree-reverse-lexicographical one, which leads to rather easy computations and the purely lexicographical one for which the Gröbner basis is more difficult to compute, but for which the information is more accessible.

We can now give a sketch of the Buchberger algorithm [2]-[4], which can be seen as a constructive definition of Gröbner bases,

Gröbner (polynomials$f_1, \ldots, f_n, <$ **a monomial ordering)**

$$Pairs = \{[f_i, f_j], 1 \leq i < j \leq n\}$$

while $Pairs \neq 0$ **do**

$$\textit{Choose and remove a pair } [f_i, f_j] \textit{ in } Pairs.$$

$$f_{n+1} = Reduce(S - pol(f_i, f_j, <), [f_1, \ldots, f_n],)$$

if $f_{n+1} \neq 0$ **then**

$$n = n + 1$$

$$Pairs = Pairs \cup \{[f_i, f_n], 1 \leq i \leq n\}$$

end if
end while
return$[f_1, \ldots, f_n]$

Definition 1: The output G of the algorithm is called a Gröbner base of I for the order $<$.

Theorem 1: G has the following properties :

1) G is an equivalent set of generators of I.
2) A polynomial p belongs to I if and only if *Reduce* $(p, G) = 0$.
3) The output of *Reduce* (p, G) does not depend on the order of the polynomials in the list. Thus, this is a canonical reduced expression modulus I, and the *Reduce* function can be used as a simplification function.
4) From G, it is easy to compute the number of complex solutions (counted with multiplicities) of the input system.
5) If $<$ is lexicographic, G has a "simple form".

Solutions of an algebraic system could be of a variety of kinds that can be classified their algebraic dimension. For example,

- finite number of isolated points, in which case we say that the dimension is 0;
- curves, where the dimension is 1;
- surfaces, where the dimension is 2.

If a system has different kinds of solutions (e.g., isolated points and curves), then the global dimension is the maximum dimension of each component.

Another meaningful interpretation of the dimension is that it corresponds to the remaining free degrees when all of the equations are satisfied.

3.3 Lexicographic Gröbner Bases

The computation time depends strongly on the monomial order that is used. In general, Gröbner bases for a lexicographic ordering are much more difficult to compute than the corresponding DRL Gröbner base. On the other hand, this computational cost is, however, worth it because the lexicographic Gröbner bases has a more or less triangular structure that is suitable for further processings. Fortunately we can compute efficiently lexicographic Gröbner bases with a different method.

3.4 Computer Algebra System

There is a Gröbner function in every computer algebra system (Maple, Mathematica, Axiom,...), but it must be emphasized that these implementations are very inefficient compared with recent software; even the specialized software (Magma, Singular, Macaulay, Asir) are unable to solve the most difficult systems. In this paper, we use FGB, an efficient C/C++ software developed by J.C Faugère [9], it includes a new generation of algorihms for solving polynomial systems.

4 An Example

Let us consider the following Ordinary Differential equation given by :

$$\dot{q}(t) = -q^2(t) \text{ with initial condition } q(0) = 1.00$$

. The theoretical solution of this ODE is :

$$q(t) = \frac{1}{t+1}; \quad \dot{q}(t) = \frac{-1}{(t+1)^2}; \quad \ddot{q}(t) = \frac{2}{(t+1)^3}$$

On $[0,1]$ interval, we obtain the exact solution of this differential equation represented by Fig. 1.

An approximation of this ODE by C^k spline functions ($k = 2$) under the computation of

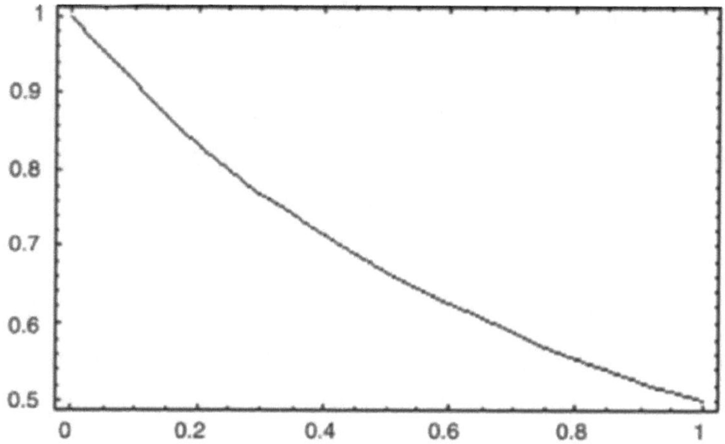

Fig. 1. Curve of the exact solution of $\dot{q}(t) = -q(t)^2$

$$\frac{\partial H(q_i^\nu)}{\partial q_i^\nu} = 0$$

gives a polynomial system (6 equations). Taking into account the initial value q(0)=1, this system reduces to 5 equations given as follow, from equation (1) to equation (5):

$$\begin{aligned}&(((6042+249\,q[0,1]^2+24\,q[0,1]^3+600\,q[0,2]+1320\,q[1,0]-1818\,q[1,0]^2-\\&2112\,q[1,1]+652\,q[1,0]\,q[1,1]-39\,q[1,1]^2+275\,q[1,2]-39\,q[1,0]\,q[1,2]+\\&\quad 2\,q[0,1]\,((3621+12\,q[0,2]+181\,q[1,0]^2-143\,q[1,1]+10\,q[1,1]^2-\\&\qquad 2q\,[1,0]\,((-330+52\,q[1,1]-5\,q[1,2]))+11\,q[1,2]))))=0 \qquad (1)\end{aligned}$$

$$\begin{aligned}&((281+600\,q[0,1]+12\,q[0,1]^2+56\,q[0,2]+330\,q[1,0]-149\,q[1,0]^2-275\,q[1,1]+\\&\qquad 39\,q[1,0]\,q[1,1]-q[1,1]^2+33\,q[1,2]-q[1,0]\,q[1,2]))=0 \qquad (2)\end{aligned}$$

$$\begin{aligned}&(((-25740)+17880\,q[1,0]+41580\,q[1,0]^2+21720\,q[1,0]^3-492\,q[1,1]-\\&7260\,q[1,0]\,q[1,1]-11196\,q[1,0]^2\,q[1,1]+660\,q[1,1]^2+2226\,q[1,0]\,q[1,1]^2-\\&\qquad 138q[1,1]^3+q[0,1]\,((1320-3636\,q[1,0]\\&+652\,q[1,1]-39\,q[1,2]))-149\,q[1,2]+660\,q[1,0]\,q[1,2]+843\,q[1,0]^2\,q[1,2]-\\&\qquad 276\,q[1,0]\,q[1,1]\,q[1,2]\\&+12q[1,1]^2\,q[1,2]+12q[1,0]\,q[1,2]^2-q[0,2]\,(((-330)+298\,q[1,0]-39\,q[1,1]\\&\quad +q[1,2]))+2\,q[0,1]^2\,((330+181\,q[1,0]-52\,q[1,1]+5\,q[1,2]))))=0 \qquad (3)\end{aligned}$$

$$((2310-492\,q[1,0]-3630\,q[1,0]^2-3732\,q[1,0]^3+2\,q[0,1]\,((-1056+326\,q[1,0]-39\,q[1,1]))+q[0,2]\,((-275+39\,q[1,0]-2\,q[1,1]))+5378\,q[1,1]+1320\,q[1,0]\,q[1,1]+2226\,q[1,0]^2\,q[1,1]-165\,q[1,1]^2-414\,q[1,0]\,q[1,1]^2+24\,q[1,1]^3+q[0,1]^2((-143-104\,q[1,0]+20\,q[1,1]))-462\,q[1,2]-138\,q[1,0]^2\,q[1,2]+24\,q[1,0]\,q[1,1]\,q[1,2]))=0 \tag{4}$$

$$((q[0,1]\,((275-39\,q[1,0]))-q[0,2]\,((-33+q[1,0]))-149\,q[1,0]+330\,q[1,0]^2+281\,q[1,0]^3+q[0,1]^2\,((11+10\,q[1,0]))-462\,q[1,1]-138\,q[1,0]^2\,q[1,1]+12\,q[1,0]\,q[1,1]^2+44\,q[1,2]+12\,q[1,0]^2\,q[1,2])))=0 \tag{5}$$

We have 5 unknowns values $q[0,1]$ (derivative of q), $q[0,2]$ (curvative of q), $q[1,0]$ (position of q), $q[1,1]$ and $q[1,2]$.

The computation of this example using Gröbner bases gives 47 square root (3 reels, 44 complexes). Only real square roots are used to obtain the global extrema (minima).

The following figure shows the performance results of the computation with FGB software . The error $e(t)$ between $q(t)$ and $\tilde{q}(t)$ on $[0,1]$ interval is given by Figure 3. Curve of the approximated solution of The computation of quadratic error of this approximation is:

$$\varepsilon = \int_0^1 (q(t)-\tilde{q}(t))^2 \; dt = 18 * 10^{-8}$$

This error shows the powerful and precise approximation using C^k spline functions and algebraic methods.

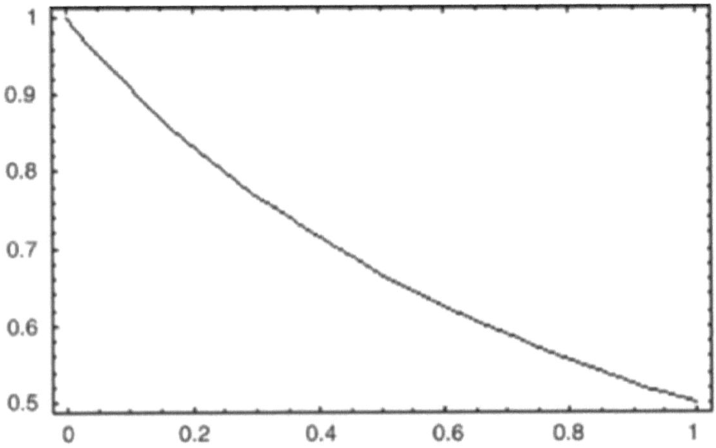

Fig. 2. Curve of the approximated solution of $\dot{q}(t) = -q(t)^2$

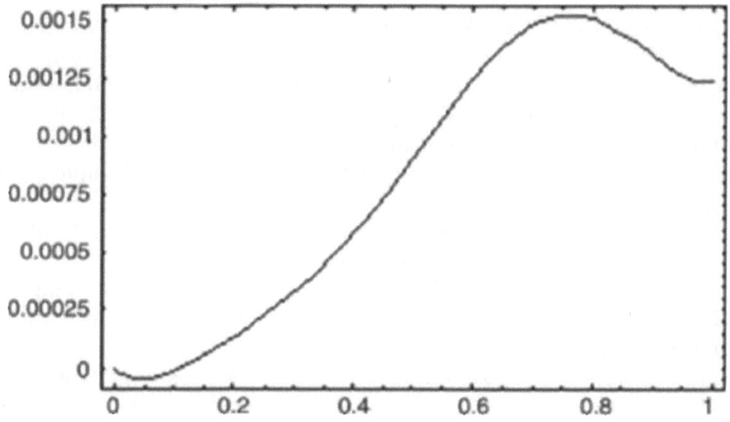

Fig. 3. Curve of $e(t) = q(t) - \tilde{q}(t)$ on [0,1]

5 Conclusion

We have presented in this paper a method to solve the Ordinary Differential Equations (ODE) with C^k spline functions and algebra computation. C^k spline functions on combination with Gröbner bases computation seem to be a new powerful tool in the investigation of nonlinear differential problems.

6 Acknowledgment

The authors would like to thank Dr. Jean Charles Faugère (Laboratoire d'Informatique de Paris 6) for his helpfull on solving these systems with FGB and for the fruitfull discussions we had.

References

1. M. Rouff (1992) The Computation of Ck spline function, Computers and Mathematics with Applications, 23(1):103-110.
2. M. Rouff (1996) "Ck spline functions in applied differential problem", Systems Analysis Modelling and Simulations, Gordon & Breach Science Publishers, 26:197-205.
3. M. Rouff and M. Alaoui (1996) "Computation of dynamical electromagnetic problems using Lagrangian formalism and multidimensional Ck spline functions", Zeitschrift für Angewandte Mathematik und Mechanik, Academie Verlag Berlin, 76(1):513-514.
4. M. Rouff and W.L. Zhang (1994) "Ck spline functions and linear operators", Computers and Mathematics with Applications, 28(4):51-59.
5. B. Buchberger (1965) Ein Algorithmus zum Auffinden der Basiselemente des Restklassenringes nach einem nulldimensionalen Polynomideal. (PhD thesis, Innsbruck).

6. B. Buchberger (1970) "An Algorithmical Criterion for the Solvability of Algebraic Systems". Aequationes Mathematicae 4(3):374-383. (German).
7. B. Buchberger (1979) "A Criterion for Detecting Unnecessary Reductions in the Construction of Gröbner Basis". In Proc. EUROSAM 79, vol. 72 of Lect. Notes in Comp. Sci., Springer Verlag, 3-21.
8. J.C Faugère (1994) Résolution des systèmes d'équations algébriques. (PhD thesis, Université Paris 6).
9. J.C Faugère (1999) "A new efficient algorithm for computing Grôbner bases (F4)". Journal of Pure and Applied Algebra 139:61-88.
10. Faugère J.C., Gianni P., Lazard D., Mora T. (1993) "Efficient Computation of Zero-Dimensional Gröbner Basis by Change of Ordering". Journal of Symbolic Computation 16(4):329-344.
11. D. Lazrad (1992) "Solving zero-dimensional algebraic systems". Journal of Symbolic Computation 13(2):117-132. available by anonymous ftp posso.lip6.fr.

Part VI

Control

Decoupling Partially Unknown Dynamical Systems by State Feedback Equations

Philippe Makany[1], Marc Rouff[21], and Zakaria Lakhdari[1*]

[1] Laboratoire des Sciences Appliquées de Cherbourg (LUSAC)
BP 78, rue Louis Aragon
50130 Cherbourg Octeville, France
[2] Laboratoire d'Ingénierie des Systèmes de Versailles (LISV)
Centre Universitaire de Technologie, 10/12 Avenue de l'Europe
78140 Velizy, France

Summary. One problem for efficient control of complex systems is to take into account the influence of unknown dynamics in interaction with a fixed defined one by a set of defined time varying parameters behaviour coming from coupling partially unknown dynamics. Under the assumption of robust model, nonlinear state feedback decoupling methods can be used for the control of such systems. We show in our paper, that if these parameters are slowly varying in time before the main dynamics (the exact mathematical meaning of this notion is given in the article), the classical decoupling methods remain valid with parameterized laws, but if parameters dynamic are non negligible before the main dynamic, we are in the fast varying parameters case, and we show that in this case decoupling methods operators must be changed in order to include these dynamical effects. We show also that in this case the use of classical decoupling methods, leads to non effcient control and multiple spurious effects. The computation of the static and dynamic feedback laws, in case of fast varying parameters, are given, for nonlinear decoupling methods, linearization and rejection of perturbations, as the research of the functional invariants of the output. A robust example in robotics is given, and the contribution of parameters dynamic is shown.

Key words: complex systems control, feedback laws, decoupling methods, robotics

1 Introduction

Many complex systems must be caracterised also by their interactions, through time varying parameters, with many other dynamics well or partially known, as shown in the following figure, and the question is, how can we control D1

* Thanks to FEDER european agency for funding.

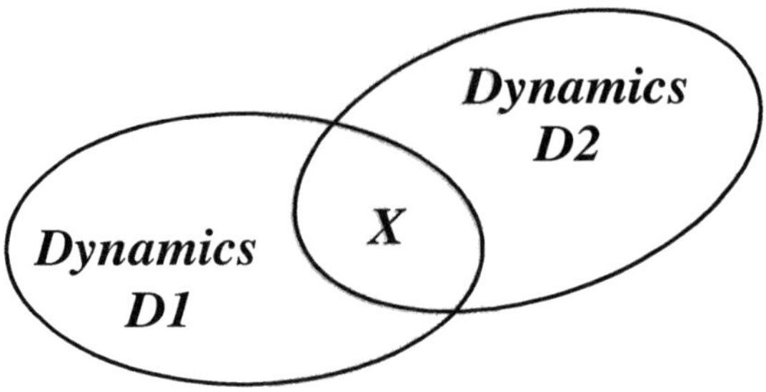

D1: known dynamics

D2: unknown dynamics

X: coupling parameter

Fig. 1. Dynamics in interaction with parameter x

a well known dynamics, knowing the interactions with an unknown dynamics D2 interacting with D1 with measurable time varying parameters? Under some conditions of robustness [1], [2], [3], [4], nonlinear state feedback methods [5], [6], are well known and well efficient methods for many nonlinear dynamical problems. However, in many applications, a complete dynamical knowledge of the system is not really needed everywhere in the state space, but only in the neighbourhood of critical behaviours characterizing the process. So one may expect the nonlinear state feedback methods to be finally applicable to a wider class of problems.

Practically, in many cases, this partial decoupling approach involved partial state space equations which generally must be parametrized by time functions in order to include the non state space behaviours of the system. If these time varying parameters are considered as constants or slowly varying compared to the dynamical behaviour of the partial state space equation (see section 2 for a precise meaning of this statement) the usual methods of computation of the feedback laws remain valid [7] by assuming that these laws are parametrised. On contrary if these time varying parameters are quickly varying compared to the dynamical behaviour of the partial state space equation, new dynamical terms appear in the computation of the feedback laws which modify drastically the behaviour of the system

We propose in this article the computation of the static and dynamic feedback laws for quickly varying parameters in case of the decoupling and/or linearization method and rejection of perturbations. Section 2 presents the basic idea and algebraic properties of these computations for static decoupling and linearization. The extension to dynamic feedback decoupling and linearization are given in section 3. Section 4 and section 5 give respectivelly, the static and dynamic state feedback laws in the case of rejection of mesurable perturbations. An example in robotics is given in section 6, and we consider here a one degree of freedom robot manipulator with its associated one degree of freedom effector, as partial state space equation we consider the dynamic of the grand motion axe, and the time varying parameter is the position of the effector. The static feedback laws are given in this example and the contribution of these new dynamical terms are discussed.

2 Static State Feedback Decoupling Method with Quickly Varying Parameters

Let us consider the following parametrised nonlinear affine system,

$$\dot{\mathbf{q}}(t) = \mathbf{A}^o(\mathbf{q}(t), \mathbf{x}(\mathbf{t})) + \sum_{j=1}^{m} u_j \, \mathbf{A}^j(\mathbf{q}(t), \mathbf{x}(t)) \tag{1}$$

$$\mathbf{y} = \mathbf{h}(\mathbf{q}(t), \mathbf{x}(t))$$

where $\mathbf{q}(t)$ is the state vector of dimension N, $\mathbf{x}(t)$ is the parameters vector of dimension o, both $\mathbf{q}$ and $\mathbf{x}$ are analytical functions of the time variable t. $\mathbf{A}^j$, $j \in [0, m]$ are $m+1$ nonlinear analytical vectorial functions of the variables $\mathbf{q}$ and $\mathbf{x}$ and of dimension N, u_j is the jth component of the control vector $\mathbf{u}$ of dimension m. $\mathbf{y}$ is the output vector of dimension r and $\mathbf{h}$ is a nonlinear vectorial analytical function of the dynamical variables $\mathbf{q}$ and $\mathbf{x}$, of dimension r. Let us now introduce the following differential operators,

$$\mathcal{A}^j = \sum_{k=1}^{N} \left[\mathbf{A}^j\right]^k \frac{\partial}{\partial q_k}$$

where $\left[\mathbf{A}^j\right]^k$ is the kth component of the vectorial function $\mathbf{A}^j(\mathbf{q}(t), \mathbf{x}(t))$ and q_k is the kth component of the state vector $\mathbf{q}$,

$$\mathcal{X}^{(n)} = \sum_{k=1}^{o} x_k^{(n)} \frac{\partial}{\partial x_k}$$

where $x_k^{(n)}$ is the nth time derivative of the kth component of the parameters vector $\mathbf{x}$ and,

$$\frac{\partial}{\partial t}$$

which is the partial time differential operator acting only on elementary time functions as $\mathbf{x}(t)$, $\mathbf{u}(t)$ or $\mathbf{p}(t)$, see the following paragraphs

Let us consider the $\phi_s + 1$ successive time derivatives of the output y_s, where ϕ_s is the characteristic number of the sth output [20], $s \in [1, r]$, define as the smallest integer as,

$$\forall j, j \in [1, m], \forall \nu \in [0, \phi_s - 1], \mathcal{A}^j \mathcal{A}^{o\nu} h_s = 0$$

and

$$\exists \ell,\, \ell \in [1,m], / \mathcal{A}^{\ell}\, \mathcal{A}^{o\phi_s} h_s \neq 0$$

We have,

$$y_s = h_s(\mathbf{q},\mathbf{x})$$

$$\dot{y}_s = \left(\mathcal{A}^o + \mathcal{X}^{(1)} + \frac{\partial}{\partial t}\right) h_s$$

$$\ddot{y}_s \equiv y_s^{(2)} = \left(\mathcal{A}^o + \mathcal{X}^{(1)} + \frac{\partial}{\partial t}\right)^2 h_s$$

$$y_s^{(3)} = \left(\mathcal{A}^o + \mathcal{X}^{(1)} + \frac{\partial}{\partial t}\right)^3 h_s$$

$$\vdots$$

$$y_s^{(\phi_s+1)} = \left(\mathcal{A}^o + \mathcal{X}^{(1)} + \frac{\partial}{\partial t}\right)^{\phi_s+1} h_s + \sum_{j=1}^{m} u_j(t)\mathcal{A}^j \mathcal{A}^{o\phi_s} h_s$$

Then, the vector $\bar{y}^{\phi+1} \equiv col(y_1^{(\phi_1+1)}, y_2^{(\phi_2+1)}, \ldots, y_r^{(\phi_r+1)})$ can be written as,

$$\bar{y}^{\phi+1} = \overline{\left(\mathcal{A}^o + \mathcal{X}^{(1)} + \frac{\partial}{\partial t}\right)^{\phi+1}\mathbf{h}} + \boldsymbol{\Omega}\,\mathbf{u}$$

where the vector,

$$\overline{\left(\mathcal{A}^o + \mathcal{X}^{(1)} + \frac{\partial}{\partial t}\right)^{\phi+1}}\mathbf{h}(\mathbf{q},\mathbf{x}) \equiv col\left(\left(\mathcal{A}^o + \mathcal{X}^{(1)} + \frac{\partial}{\partial t}\right)^{\phi_1+1} h_1(\mathbf{q},\mathbf{x}),\right.$$

$$\left.\ldots, \left(\mathcal{A}^o + \mathcal{X}^{(1)} + \frac{\partial}{\partial t}\right)^{\phi_r+1} h_r(\mathbf{q},\mathbf{x})\right)$$

and the matrix,

$$\boldsymbol{\Omega}(\mathbf{q},\mathbf{x}) \equiv [\mathcal{A}^j \mathcal{A}^{o\phi_s} h_s]$$

where s and j are respectively the row and column indexes

If we take control laws of type,

$$\mathbf{u} = \boldsymbol{\alpha}(\mathbf{q},\mathbf{x}) + \boldsymbol{\beta}(\mathbf{q},\mathbf{x})\,\mathbf{v} \tag{2}$$

where $\boldsymbol{\alpha}$ and $\boldsymbol{\beta}$ are respectively m and $m \times m$ nonlinear analytic functions of $\mathbf{q}$ and $\mathbf{x}$, $\mathbf{v}$ is the new control vector of dimension m, the dynamical system (1) can be decoupled and/or linearized through its input-output behaviour by assuming that $r = m$ and that $\boldsymbol{\Omega}$ is invertible. If

$$\boldsymbol{\Gamma}^o = col\left(\ldots, -\left(\mathcal{A}^o + \mathcal{X}^{(1)} + \frac{\partial}{\partial t}\right)^{\phi_s+1} h_s + \right.$$

$$\left. \sum_{j=1}^{r}\sum_{k=0}^{\phi_j} \sigma_{sjk}\left(\mathcal{A}^o + \mathcal{X}^{(1)} + \frac{\partial}{\partial t}\right)^k h_j, \ldots\right)$$

with $s \in [1, r]$, and

$$\boldsymbol{\Gamma}^1 = \left[g_i^j\right], \quad (i,j) \in [1,m]^2$$

where the σ_{sjk}, g_i^j are arbitrary constants choosing by the designer, the feedback laws linearizing the system are obtained by,

$$\boldsymbol{\alpha}(\mathbf{q},\mathbf{x}) = \boldsymbol{\Omega}^{-1}\,\boldsymbol{\Gamma}^o$$

$$\boldsymbol{\beta}(\mathbf{q},\mathbf{x}) = \boldsymbol{\Omega}^{-1}\,\boldsymbol{\Gamma}^1$$

Through these new feedback laws the input-output dynamical behaviour of the system can be written as,

$$y_s^{(\phi_s+1)} = \sum_{j=1}^{r}\sum_{k=0}^{\phi_j} \sigma_{sjk}\, y_j^{(k)} + \sum_{j=1}^{m} g_s^j\, v_j \tag{3}$$

$s \in [1, r]$, which are linear differential equations of the output $\mathbf{y}$ and of the input $\mathbf{v} \equiv col(v_1, \ldots, v_m)$

The decoupling conditions are given by,

$$\sigma_{sjk} = \delta[s-j]\,\sigma_{ssk}$$

$$g_i^j = \delta[i-j]\,g_i^i$$

where $\delta[.]$ is the Kronecker symbol.

We can now easily see the influence of fast varying parameters. The slowly varying parameters assumption leads to neglect $\left(\mathcal{X}^{(1)} + \frac{\partial}{\partial t}\right)$ before $\mathcal{A}^o$.

3 The Dynamic State Feedback Approach

If systems of type (1) could not be linearized and/or decoupled by static state feedback laws, i.e. ϕ_s does not exist, or $r \neq m$, or $\boldsymbol{\Omega}$ is not invertible, or if we consider systems of type (4),

$$\dot{\mathbf{q}}(t) = \mathbf{F}(\mathbf{q}(t), \mathbf{x}(t), \mathbf{u}(t)) \tag{4}$$

$$\mathbf{y}(t) = \mathbf{h}(\mathbf{q}(t), \mathbf{x}(t))$$

dynamic state feedback laws of type ,

$$\begin{aligned} \mathbf{u} &= \boldsymbol{\alpha}(\mathbf{q},\mathbf{x},\boldsymbol{\xi}) + \textstyle\sum_{j=1}^{m} \boldsymbol{\beta}_j(\mathbf{q},\mathbf{x},\boldsymbol{\xi})\, v_j(t) \\ \dot{\boldsymbol{\xi}} &= \boldsymbol{\gamma}(\mathbf{q},\mathbf{x},\boldsymbol{\xi}) + \textstyle\sum_{j=1}^{m} \boldsymbol{\delta}_j(\mathbf{q},\mathbf{x},\boldsymbol{\xi})\, v_j(t) \end{aligned} \tag{5}$$

can be used to "repear" the system and to transform it in a nonlinear affine system of type (1) linearizable and/or decouplable by static state feedback laws of type (2). The basic idea of this method is to enhance the state vector of (1) or (4) by the dynamic state feedback laws in order to create a new nonlinear affine system of type (1) with appropriate properties. In this way (1) can be written as,

$$\dot{\mathbf{q}}(t) = \mathbf{A}^o(\mathbf{q},\mathbf{x}) + \sum_{j=1}^{m} \left(\alpha_j(\mathbf{q},\mathbf{x},\boldsymbol{\xi}) + \sum_{i=1}^{m} \beta_{ji}(\mathbf{q},\mathbf{x},\boldsymbol{\xi})\, v_i \right) \mathbf{A}^j(\mathbf{q},\mathbf{x})$$

$$\dot{\boldsymbol{\xi}}(t) = \boldsymbol{\gamma}(\mathbf{q},\mathbf{x}) + \sum_{j=1}^{m} \boldsymbol{\delta}_j(\mathbf{q},\mathbf{x},\boldsymbol{\xi})\, v_j$$

$$\mathbf{y}(t) = \mathbf{h}(\mathbf{q},\mathbf{x})$$

So with $\tilde{\mathbf{q}} = col(\mathbf{q},\boldsymbol{\xi})$, (1) can be written as,

$$\dot{\tilde{\mathbf{q}}} = \tilde{\mathbf{A}}^{\mathbf{o}}(\tilde{\mathbf{q}},\mathbf{x}) + \sum_{\mathbf{j}=\mathbf{1}}^{\mathbf{m}} \tilde{\mathbf{A}}^{\mathbf{j}}(\tilde{\mathbf{q}},\mathbf{x})\, \mathbf{v_j}$$

$$\mathbf{y} = \mathbf{h}(\mathbf{q},\mathbf{x})$$

with,

$$\tilde{\mathbf{A}}^o(\tilde{\mathbf{q}},\mathbf{x}) = \begin{pmatrix} \mathbf{A}^o(\mathbf{q},\mathbf{x}) + \sum_{j=1}^{m} \alpha_j(\mathbf{q},\mathbf{x},\boldsymbol{\xi}) \mathbf{A}^j(\mathbf{q},\mathbf{x}) \\ \boldsymbol{\gamma}(\mathbf{q},\mathbf{x},\boldsymbol{\xi}) \end{pmatrix}$$

$$\tilde{\mathbf{A}}^j(\tilde{\mathbf{q}},\mathbf{x}) = \begin{pmatrix} \sum_{i=1}^{m} \beta_{ij}(\mathbf{q},\mathbf{x},\boldsymbol{\xi})\, \mathbf{A}^i(\mathbf{q},\mathbf{x}) \\ \boldsymbol{\delta}_j(\mathbf{q},\mathbf{x},\boldsymbol{\xi}) \end{pmatrix}$$

By the same way (4) can be written as,

$$\dot{\mathbf{q}}(t) = \mathbf{F}\left(\mathbf{q},\mathbf{x},\boldsymbol{\alpha}(\mathbf{q},\mathbf{x},\boldsymbol{\xi})\right)$$

$$\dot{\boldsymbol{\xi}} = \boldsymbol{\gamma}(\mathbf{q},\mathbf{x},\boldsymbol{\xi}) + \sum_{j=1}^{m} \mathbf{q},\mathbf{x},\boldsymbol{\xi})\, v_j$$

$$\mathbf{y} = \mathbf{h}(\mathbf{q},\mathbf{x})$$

so with $\tilde{\mathbf{q}} = col(\mathbf{q},\boldsymbol{\xi})$, (4) can be written as,

$$\dot{\tilde{\mathbf{q}}} = \tilde{\mathbf{A}}^{\mathbf{o}}(\tilde{\mathbf{q}},\mathbf{x}) + \sum_{\mathbf{j}=\mathbf{1}}^{\mathbf{m}} \tilde{\mathbf{A}}^{\mathbf{j}}(\tilde{\mathbf{q}},\mathbf{x})\, \mathbf{v_j}$$

$$\mathbf{y} = \mathbf{h}(\mathbf{q},\mathbf{x})$$

with

$$\tilde{\mathbf{A}}^o(\tilde{\mathbf{q}},\mathbf{x}) = \begin{pmatrix} \mathbf{F}\left(\mathbf{q},\mathbf{x},\boldsymbol{\alpha}(\mathbf{q},\mathbf{x},\boldsymbol{\xi})\right) \\ \boldsymbol{\gamma}(\mathbf{q},\mathbf{x},\boldsymbol{\xi}) \end{pmatrix}$$

$$\tilde{\mathbf{A}}^j(\tilde{\mathbf{q}},\mathbf{x}) = \begin{pmatrix} 0 \\ \boldsymbol{\delta}_j(\mathbf{q},\mathbf{x},\boldsymbol{\xi}) \end{pmatrix}$$

4 Rejection of Mesurable Perturbations by Static State Feedback Laws

Let us consider nonlinear affine dynamical systems of the following type,

$$\dot{\mathbf{q}} = \mathbf{A}^o(\mathbf{q},\mathbf{x}) + \sum_{i=1}^{m} u_i(t)\,\mathbf{A}^i(\mathbf{q},\mathbf{x}) + \sum_{j=1}^{p} p_j(t)\,\mathbf{P}(\mathbf{q},\mathbf{x}) \tag{6}$$

$$\mathbf{y} = \mathbf{h}(\mathbf{q},\mathbf{x})$$

and let us introduce the characteristic numbers ϕ_s and ϕ_{ps}. ϕ_s is the characteristic number associated to the control vector $\mathbf{u}$ and to the sth output y_s. ϕ_s is defined as in paragraph **2** as the smallest integer as,

$$\forall j,\, j \in [1,m],\, \forall \nu \in [0,\phi_s - 1],\, \mathcal{A}^j \mathcal{A}^{o\nu} h_s = 0$$

and

$$\exists \ell,\, \ell \in [1,m], / \mathcal{A}^\ell\, \mathcal{A}^{o\phi_s} h_s \neq 0$$

ϕ_{ps} is the characteristic number associated to the perturbation vector $\mathbf{p}$ and to the sth output defined as the smallest integer as,

$$\forall j,\, j \in [1,p],\, \forall \nu \in [0,\phi_{ps} - 1],\, \mathcal{P}^j \mathcal{A}^{o\nu} h_s = 0$$

and

$$\exists \ell,\, \ell \in [1,p], / \mathcal{P}^\ell\, \mathcal{A}^{o\phi_{ps}} h_s \neq 0$$

Clearly the computation of the successive time derivatives of the output vector depends on the three cases,

$$\phi_{ps} = \phi_s, \quad \phi_{ps} > \phi_s, \quad \phi_{ps} < \phi_s$$

First if $\phi_{ps} = \phi_s$ the $y_s^{(n)}$ can be written as,

$$y_s = h_s(\mathbf{q},\mathbf{x})$$

$$\dot{y}_s = \left(\mathcal{A}^o + \mathcal{X}^{(1)} + \frac{\partial}{\partial t}\right) h_s$$

$$\ddot{y}_s \equiv y_s^{(2)} = \left(\mathcal{A}^o + \mathcal{X}^{(1)} + \frac{\partial}{\partial t}\right)^2 h_s$$

$$y_s^{(3)} = \left(\mathcal{A}^o + \mathcal{X}^{(1)} + \frac{\partial}{\partial t}\right)^3 h_s$$

$$\vdots$$

$$y_s^{(\phi_s+1)} = \left(\mathcal{A}^o + \mathcal{X}^{(1)} + \frac{\partial}{\partial t}\right)^{(\phi_s+1)} h_s + \sum_{j=1}^{m} u_j(t)\mathcal{A}^j \mathcal{A}^{o\phi_s} h_s + \sum_{j=1}^{p} p_j(t)\mathcal{P}^j \mathcal{A}^{o\phi_s}$$

where $\mathcal{P}^j = \sum_{k=1}^{N} [\mathbf{P}^j]^k \frac{\partial}{\partial q_k}$ and $[\mathbf{P}^j]^k$ is the kth component of the vectorial function $\mathbf{P}^j(\mathbf{q}, \mathbf{x})$ defined in (4)

If we take,

$$\boldsymbol{\Gamma}^o \equiv col(\ldots, -\left(\mathcal{A}^o + \mathcal{X}^{(1)} + \frac{\partial}{\partial t}\right)^{\phi_s+1} h_s$$

$$+ \sum_{j=1}^{r}\sum_{k=0}^{\phi_j} \sigma_{sjk} \left(\mathcal{A}^o + \mathcal{X}^{(1)} + \frac{\partial}{\partial t}\right)^k h_j - \sum_{j=1}^{p} p_j(t)\mathcal{P}^j \mathcal{A}^{o\phi_s} h_s, \ldots)$$

and

$$\boldsymbol{\Gamma}^1 \equiv [g_i^j], \quad (i,j) \in [1,m]^2$$

where the σ_{sjk} and g_i^j are arbitrary constants choosen by the designer. We can easily see that static feedback laws of type,

$$\mathbf{u} = \boldsymbol{\alpha}(\mathbf{q}, \mathbf{x}, \mathbf{p}) + \boldsymbol{\beta}(\mathbf{q}, \mathbf{x})\mathbf{v}$$

linearize the system et reject the mesurable perturbations $\mathbf{p}$ with,

$$\boldsymbol{\alpha}(\mathbf{q}, \mathbf{x}, \mathbf{p}) = \boldsymbol{\Omega}^{-1}\boldsymbol{\Gamma}^o$$

$$\boldsymbol{\beta}(\mathbf{q}, \mathbf{x}) = \boldsymbol{\Omega}^{-1}\boldsymbol{\Gamma}^1$$

Secondly let us consider the case $\phi_{ps} > \phi_s$, we have then,

$$y_s = h_s(\mathbf{q}, \mathbf{x})$$

$$\dot{y}_s = \left(\mathcal{A}^o + \mathcal{X}^{(1)} + \frac{\partial}{\partial t}\right) h_s$$

$$\ddot{y}_s \equiv y_s^{(2)} = \left(\mathcal{A}^o + \mathcal{X}^{(1)} + \frac{\partial}{\partial t}\right)^2 h_s$$

$$\vdots$$

$$y_s^{(\phi_s+1)} = \left(\mathcal{A}^o + \mathcal{X}^{(1)} + \frac{\partial}{\partial t}\right)^{\phi_s+1} h_s + \sum_{j=1}^{m} u_j(t)\mathcal{A}^j \mathcal{A}^{o\phi_s} h_s$$

Clearly if we take,

$$\boldsymbol{\Gamma}^o = col(\ldots, -\left(\mathcal{A}^o + \mathcal{X}^{(1)} + \frac{\partial}{\partial t}\right)^{\phi_s+1} h_s +$$

$$\sum_{j=1}^{r}\sum_{k=1}^{\phi_j} \sigma_{sjk} \left(\mathcal{A}^o + \mathcal{X}^{(1)} + \frac{\partial}{\partial t}\right)^{k} h_j, \ldots)$$

and,

$$\boldsymbol{\Gamma}^1 = [g_i^j], \quad (i,j) \in [1,m]^2$$

where σ_{sjk} and g_i^j are the usual arbitrary constants, the linearizing and rejecting laws are given by,

$$\mathbf{u} = \boldsymbol{\alpha}(\mathbf{q},\mathbf{x}) + \boldsymbol{\beta}(\mathbf{q},\mathbf{x})\mathbf{v}$$

with

$$\boldsymbol{\alpha}(\mathbf{q},\mathbf{x}) = \boldsymbol{\Omega}^{-1}\,\boldsymbol{\Gamma}^o$$
$$\boldsymbol{\beta}(\mathbf{q},\mathbf{x}) = \boldsymbol{\Omega}^{-1}\,\boldsymbol{\Gamma}^1$$

The perturbations $\mathbf{p}$ are rejected without the knowledge of $\mathbf{p}$.

Finally we consider the case $\phi_s > \phi_{ps}$, we have,

$$y_s = h_s(\mathbf{q},\mathbf{x})$$

$$\vdots$$

$$y_s^{(\phi_{ps}+1)} = \left(\mathcal{A}^o + \mathcal{X}^{(1)} + \frac{\partial}{\partial t}\right)^{\phi_{ps}+1} h_s + \sum_{j=1}^{p} p_j(t)\mathcal{P}^j\mathcal{A}^{o\phi_{ps}} hs$$

$$\vdots$$

$$y_s^{(\phi_s+1)} = \left(\mathcal{A}^o + \sum_{j=1}^{p} p_j(t)\mathcal{P}^j + \mathcal{X}^{(1)} + \frac{\partial}{\partial t}\right)^{\phi_s-\phi_{ps}}$$

$$\left\{\left(\mathcal{A}^o + \mathcal{X}^{(1)} + \frac{\partial}{\partial t}\right)^{\phi_{ps}+1} h_s + \sum_{j=1}^{p} p_j(t)\mathcal{P}^j\mathcal{A}^{o\phi_{ps}} h_s\right\} +$$

$$\sum_{j=1}^{m} u_j(t)\mathcal{A}^j\mathcal{A}^{o\phi_s} hs$$

By assuming that,

$$\boldsymbol{\Gamma}^o = col(\ldots, -\left(\mathcal{A}^o + \sum_{j=1}^{p} p_j(t)\mathcal{P}^j + \mathcal{X}^{(1)} + \frac{\partial}{\partial t}\right)^{\phi_s-\phi_{ps}}$$

$$\left(\left(\mathcal{A}^o + \mathcal{X}^{(1)} + \frac{\partial}{\partial t}\right)^{\phi_{ps}+1} h_s + \sum_{j=1}^{p} p_j(t)\mathcal{P}^j \mathcal{A}^{o\phi_{ps}} h_s\right)$$

$$+\sum_{j=1}^{r}\sum_{k=0}^{\phi_j} \sigma_{sjk}\left(\mathcal{A}^o + \mathcal{X}^{(1)} + \frac{\partial}{\partial t}\right)^k h_j, \ldots)$$

$$\boldsymbol{\Gamma}^1 = [g_i^j], \quad (i,j) \in [1,m]^2$$

with the σ_{sjk}, g_i^j the usual arbitrary constants, we can easily see that static feedback laws of type,

$$\mathbf{u}(t) = \boldsymbol{\alpha}(\mathbf{q}, \mathbf{x}, \mathbf{p}, \ldots, \mathbf{p}^{(\phi_s - \phi_{ps})}) + \boldsymbol{\beta}(\mathbf{q}, \mathbf{x})\,\mathbf{v}(t)$$

reject the mesurable perturbations and linearize the system with,

$$\boldsymbol{\alpha}(\mathbf{q}, \mathbf{x}, \mathbf{p}, \ldots, \mathbf{p}^{(\phi_s - \phi_{ps})}) = \boldsymbol{\Omega}^{-1}\boldsymbol{\Gamma}^o$$

$$\boldsymbol{\beta}(\mathbf{q}, \mathbf{x}) = \boldsymbol{\Omega}^{-1}\boldsymbol{\Gamma}^1$$

5 Rejection of Mesurable Perturbations by Dynamic State Feedback Laws

Like in paragraph **3**, if systems of type (6) could not be decoupled and/or linearized by static state feedback laws, or if we consider systems of the following type,

$$\dot{\mathbf{q}} = \mathbf{F}(\mathbf{q}, \mathbf{x}, \mathbf{u}) + \sum_{j=1}^{p} p_j(t)\mathbf{P}^j(\mathbf{q}, \mathbf{x})$$

$$\mathbf{y} = \mathbf{h}(\mathbf{q}, \mathbf{x})$$

dynamic state feedback laws of type (5) can be used to "repear" the system, i.e. in order to create appropriate characteristic numbers, and/or make $\boldsymbol{\Omega}$ invertible, and/or make the system nonlinear affine, and/or modify ϕ_s before ϕ_{ps}.

6 Example in Robotics

As example of this appoach we will consider the control by linearization of a one degree of freedom motorized robot manipulator with compliance [3] and realistic friction model [4]. As in many cases in robotics the parameter of this dynamical model is the position of the effector. See figure 2 for a schematic representation of this structure

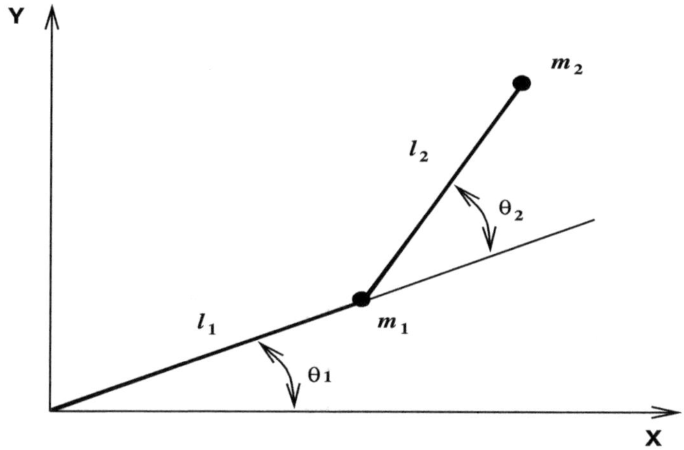

Fig. 2. Schematic representation of a one degree of freedom robot manipulator (θ_1), with its associated one degree of freedom effector (θ_2). Masses m_1 and m_2 are localized in the extremities of these two arms of length ℓ_1 and ℓ_2 respectively.

The Lagrangian appoach of the mechanical structure leads to the following dynamical equation,

$$g\ell_1 m_1 cos(\theta_1) - m_2 g(\ell_1 cos(\theta_1) + \ell_2 cos(\theta_1 + \theta_1)) + 2\ell_1\ell_2 m_2 sin(\theta_2)\dot{\theta}_1\dot{\theta}_2 +$$
$$\ell_1\ell_2 m_2 sin(\theta_2)\dot{\theta}_2^2 + (-\ell_1^2 m_1 - m_2(\ell_1^2 + \ell_2^2 + 2\ell_1\ell_2 cos(\theta_2)))\ddot{\theta}_1 -$$
$$\ell_2 m_2(\ell_2 + \ell_1 cos(\theta_2))\ddot{\theta}_2 - \gamma + f = 0$$

where θ_1, θ_2, ℓ_1, ℓ_2, m_1, m_2 are respectively the angular position, the length of the arm, and the localized mass, for the robot subscript 1 and for the effector subscript 2. g is the gravity constant, γ is the applied torque, and f is the friction torque

The dynamical behaviour of the motor is given by the following equation,

$$u = Ri + L\frac{di}{dt} + k\dot{\theta}_m$$

$$\gamma = k''i$$

where u, i, R, L, θ_m are respectively the voltage control of the motor, the induction current, the resistance of the inductor, the inductance of the inductor, and the angular position of the motor. k and k'' are two positive constants

The dynamical behaviour of the compliance is given by the following equation [3],

$$\gamma = k'(\theta_m - \theta_1) + c(\dot{\theta}_m - \dot{\theta}_1)$$

where γ is the motor torque and k', and c are two positive real constants

The dynamical behaviour of the friction is given by the following equations [4],

$$f = \tilde{\sigma}_0 z + \tilde{\sigma}_1 \dot{z} + \tilde{\alpha}_2 \dot{\theta}_1$$

$$\dot{z} = \dot{\theta}_1 - \frac{\tilde{\sigma}_0}{(\tilde{\alpha}_0 + \tilde{\alpha}_1 e^{-(\dot{\theta}_1/v_0)^2})} z|\dot{\theta}_1|$$

where $\tilde{\sigma}_0$, $\tilde{\sigma}_1$, $\tilde{\alpha}_0$, $\tilde{\alpha}_1$, $\tilde{\alpha}_2$, v_0 are real positive constants, e is the exponential function

By assuming that, $i = q_1$, $\theta_m = q_2$, $\theta_1 = q_3$, $\dot{\theta}_1 = q_4$, $z = q_5$, and $\theta_2 = x$, the dynamical behaviour of the whole system can be written in the form (1), with $\mathbf{q} \equiv col(q_1, q_2, q_3, q_4, q_5)$, $\mathbf{x} \equiv col(x, x^{(1)}, x^{(2)})$ and,

$$\mathbf{h} \equiv h = \ell_1 cos(q_3) + \ell_2 cos(q_3 + x)$$

the output $\mathbf{y}$ is here the projection of the end of the effector on the $\mathbf{X}$ axis (see figure 1), and we have $N = 5$, $o = 3$, $m = 1$, $r = 1$, and,

$$\mathbf{A}^o = \begin{pmatrix} -\frac{k}{L}(\frac{k'' q_1}{c} - \frac{k'}{c}(q_2 - q_3) + q_4) + \frac{u}{L} - \frac{R}{L} q_1 \\ q_4 - \frac{k'}{c}(q_2 - q_3) + \frac{k''}{c} q_1 \\ q_4 \\ -\frac{\sigma_1}{D}(q_4 - \frac{\sigma_0 q_5 |q_4|}{(\alpha_0 + \alpha_1 e^{-(q_4/v_0)^2})}) + \tilde{f}(q_1, q_3, q_4, q_5, x, x^{(1)}, x^{(2)})/D \\ q_4 - \frac{\sigma_0}{(\alpha_0 + \alpha_1 e^{-(q_4/v_0)^2})} q_5 |q_4| \end{pmatrix}$$

$$\mathbf{A}^1 = \begin{pmatrix} 1/L \\ 0 \\ 0 \\ 0 \\ 0 \end{pmatrix}$$

with,

$$\tilde{f}(q_1, q_3, q_4, q_5, x, x^{(1)}, x^{(2)}) = -g\ell_1 m_1 cos(q_3) - m_2 g(\ell_1 cos(q_3) + \ell_2 cos(q_3 + x))$$

$$+2\ell_1 \ell_2 m_2 sin(q_3) q_4 x^{(1)} + \ell_1 \ell_2 m_2 sin(x) x^{(1)2} - \ell_2 m_2(\ell_2 + \ell_1 cos(x)) x^{(2)} + k'' q_1 -$$

$$\sigma_0 q_5 - \alpha_2 q_4$$

$$D = \ell_1^2 m_1 + m_2(\ell_1^2 + \ell_2^2 + 2\ell_1 \ell_2 cos(x))$$

Let us first compute the linearizing feedback law by the classical method [7], i.e. by assuming that $\mathbf{A}^o$ is parametred by $\mathbf{x}$ without any consideration of the successive derivatives of $\mathbf{x}$

Then the computation of the characteristic number leads to,

$$\mathcal{A}^o h = \frac{\partial h}{\partial q_3} q_4$$

$$\mathcal{A}^{o2} h = q_4^2 \frac{\partial^2 h}{\partial q_3^2} + [\mathbf{A}^o]^4 \frac{\partial h}{\partial q_3}$$

$$\mathcal{A}^{o3} h = \frac{\partial h}{\partial q_3}([\mathbf{A}^o]^1 \frac{k''}{D} + [\mathbf{A}^o]^4 \frac{\partial}{\partial q_4}[\mathbf{A}^o]^4 + [\mathbf{A}^o]^5 \frac{\partial}{\partial q_5}[\mathbf{A}^o]^4) + q_4^3 \frac{\partial^3 h}{\partial q_3^3}$$
$$+ q_4 \frac{\partial [\mathbf{A}^o]^4}{\partial q_3} \frac{\partial h}{\partial q_3} + 3[\mathbf{A}^o]^4 q_4 \frac{\partial^2 h}{\partial q_3^2}$$

$$\mathcal{A}^1 h \equiv 0, \quad \mathcal{A}^1 \mathcal{A}^o h \equiv 0$$

$$\mathcal{A}^1 \mathcal{A}^{o2} h = \frac{k''}{L\,D} \frac{\partial h}{\partial q_3}$$

with $[\mathbf{A}^o]^i$ the ith component of the vector $\mathbf{A}^o$. The characteristic number is then 2, this leads to the following $\boldsymbol{\Omega}$ matrix and to the following linearizing feedback law,

$$\boldsymbol{\Omega} = \frac{k''}{L\,D} \frac{\partial h}{\partial q_3}$$

$$\boldsymbol{\alpha} = \frac{L\,D}{k'' \frac{\partial h}{\partial q_3}} (-\mathcal{A}^{o3} h + \sigma_0 h + \sigma_1 \mathcal{A}^o h + \sigma_2 \mathcal{A}^{o2} h)$$

$$\boldsymbol{\beta} = \frac{L\,D}{k'' \frac{\partial h}{\partial q_3}}$$

where σ_0, σ_1, σ_2 are arbitrary constants, see the σ_{sjk} in paragraph **2**

Introducing the control law of type (2) into (1), leads to the following systems,

$$\dot{q}_1 = \frac{D}{k'' \frac{\partial h}{\partial q_3}} \Bigg((-[\mathbf{A}^o]^4 \frac{\partial}{\partial q_4}[\mathbf{A}^o]^4 - [\mathbf{A}^o]^5 \frac{\partial}{\partial q_5}[\mathbf{A}^o]^4) \frac{\partial h}{\partial q_3} - q_4^3 \frac{\partial^3 h}{\partial q_3^3} - q_4 \frac{\partial [\mathbf{A}^o]^4}{\partial q_3} \frac{\partial h}{\partial q_3}$$
$$-3[\mathbf{A}^o]^4 q_4 \frac{\partial^2 h}{\partial q_3^2} + \sigma_0 h + \sigma_1 \mathcal{A}^o h + \sigma_2 \mathcal{A}^{o2} h + v \Bigg) \tag{7}$$

$$\dot{q}_2 = [\mathbf{A}^o]^2$$

$$\dot{q}_3 = [\mathbf{A}^o]^3$$

$$\dot{q}_4 = [\mathbf{A}^o]^4$$

$$\dot{q}_5 = [\mathbf{A}^o]^5$$

The successive time derivations up to 3 $(\phi + 1)$, of the output y leads to the following expressions,

$$y = h$$

$$\dot{y} = \frac{\partial h}{\partial q_3} q_4 - \ell_2 sin(q_3 + x) x^{(1)}$$

$$\ddot{y} \equiv y^{(2)} = \frac{\partial^2 h}{\partial q_3^2} q_4^2 - 2\ell_2 cos(q_3 + x) q_4 x^{(1)} - \ell_2 cos(q_3 + x) x^{(1)2} + \frac{\partial h}{\partial q_3} [\mathbf{A}^o]^4 -$$
$$\ell_2 sin(q_3 + x) x^{(2)}$$

$$y^{(3)} = \frac{\partial^3 h}{\partial q_3^3} q_4^3 + \ell_2 sin(q_3 + x)(3q_4^2 x^{(1)} + 3q_4 x^{(1)2} + x^{(1)3}) + 3\frac{\partial^2 h}{\partial q_3^2} q_4 [\mathbf{A}^o]^4$$
$$-3cos(q_3 + x)(q_4 x^{(2)} + x^{(1)}[\mathbf{A}^o]^4 + x^{(1)} x^{(2)}) + \frac{\partial h}{\partial q_3}(\frac{\partial}{\partial q_1}[\mathbf{A}^o]^4 \dot{q}_1 + \frac{\partial [\mathbf{A}^o]^4}{\partial q_3}[\mathbf{A}^o]^3$$
$$+\frac{\partial [\mathbf{A}^o]^4}{\partial q_4}[\mathbf{A}^o]^4 + \frac{\partial [\mathbf{A}^o]^4}{\partial q_5}[\mathbf{A}^o]^5 + \frac{\partial [\mathbf{A}^o]^4}{\partial x} x^{(1)} + \frac{\partial [\mathbf{A}^o]^4}{\partial x^{(1)}} x^{(2)} + \frac{\partial [\mathbf{A}^o]^4}{\partial x^{(2)}} x^{(3)} + x^{(3)}) \tag{8}$$

Introducing (7) into $y^{(3)}$leads to the following input-output equation,

$$y^{(3)} = \sigma_0 y + \sigma_1 y^{(1)} + \sigma^2 y^{(2)} + v + \ell_2 sin(q_3 + x)(3q_4^2 x^{(1)} + 3q_4 x^{(1)2} + x^{(1)3})$$
$$-3cos(q_3 + x)(q_4 x^{(2)} + x^{(1)}[\mathbf{A}^o]^4 + x^{(1)} x^{(2)}) + \frac{\partial h}{\partial q_3}(\frac{\partial [\mathbf{A}^o]^4}{\partial x} x^{(1)} + \frac{\partial [\mathbf{A}^o]^4}{\partial x^{(1)}} x^{(2)} +$$
$$\frac{\partial [\mathbf{A}^o]^4}{\partial x^{(2)}} x^{(3)} + x^{(3)})$$
$$+\sigma_1 \ell_2 sin(q_3 + x) x^{(1)} + \sigma_2 (2\ell_2 cos(q_3 + x) q_4 x^{(1)} + \ell_2 cos(q_3 + x) x^{(1)2} +$$
$$\ell_2 sin(q_3 + x) x^{(2)})$$

We can easily see that in the case of slowly varying parameter $x^{(i)} \simeq 0, \forall i \in [1,3]$ this system is linearized, and can be written as,

$$y^3 = \sigma_0\, y + \sigma_1\, y^{(1)} + \sigma_2\, y^{(2)} + v$$

see (3) in paragraph **2**

But if the $x^{(i)}$ is not negligeable the system is clearly nonlinear

We will now show that by the use of our approach the system is perfectly linearized for all $x^{(i)}$

In our approach the characteristic number, $\boldsymbol{\Omega}$, and $\boldsymbol{\beta}$ remain the same as the preceeding approach, and $\boldsymbol{\alpha}$ is written as,

$$\boldsymbol{\alpha} = \frac{L\,D}{k'' \frac{\partial h}{\partial q_3}}(-(\mathcal{A}^o + \mathcal{X}^{(1)} + \frac{\partial}{\partial t})^3 h + \sigma_0 h + \sigma_1 (\mathcal{A}^o + \mathcal{X}^{(1)} + \frac{\partial}{\partial t}) h + \sigma_2 (\mathcal{A}^o + \mathcal{X}^{(1)} +$$
$$\frac{\partial}{\partial t})^2 h)$$

with,

$$(\mathcal{A}^o + \mathcal{X}^{(1)} + \frac{\partial}{\partial t})h = [\mathbf{A}^o]^3 \frac{\partial h}{\partial q_3} + x^{(1)} \frac{\partial h}{\partial x}$$

$$(\mathcal{A}^o + \mathcal{X}^{(1)} + \frac{\partial}{\partial t})^2 h = [\mathbf{A}^o]^4 \frac{\partial h}{\partial q_3} + q_4^2 \frac{\partial^2 h}{\partial q_3^2} + 2q_4 x^{(1)} \frac{\partial^2 h}{\partial q_3 \partial x} + x^{(1)2} \frac{\partial^2 h}{\partial x^2} + x^{(2)} \frac{\partial h}{\partial x}$$

$$(\mathcal{A}^o + \mathcal{X}^{(1)} + \frac{\partial}{\partial t})^3 h = [\mathbf{A}^o]^1 \frac{k''}{D} \frac{\partial h}{\partial q_3} + q_4 \frac{\partial [\mathbf{A}^o]^4}{\partial q_3} \frac{\partial h}{\partial q_3} + [\mathbf{A}^o]^4 \frac{\partial [\mathbf{A}^o]^4}{\partial q_4} \frac{\partial h}{\partial q_3} +$$

$$[\mathbf{A}^o]^5 \frac{\partial [\mathbf{A}^o]^4}{\partial q_5} \frac{\partial h}{\partial q_3}$$

$$+3[\mathbf{A}^o]^4 q_4 \frac{\partial^2 h}{\partial q_3^2} + 3[\mathbf{A}^o]^4 x^{(1)} \frac{\partial^2 h}{\partial q_3 \partial x} + 3q_4^2 x^{(1)} \frac{\partial^3 h}{\partial q_3^2 \partial x} + 3q_4 x^{(1)2} \frac{\partial^3 h}{\partial q_3 \partial x^2} + x^{(1)3} \frac{\partial^3 h}{\partial x^3}$$

$$+x^{(3)} \frac{\partial h}{\partial q_3} + q_4^3 \frac{\partial^3 h}{\partial q_3^3} + 3q_4 x^{(2)} \frac{\partial^2 h}{\partial q_3 \partial x} + 3x^{(1)} x^{(2)} \frac{\partial^2 h}{\partial x^2} + \frac{\partial h}{\partial q_3} (\frac{\partial [\mathbf{A}^o]^4}{\partial x} x^{(1)} +$$

$$\frac{\partial [\mathbf{A}^o]^4}{\partial x^{(1)}} x^{(2)} + \frac{\partial [\mathbf{A}^o]^4}{\partial x^{(2)}} x^{(3)})$$

Introducing this new feedback law into the dynamical system,leads to

$$\dot{q}_1 = \frac{D}{k'' \frac{\partial h}{\partial q_3}} \left(-q_4 \frac{\partial [\mathbf{A}^o]^4}{\partial q_3} \frac{\partial h}{\partial q_3} - [\mathbf{A}^o]^4 \frac{\partial [\mathbf{A}^o]^4}{\partial q_4} \frac{\partial h}{\partial q_3} - [\mathbf{A}^o]^5 \frac{\partial [\mathbf{A}^o]^4}{\partial q_5} \frac{\partial h}{\partial q_3} \right.$$

$$-3[\mathbf{A}^o]^4 q_4 \frac{\partial^2 h}{\partial q_3^2} - 3[\mathbf{A}^o]^4 x^{(1)} \frac{\partial^2 h}{\partial q_3 \partial x} - 3q_4^2 x^{(1)} \frac{\partial^3 h}{\partial q_3^2 \partial x} - 3q_4 x^{(1)2} \frac{\partial^3 h}{\partial q_3 \partial x^2} - x^{(1)3} \frac{\partial^3 h}{\partial x^3}$$

$$-x^{(3)} \frac{\partial h}{\partial q_3} - q_4^3 \frac{\partial^3 h}{\partial q_3^3} - 3q_4 x^{(2)} \frac{\partial^2 h}{\partial q_3 \partial x} - 3x^{(1)} x^{(2)} \frac{\partial^2 h}{\partial x^2} - \frac{\partial h}{\partial q_3} (\frac{\partial [\mathbf{A}^o]^4}{\partial x} x^{(1)} + \frac{\partial [\mathbf{A}^o]^4}{\partial x^{(1)}} x^{(2)}$$

$$\left. + \frac{\partial [\mathbf{A}^o]^4}{\partial x^{(2)}} x^{(3)}) + \sigma_0 h + \sigma_1 (\mathcal{A}^o + \mathcal{X}^{(1)} + \frac{\partial}{\partial t}) h + \sigma_2 (\mathcal{A}^o + \mathcal{X}^{(1)} + \frac{\partial}{\partial t})^2 h + v \right) \tag{9}$$

$$\dot{q}_2 = [\mathbf{A}^o]^2$$

$$\dot{q}_3 = [\mathbf{A}^o]^3$$

$$\dot{q}_4 = [\mathbf{A}^o]^4$$

$$\dot{q}_5 = [\mathbf{A}^o]^5$$

Introducing (9) into (8) leads to the following input-output dynamical equation,

$$y^{(3)} = \sigma_0 y + \sigma_1 y^{(1)} + \sigma_2 y^{(2)}$$

which is valid for all $x^{(i)}, i \in [1, 3]$. Our approach leads then to an exact linearization of the input-output behaviour for all $\mathbf{x}$.

7 Conclusion

We have presented in this paper, a new algorithmic approach, for computing in the framework of the nonlinear decoupling method, the feedback laws of parametrized dynamical systems. This approach can be used with static as with dynamic state feedback laws and leads to an exact linearization and decoupling for all behaviours of the parameters as well as their derivatives. The only condition is that the parameters as well as their derivatives are mesurable. Rejection of mesurable perturbations can be also considered with the same exact approach. Our example shows clearly the important influence of the derivatives of the parameters on the input-output dynamical behaviour of the linearized system and shows that neglecting these derivatives is an important cause of disfunctionment of the decoupling method. Our approach is perfectly algorithmic and can be easily translated in a formal language computer program.

References

1. Rouff M., Tran Du Ban (1989) Contrôle des robots élastiques: Utilisation de la linéarisation statique, *Canadian Society of Mechanical Engineeering*, 13 1(3):17-21
2. Rouff M., Verdier M. (1999) Nonlinear model matching a multiclosed loop approach, *International Journal of Robotics and Automation*, 12(3):100-109
3. Nordin M., Galic J., Gutman P. (1997) New Models for backlash and Gearplay, *Int. J. of Adap. Contr. & signal Process.*, 11(1):49-63
4. Canudas de Wit C., Lischinsky P. (1997) Adaptative friction compensation with partially known dynamic friction model, *Int. J. of Adap. Contr. & signal Process.*, 11(1):65-80
5. Cotsaftis M., Robert J., Rouff M., Vibet C. (1995) Application of decoupling method to Hamiltonian systems, *IEE Proc.-Control Theory Appli.*, 142:595-602
6. Cotsaftis M., Robert J., Rouff M., Vibet C. (1998) Applications and prospect of the nonlinear decoupling method, *Comput. Methods Appl. Mech. Engrg.*, 154:163-178
7. Rouff M. (1992) Computation of feedback laws for static linearization of nonlinear dynamical systems, *Mechatronics*, 2(6):613-623

Eigenfrequencies Invariance Properties in Deformable Flexion-Torsion Loaded Bodies -1- General Properties

Marc Rouff[2,1], Zakaria Lakhdari[1], and Philippe Makany[1*]

[1] Laboratoire des Sciences Appliquées de Cherbourg (LUSAC)
BP 78, rue Louis Aragon
50130 Cherbourg Octeville, France

[2] Laboratoire d'Ingénierie des Systèmes de Versailles (LISV)
Centre Universitaire de Technologie, 10/12 Avenue de l'Europe
78140 Velizy, France

Summary. We present here, in the framework of resonnant behaviours, an eigenfrequency invariance properties for loaded flexural and torsional deformable bodies. We found that under a preload condition each eigenfrequency is invariant versus the load m_N. For the mode 7 and beyond the preload conditions are less than $m_N = 1$, and the seven first modes (form 0 to 6) can be easily modelled by polynomial functions of m_N in the neighbourhood of useful m_N. These properties open the way to new finite dimension controllers.

Key words: resonnant behavior, eigenfrequency invariante properties, flexion-torsion bodies, controller

1 Introduction

The research of higher performances in speed and precision for industrial systems has been leading to lightweight mechanical structures [1]. Such structures cannot be maintained as rigid bodies and deformations take place which considerably modify their behaviour. As the system becomes now infinite dimensional, the control to be applied for stabilizing its trajectory becomes itself infinite dimensional, which fundamentally changes its nature. First sensing the modes and computing the corresponding control rapidly reaches the limits of existing technology, and second actuators themselves cannot deliver the required power in the high frequency range corresponding to these modes. It is very important to research a way to approach differently the problem. In a first step, it is interesting to identify system properties allowing to determine

* Thanks to FEDER european agency for funding.

in which cases, if any, the infinite dimensional deformable system reduces to a finite dimensional one controlled by a finite dimensional controller. Here specializing to the most penalizing case of initially rigid one link system, it is shown that past a fixed order (typically 7) the modes are totally invariant as a fonction of the applied tip load, and for lower order (< 7), are exhibiting a smooth behavior easily modelled by simple rational function of the load.

2 The Equation of Degeneration

Applying Lagrangian Formalism the complete equations of one link deformable mechanical system, is in the following form,

$$\rho A(x\frac{d^2\theta}{dt^2} + \frac{\partial^2 u(t,x)}{\partial t^2}) = -\frac{\partial^2}{\partial x^2}EI\frac{\partial^2 u(t,x)}{\partial x^2} \tag{1}$$

$$\rho K^2\frac{\partial^2 \gamma(t,x)}{\partial t^2} = \frac{\partial}{\partial x}GJ\frac{\partial \gamma(t,x)}{\partial x} \tag{2}$$

$$m\frac{\partial^2 X}{\partial t^2} = \frac{\partial}{\partial x}EI\left.\frac{\partial^2 u(t,x)}{\partial x^2}\right|_{x=L} \tag{3}$$

$$ml_f\frac{\partial^2 X}{\partial t^2} + J_f\left.(x\frac{d^2\theta}{dt^2} + \frac{\partial^3 u(t,x)}{\partial t^2 \partial x})\right|_{x=L} = -EI\left.\frac{\partial^2 u(t,x)}{\partial x^2}\right|_{x=L} \tag{4}$$

$$ml_t\frac{\partial^2 X}{\partial t^2} + J_t\left.\frac{\partial^2 \gamma(t,x)}{\partial t^2}\right|_{x=L} = -GJ\left.\frac{\partial \gamma(t,x)}{\partial x}\right|_{x=L} \tag{5}$$

with θ, $u(t,x)$, $\gamma(t,x)$ respectively the articular variable, and the deformation, flexion and torsion, variables, $(l_f,\, l_t)$ the coordinates of the tip mass m with respect to the end of the link, L is the length of the body, and the various other coefficients characterizing the beam are as usual within Euler-Bernouilli approximation [2]. Boundary Conditions are given by eqns(3,4,5), and,

$$X = (L+l_f)\theta + l_f\left.\frac{\partial u(t,x)}{\partial x}\right|_{x=L} + l_t\gamma(t,x)\Big|_{x=L} + u(t,x)\Big|_{x=L} \tag{6}$$

We want to study the resonant behaviour of this non dissipative system, and we pose for that,$\dot\theta \equiv 0$, we suppose also for reasons of simplicity that EI and GJ do not depend on x. By assuming that, $\zeta = x/L$, $\lambda_f = l_f/L$, $\lambda_t = l_t/L$, $\tilde{u} = u/L$, $\omega_f^2 = EI/\rho AL^4$, $\omega_t^2 = GJ/\rho K^2L^2$, $\lambda_\omega = \omega_f/\omega_t$, $\lambda_m = K^2/AL^2$, $m_N = m/\rho AL$, $J_{fN} = J_f/\rho AL^3$, $J_{tN} = J_t/\rho K^2L$, $\tau = t\omega_f$, we found the following normalized equations of motion,

$$\frac{\partial^2 \tilde{u}}{\partial \tau^2} + \frac{\partial^4 \tilde{u}}{\partial \zeta^4} = 0 \tag{7}$$

$$\frac{\partial^2\gamma}{\partial\tau^2} - (\frac{1}{\lambda_\omega})^2\frac{\partial^2\gamma}{\partial\zeta^2} = 0 \tag{8}$$

with, for all τ,the following normalized boundary conditions,

$$\tilde{u}(0,\tau) = 0 \quad (9), \quad \tilde{u}'(0,\tau) \equiv \left.\frac{\partial\tilde{u}}{\partial\zeta}\right|_{\zeta=0} = 0 \quad (10)$$

$$\gamma(0,\tau) = 0 \quad (11)$$

$$\tilde{u}''(1,\tau) \equiv \left.\frac{\partial^2\tilde{u}}{\partial\zeta^2}\right|_{\zeta=1}$$

$$= -\lambda_f\tilde{u}'''(1,\tau) - J_{fN}\left.\frac{\partial^3\tilde{u}}{\partial\tau^2\partial\zeta}\right|_{\zeta=1} \tag{12}$$

$$u'''(1,\tau) \equiv \left.\frac{\partial^3\tilde{u}}{\partial\zeta^3}\right|_{\zeta=1}$$

$$= m_N\left(\frac{\partial^2\tilde{u}}{\partial\tau^2} + \lambda_f\frac{\partial^3\tilde{u}}{\partial\zeta\partial\tau^2} + \lambda_t\frac{\partial^2\gamma}{\partial\tau^2}\right)_{\zeta=1} \tag{13}$$

$$(\frac{1}{\lambda_\omega})^2\gamma'(1,\tau) \equiv (\frac{1}{\lambda_\omega})^2\left.\frac{\partial\gamma}{\partial\zeta}\right|_{\zeta=1}$$

$$= -(\frac{\lambda_t}{\lambda_m})u'''(1,\tau) - J_{tN}\left.\frac{\partial^2\gamma}{\partial\tau^2}\right|_{\zeta=1} \tag{14}$$

In the case of stationary resonant behaviours $\tilde{u}(\zeta,\tau)$, solution of (7), can be written as,

$$\tilde{u}(\zeta,\tau) = \tilde{u}_1(\zeta)\,cos(\omega_N\,\tau) \tag{15}$$

with $\omega_N = \omega/\omega_f$, where ω is the frequency variable, and,

$$\tilde{u}_1(\zeta) = a\,cosh(\omega_N^{1/2}\zeta) + b\,sinh(\omega_N^{1/2}\zeta)$$

$$+c\,sin(\omega_N^{1/2}\zeta) + d\,cos(\omega_N^{1/2}\zeta) \tag{16}$$

$\gamma(\zeta,\tau)$, solution of (8), can be written as,

$$\gamma(\zeta,\tau) = \gamma_1(\zeta)\,cos(\omega_N\tau) \tag{17}$$

with,

$$\gamma_1(\zeta) = \bar{a}\,cos(\omega_N\,\lambda_\omega\,\zeta) + \bar{b}\,sin(\omega_N\,\lambda_\omega\,\zeta) \tag{18}$$

Introducing (15) and (17) into the six boundary conditions of the problem, equations (9) to (14), leads by the elimination of $cos(\omega_N\tau)$ to a set of 6 linear equations of the six initial values $(a,b,c,d,\bar{a},\bar{b})$. While this linear system is degenerated the only condition of existence of resonant mode is that the determinant of this linear system equals zero. This leads to the following equation of degeneration,

$$cos(\lambda_\omega \omega_N)\left(-\frac{2\omega_N^4}{\lambda_\omega}-\frac{2J_{fN}m_N\omega_n^6}{\lambda_\omega}\right)$$

$$+sin(\lambda_\omega \omega_N)\left(2J_{tN}\omega^5+\frac{2\lambda_t^2\omega^5}{\lambda_m}+2J_{fN}J_{tN}m_N\omega_N^7\right)$$

$$+cos(\omega_N^{1/2}-\lambda_\omega\omega_N)cosh(\omega_N^{1/2})$$

$$\left(-\frac{\omega_N^4}{\lambda_\omega}-J_{tN}m_N\omega_N^{11/2}+\frac{J_{fN}m_N\omega_N^6}{\lambda_\omega}-J_{fN}J_{tN}\omega_N^{13/2}-J_{tN}\lambda_f^2\omega_N^{13/2}\right.$$

$$\left.-\frac{J_{fN}\lambda_t^2\omega_N^{13/2}}{\lambda_m}\right)+sin(\omega_N^{1/2}-\lambda_\omega\omega_N)sinh(\omega_N^{1/2})$$

$$\left(\frac{\lambda_f\omega_N^5}{\lambda_\omega}+\frac{\lambda_f m_N\omega_N^5}{\lambda_\omega}-J_{tN}m_N\omega_N^{11/2}+J_{fN}J_{tN}\omega_N^{13/2}+J_{tN}\lambda_f^2\omega_N^{13/2}\right.$$

$$\left.+\frac{J_{fN}\lambda_t^2\omega_N^{13/2}}{\lambda_m}\right)+cos(\omega_N^{1/2}+\lambda_\omega\omega_N)cosh(\omega_N^{1/2})$$

$$\left(-\frac{\omega_N^4}{\lambda_\omega}+J_{tN}m_N\omega_N^{11/2}+\frac{J_{fN}m_N\omega_N^6}{\lambda_\omega}+J_{fN}J_{tN}\omega_N^{13/2}+J_{tN}\lambda_f^2\omega_N^{13/2}\right.$$

$$\left.+\frac{J_{fN}\lambda_t^2\omega_N^{13/2}}{\lambda_m}\right)+sin(\omega_N^{1/2}+\lambda_\omega\omega_N)sinh(\omega_N^{1/2})$$

$$\left(\frac{\lambda_f\omega_N^5}{\lambda_\omega}+\frac{\lambda_f m_N\omega_N^5}{\lambda_\omega}+J_{tN}m_N\omega_N^{11/2}-J_{fN}J_{tN}\omega_N^{13/2}-J_{tN}\lambda_f^2\omega_N^{13/2}\right.$$

$$\left.-\frac{J_{fN}\lambda_t^2\omega_N^{13/2}}{\lambda_m}\right)+cos(\omega_N^{1/2}-\lambda_\omega\omega_N)sinh(\omega_N^{1/2})$$

$$\left(-\frac{m_N\omega_N^{9/2}}{\lambda_\omega}+\frac{J_{fN}\omega_N^{11/2}}{\lambda_\omega}+\frac{\lambda_f^2\omega_N^{11/2}}{\lambda_\omega}-J_{tN}\lambda_f\omega_N^6-J_{tN}\lambda_f m_N\omega_N^6\right)$$

$$+sin(\omega_N^{1/2}-\lambda_\omega\omega_N)cosh(\omega_N^{1/2})$$

$$\left(\frac{m_N\omega_N^{9/2}}{\lambda_\omega}-J_{tN}\omega_N^5-\frac{\lambda_t^2\omega_N^5}{\lambda_m}+\frac{J_{fN}\omega_N^{11/2}}{\lambda_\omega}+\frac{\lambda_f^2\omega_N^{11/2}}{\lambda_\omega}+J_{fN}J_{tN}m_N\omega_N^7\right)$$

$$+cos(\omega_N^{1/2}+\lambda_\omega\omega_N)sinh(\omega_N^{1/2})$$

$$\left(-\frac{m_N\omega_N^{9/2}}{\lambda_\omega}+\frac{J_{fN}\omega_N^{11/2}}{\lambda_\omega}+\frac{\lambda_f^2\omega_N^{11/2}}{\lambda_\omega}+J_{tN}\lambda_f\omega_N^6-J_{tN}\lambda_f m_N\omega_N^6\right)$$

$$+sin(\omega_N^{1/2} + \lambda_\omega \omega_N)cosh(\omega_N^{1/2})$$

$$\left(\frac{m_N \omega_N^{9/2}}{\lambda_\omega} + J_{tN}\omega_N^5 + \frac{\lambda_t^2 \omega_N^5}{\lambda_m} + \frac{J_{fN}\omega_N^{11/2}}{\lambda_\omega} + \frac{\lambda_f^2 \omega_N^{11/2}}{\lambda_\omega} - J_{fN} J_{tN} m_N \omega_N^7 \right) = 0 \tag{19}$$

This equation gives implicitly, for a given geometry, the eigenfrequency ω_N versus the load m_N, thus permitting the study of the resonnant mode versus the load.

3 Invariance Properties

The study of (19) in the limit of large ω_N gives two interesting results. First the case of large m_N leads to the equation

$$\omega_N = \frac{Arctg}{\lambda_\omega} \left(\frac{\frac{-2\omega_N^{9/2}}{\lambda_\omega} + \frac{2J_{fN}\omega_N^6}{\lambda_\omega}}{2J_{fN}J_{tN}\omega_N^7 - 2J_{tN}\omega_N^{11/2}} \right) + \frac{k\pi}{\lambda_\omega}$$

where $k \in \mathbb{N}$, is the mode index, for large values of ω_N we have,

$$\omega_N = \frac{k\pi}{\lambda_\omega}$$

Secondly the case $m_N = 0$ leads for large ω_N to the equation,

$$\omega_N = \frac{Arctg}{\lambda_\omega}$$

$$\left(\frac{\frac{2\omega_N^4}{\lambda_\omega} + \frac{2J_{fN}\omega_N^{11/2}}{\lambda_\omega} - \frac{\lambda_t^2\omega_N^{11/2}}{\lambda_\omega}}{2j_{tN}\omega_N^5 + 2\frac{\lambda_t^2\omega_N^5}{\lambda_m} - 2J_{fN}J_{tN}\omega_N^{13/2} - 2J_{tN}\lambda_t^2\omega_N^{13/2} - 2\frac{J_{tN}\lambda_t^2\omega_N^{13/2}}{\lambda_\omega}} \right) + \frac{k\pi}{\lambda_\omega}$$

i.e. for large values of ω_N,

$$\omega_N = \frac{k\pi}{\lambda_\omega}$$

We found then that $\omega_N = f(m_N)$ has a constant behaviour (invariance property) for large values of m_N, and that each mode is separated from his neighbour by $\Delta\omega_N = \pi/\lambda_\omega$. We see also that for large values of ω_N this invariance property do not depends on the geometry of the system but only on λ_ω. For a more precise approach of these phenomena we have computed using (19) the curves $\omega_N = f(m_N)$ for the eight first modes see Figures 1 to 6. We see that as predicted theoretically every mode after preload condition converges to a constant value (invariance). For the mode $k = 0$ this mode converges to 0 for

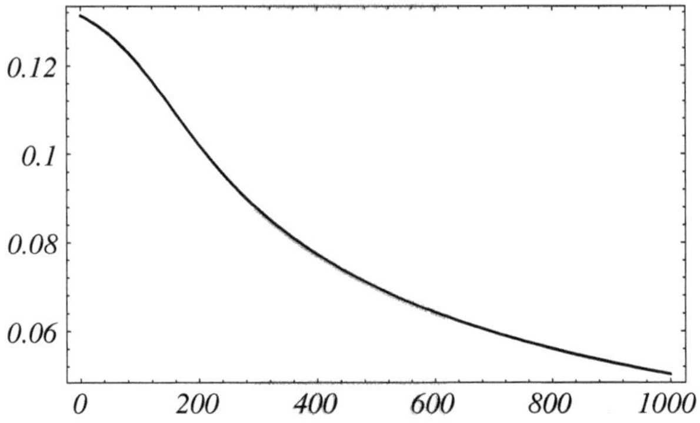

Fig. 1. mode $k = 0$, $m_N \in [0, 1000]$

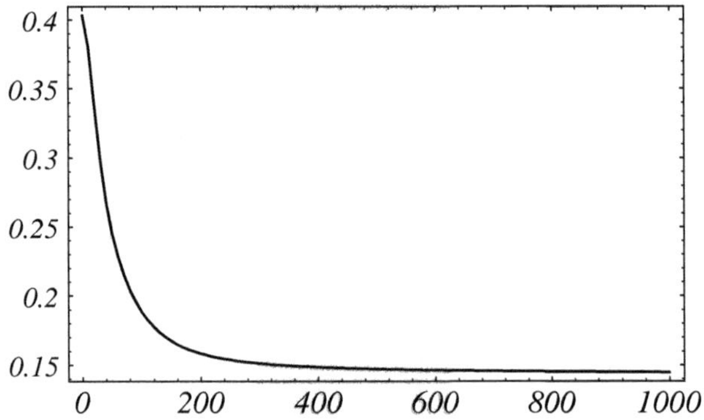

Fig. 2. mode $k = 1$, $m_N \in [0, 1000]$.

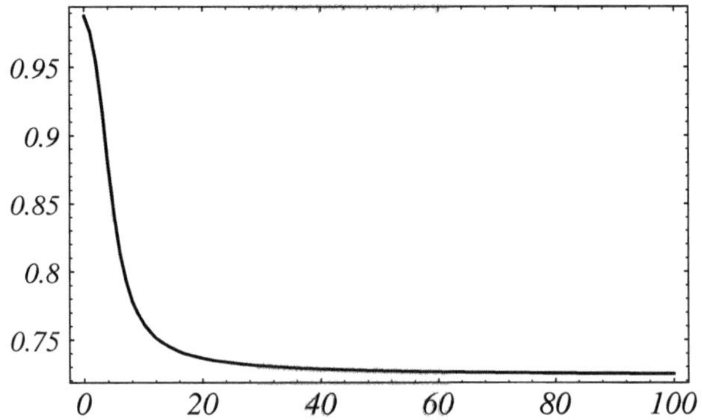

Fig. 3. mode $k = 3$, $m_N \in [0, 100]$

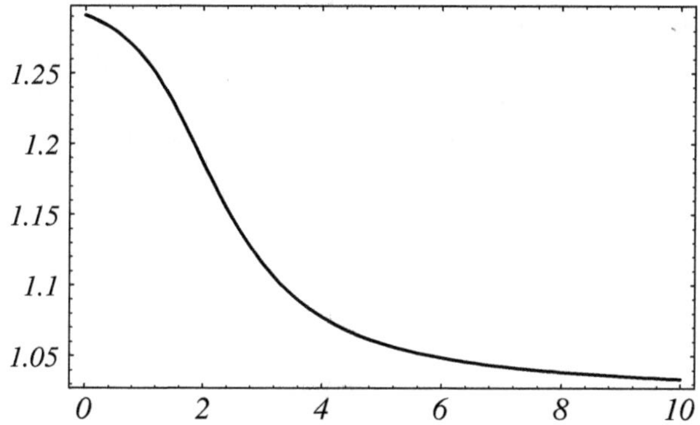

Fig. 4. mode $k = 4$, $m_N \in [0, 10]$

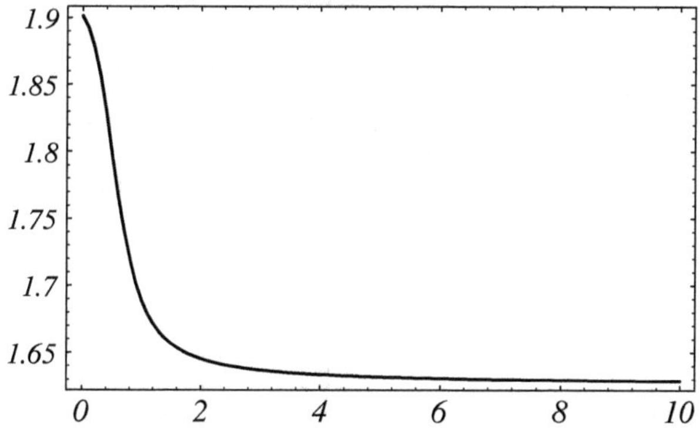

Fig. 5. mode $k = 6$, $m_N \in [0, 10]$

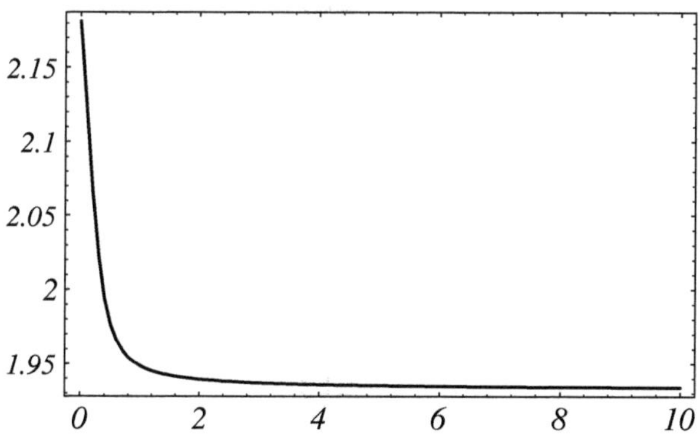

Fig. 6. mode $k = 7$, $m_N \in [0, 10]$

only $m_N > 10000$ not presented in our figure. We see also that for $k \geq 7$ the preload condition is less than $m_N = 1$, so the infinite subset of the eigenfrequnecies $k \in [7, +\infty]$ can be represented by precalculated constants opening the way to new finite dimension controllers. The seven first mode $k \in [0, 6]$ can be easily modelled by polynomial functions of m_N in the neighbourhood of the usefull m_N.

4 Conclusion

Analysis of resonant solutions of one link deformable system shows that, as a function of the applied tip load, the resonant frequencies are distributed in equally spaced layers splitted by a distance $\pi\omega_t/\omega_f$, where ω_t and ω_f are the characteristic torsion and flexion frequencies of the system. Furthermore, the modes of order $k > 7$ are typically converging to the layer boundary $k\pi\omega_t/\omega_f$ for normalized tip load equal to 1, and have for $k < 7$ a very smooth behavior representable by simple rational fraction of tip load, allowing to easily predict the location of resonances for well defined system parameters, and to determine the domains in which they are invariant, a property which will be used later for control.

References

1. W.J. Book (1984) "New concepts in leightweight arms",*Proceedings of the 2nd International Symposium on Robotics research*, Kyoto, Japan, 203-205
2. R.W. Clough and J. Penzien (1975) "Dynamics of structures", *Mc Graw Hill*, New York

Eigenfrequencies Invariance Properties in Deformable Flexion-Torsion Loaded Bodies -2- The Compliant Case

Marc Rouff[2,1], Zakaria Lakhdari[1], and Philippe Makany[1*]

[1] Laboratoire des Sciences Appliquées de Cherbourg (LUSAC)
BP 78, rue Louis Aragon
50130 Cherbourg Octeville, France

[2] Laboratoire d'Ingénierie des Systèmes de Versailles (LISV)
Centre Universitaire de Technologie, 10/12 Avenue de l'Europe
78140 Velizy, France

Summary. We present here in the framework of resonnant behaviours, eigenfrequencies invariance properties for compliant loaded flexural and torsional deformable bodies. We will see that compliant behaviours create a new order on the system which leads to a large invariance property in the $(\omega_N \times m_N \times \lambda_0)$ space. As in the preceeding paper the notion of preload condition can be used and we see that compliance diminishes drastically the preload condition.

Key words: resonnant behavior, eigenfrequency invariante properties, flexion-torsion bodies, controller

1 Introduction

Previous analysis of deformation resonance modes of one link system was showing interesting invariance properties of resonance frequencies versus applied tip load, allowing to construct finite dimensional controllers with arbitrary prescribed preciseness. However to further improve this result, the role of passive system parameters has ben studied. It is shown that addition of compliance effect at link origin changes the structure of equations as a consequence of modification of power distribution between resonants modes. In this enlarged system, non compliant mode structure appears to be singular. The most significant results of this compliant approach are that there exits a threshold compliance value above which all resonance frequencies are invariant versus the applied tip load. This new invariance has not the same behavior as in the non compliant case, as shows a lower threshold value.

* Thanks to FEDER european agency for funding.

2 The Equation of Degeneration

If we consider as in the preceeding paper the dynamical equation of a one link deformable mechanical system, with the same notation and with the same normalized variables, we found the following normalized equations of motion,

$$\frac{\partial^2 \tilde{u}}{\partial \tau^2} + \frac{\partial^4 \tilde{u}}{\partial \zeta^4} = 0 \tag{1}$$

$$\frac{\partial^2 \gamma}{\partial \tau^2} - (\frac{1}{\lambda_\omega})^2 \frac{\partial^2 \gamma}{\partial \zeta^2} = 0 \tag{2}$$

with, for all τ,the following normalized boundary conditions,

$$\tilde{u}(0,\tau) = 0 \quad (3), \quad \gamma(0,\tau) = 0 \quad (4)$$

$$\lambda_0 \tilde{u}''(0,\tau) + (1-\lambda_0)\tilde{u}'(0,\tau) \equiv \lambda_0 \left.\frac{\partial^2 \tilde{u}}{\partial \zeta^2}\right|_{\zeta=0} + (1-\lambda_0) \left.\frac{\partial \tilde{u}}{\partial \zeta}\right|_{\zeta=0} = 0 \tag{5}$$

$$\tilde{u}''(1,\tau) \equiv \left.\frac{\partial^2 \tilde{u}}{\partial \zeta^2}\right|_{\zeta=1} = -\lambda_f \tilde{u}'''(1,\tau) - J_{fN} \left.\frac{\partial^3 \tilde{u}}{\partial \tau^2 \partial \zeta}\right|_{\zeta=1} \tag{6}$$

$$u'''(1,\tau) \equiv \left.\frac{\partial^3 \tilde{u}}{\partial \zeta^3}\right|_{\zeta=1} = m_N \left(\frac{\partial^2 \tilde{u}}{\partial \tau^2} + \lambda_f \frac{\partial^3 \tilde{u}}{\partial \zeta \partial \tau^2} + \lambda_t \frac{\partial^2 \gamma}{\partial \tau^2} \right)_{\zeta=1} \tag{7}$$

$$(\frac{1}{\lambda_\omega})^2 \gamma'(1,\tau) \equiv (\frac{1}{\lambda_\omega})^2 \left.\frac{\partial \gamma}{\partial \zeta}\right|_{\zeta=1} = -(\frac{\lambda_t}{\lambda_m}) u'''(1,\tau) - J_{tN} \left.\frac{\partial^2 \gamma}{\partial \tau^2}\right|_{\zeta=1} \tag{8}$$

Equation (5) is the compliant boundary condition that we supposed to be applied only on $\zeta = 0$, λ_0 is the compliant parameter, $\lambda \in [0,1]$, $\lambda = 0$ no compliance, $\lambda = 1$ full compliance.In the case of stationary resonnant behaviours $\tilde{u}(\zeta,\tau)$, solution of (1), can be written as,

$$\tilde{u}(\zeta,\tau) = \tilde{u}_1(\zeta)\, cos(\omega_N\, \tau) \tag{9}$$

with $\omega_N = \omega/\omega_f$, where ω is the frequency variable, and,

$$\begin{aligned} \tilde{u}_1(\zeta) &= a\, cosh(\omega_N^{1/2}\zeta) + b\, sinh(\omega_N^{1/2}\zeta) \\ &+ c\, sin(\omega_N^{1/2}\zeta) + d\, cos(\omega_N^{1/2}\zeta) \end{aligned} \tag{10}$$

$\gamma(\zeta,\tau)$, solution of (2), can be written as,

$$\gamma(\zeta,\tau) = \gamma_1(\zeta)\, cos(\omega_N \tau) \tag{11}$$

with,

$$\gamma_1(\zeta) = \bar{a}\, cos(\omega_N\, \lambda_\omega\, \zeta) + \bar{b}\, sin(\omega_N\, \lambda_\omega\, \zeta) \tag{12}$$

Introducing (9) and (11) into the six boundary conditions of the problem, equations (3) to (8), leads by the elimination of $cos(\omega_N \tau)$ to a set of 6 linear equations of the six intial values $(a, b, c, d, \bar{a}, \bar{b})$. While this linear system is degenerated the only condition of existance of the resonnant mode is determinant of this linear system equals zero. This leads to the following equation of degeneration,

$$
\begin{aligned}
&\frac{F_1(\omega_N)}{\lambda_\omega}(1-\lambda_0)\omega_N^4\left(-1+\lambda_f\omega_N(1-m_N)\right.\\
&\qquad\left.-J_{fN}m_N\omega_N^2\right)\\
&+\frac{F_2(\omega_N)}{\lambda_\omega}\left(2(1-\lambda_0)\omega_N^4\left(-1+J_{fN}m_N\omega_N^2\right)\right.\\
&\qquad\left.-2\lambda_0\omega_N^6\left(J_{fN}+\lambda_f^2\right)\right)\\
&+\frac{F_3(\omega_N)}{\lambda_\omega}(1-\lambda_0)\omega_N^4\left(-1-\lambda_f\omega_N(1-m_N)\right.\\
&\qquad\left.-J_{fN}m_N\omega_N^2\right)\\
&+\frac{F_4(\omega_N)}{\lambda_\omega}\Big(2(1-\lambda_0)\omega_N^{9/2}\left(m_N+J_{fN}\omega_N\right.\\
&\qquad\left.+\lambda_f\omega_N\right)\\
&+2\lambda_0\omega_N^{9/2}\left(-1-\lambda_f\omega_N(1+m_N)+J_{fN}m_N\omega_N^2\right)\Big)\\
&+\frac{F_5(\omega_N)}{\lambda_\omega}(1-\lambda_0)\omega_N^4\left(-1+\lambda_f\omega_N(1-m_N)\right.\\
&\qquad\left.-J_{fN}m_N\omega_N^2\right)\\
&+F_6(\omega_N)(1-\lambda_0)\omega_N^5\left(J_{tN}+\lambda_t^2/\lambda_m-J_{tN}\lambda_f\omega_N(1-m_N)\right.\\
&\qquad\left.+J_{fN}J_{tN}m_N\omega_N^2\right)\\
&+F_7(\omega_N)\left((1-\lambda_0)\omega_N^5\left(2J_{tN}+2\lambda_t^2/\lambda_m-2J_{tN}J_{fN}m_N\omega_N^2\right)\right.\\
&\qquad\left.+4\lambda_0\omega_N^7\left(J_{fN}J_{tN}+J_{tN}\lambda_f^2+J_{fN}\lambda_t^2/\lambda_m\right)\right)\\
&+F_8(\omega_N)(1-\lambda_0)\omega_N^5\left(J_{tN}+\lambda_t^2/\lambda_m+J_{tN}\lambda_f\omega_N(1-m_N)\right.\\
&\qquad\left.+J_{fN}J_{tN}m_N\omega_N^2\right)\\
&+F_9(\omega_N)\Big((1-\lambda_0)\omega_N^{11/2}\left(-2J_{tN}m_N-2J_{tN}J_{fN}\omega_N\right.\\
&\qquad\left.-2J_{tN}\lambda_f^2\omega_N-2J_{fN}\lambda_t^2/\lambda_m\omega_N\right)\\
&+2\lambda_0\omega_N^{11/2}\left(J_{tN}+\lambda_t^2/\lambda_m+J_{tN}\lambda_f\omega_N(1+m_N)\right.\\
&\qquad\left.-J_{fN}J_{tN}m_N\omega_N^2\right)\Big)
\end{aligned}
$$

$$
\begin{aligned}
&+F_{10}(\omega_N)(1-\lambda_0)\omega_N^5\left(J_{tN}+\lambda_t^2/\lambda_m-J_{tN}\lambda_f\omega_N(1-m_N)\right.\\
&\qquad\left.+J_{tN}J_{fN}m_N\omega_N^2\right)\\
&+\frac{F_{11}(\omega_N)}{\lambda_\omega}\Big(2(1-\lambda_0)\omega_N^{9/2}\left(-m_N+J_{fN}\omega_N+\right.\\
&\qquad\left.\lambda_f^2\omega_N\right)\\
&+2\lambda_0\omega_N^{9/2}\left(1-\lambda_f\omega_N(1+m_N)-J_{fN}m_N\omega_N^2\right)\Big)\\
&+\frac{F_{12}(\omega_N)}{\lambda_\omega}\left(2(1-\lambda_0)\omega_N^5\left(\lambda_f(1+m_N)\right)\right.\\
&\qquad\left.-4\lambda_0\omega_N^5m_N\right)\\
&+F_{13}(\omega_N)\Big(2(1-\lambda_0)\omega_N^{11/2}\left(J_{tN}m_N-J_{tN}J_{fN}\omega_N\right.\\
&\qquad\left.-J_{tN}\lambda_f^2\omega_N-J_{fN}\lambda_t^2/\lambda_m\omega_N\right)\\
&+2\lambda_0\omega_N^{11/2}\left(-J_{tN}-\lambda_t^2/\lambda_m+J_{tN}\lambda_f\omega_N(1+m_N)\right.\\
&\qquad\left.+J_{fN}J_{tN}m_N\omega_N^2\right)\Big)\\
&+F_{14}(\omega_N)\omega_N^6\left(2(1-\lambda_0)\left(-J_{tN}\lambda_f(1+m_N)\right)+4\lambda_0J_{tN}m_N\right)\\
&+\frac{F_{15}(\omega_N)}{\lambda_\omega}(1-\lambda_0)\omega_N^4\left(1+\lambda_f\omega_N(1-m_N)+J_{fN}m_N\omega_N^2\right)\\
&+F_{16}(\omega_N)(1-\lambda_0)\omega_N^5\left(-J_{tN}-\lambda_t^2/\lambda_m-J_{tN}\lambda_f\omega_N(1-m_N)\right.\\
&\qquad\left.-J_{tN}J_{fN}m_N\omega_N^2\right)=0
\end{aligned}
\tag{13}
$$

with

$$
\begin{aligned}
F_1(\omega_N)&=cos^2(\omega_N^{1/2})cos(\lambda_\omega\omega_N)\\
F_2(\omega_N)&=cos(\omega_N^{1/2})cos(\lambda_\omega\omega_N)cosh(\omega_N^{1/2})\\
F_3(\omega_N)&=cosh^2(\omega_N^{1/2})cos(\lambda_\omega\omega_N)\\
F_4(\omega_N)&=cosh(\omega_N^{1/2})cos(\lambda_\omega\omega_N)sin(\omega_N^{1/2})\\
F_5(\omega_N)&=sin^2(\omega_N^{1/2})cos(\lambda_\omega\omega_N)\\
F_6(\omega_N)&=cos^2(\omega_N^{1/2})sin(\lambda_\omega\omega_N)\\
F_7(\omega_N)&=cos(\omega_N^{1/2})sin(\lambda_\omega\omega_N)cosh(\omega_N^{1/2})\\
F_8(\omega_N)&=cosh^2(\omega_N^{1/2})sin(\lambda_\omega\omega_N)\\
F_9(\omega_N)&=cosh(\omega_N^{1/2})sin(\lambda_\omega\omega_N)sin(\omega_N^{1/2})\\
F_{10}(\omega_N)&=sin^2(\omega_N^{1/2})sin(\lambda_\omega\omega_N)
\end{aligned}
$$

$$F_{11}(\omega_N) = cos(\omega_N^{1/2})cos(\lambda_\omega \omega_N)sinh(\omega_N^{1/2})$$

$$F_{12}(\omega_N) = sin(\omega_N^{1/2})cos(\lambda_\omega \omega_N)sinh(\omega_N^{1/2})$$

$$F_{13}(\omega_N) = cos(\omega_N^{1/2})sin(\lambda_\omega \omega_N)sinh(\omega_N^{1/2})$$

$$F_{14}(\omega_N) = sin(\omega_N^{1/2})sin(\lambda_\omega \omega_N)sinh(\omega_N^{1/2})$$

$$F_{15}(\omega_N) = cos(\lambda_\omega \omega_N)sinh^2(\omega_N^{1/2})$$

$$F_{16}(\omega_N) = sin(\lambda_\omega \omega_N)sinh^2(\omega_N^{1/2})$$

Equation (13) gives implicitly, for a given geometry, the eigenfrequencies ω_N versus the load m_N, thus permitting the study of the resonnant mode versus the load.

3 Invariance Properties

The study of (13) in the compliant case $\lambda_0 = 1$ and in the limit of large ω_N gives two interesting results. First, the case of large ω_N leads to the equation,

$$\omega_N^{1/2} = Artg\left(\frac{1}{m_N}\right) + k\pi$$

i.e. , when $m_N \to 0$,

$$\omega_N^{1/2} = \frac{\pi}{2} + k\pi$$

when $m_N \to +\infty$,

$$\omega_N^{1/2} = k\pi$$

Secondly, we see that these asymptotic behaviours are different as the non compliant case. Here $\Delta\omega_N = \pi^2$ in place of π/λ_ω and the analytical approach of (19) show that the equation of compliance are completely different in its structure as the non compliant case. Fore a more precise approach of these phenomena we have computed using (13) the curves $\omega_N = f(m_N, \lambda_0)$, where λ_0 is the compliant variable, for the eight first modes see figure 1. We see as predicted theoretically that in the $(\omega_N, m_N, \lambda_0)$ space there are mainly two behaviors, the non compliant one wich is singular and defined for $\lambda_0 < 0.3$ and the compliant one which stays in the rest of the space and leads to a large invariant behaviour. In the framework of the compliant behavior we see easily even for the mode $k = 0$ that the threshold value of invariance is drastically reduced and that the invariance is more precise and stable than without compliance. For example the threshold condition for the mode $k = 6$ is with compliance $m_N = 0.5$ and $m_N = 2$ without compliance.

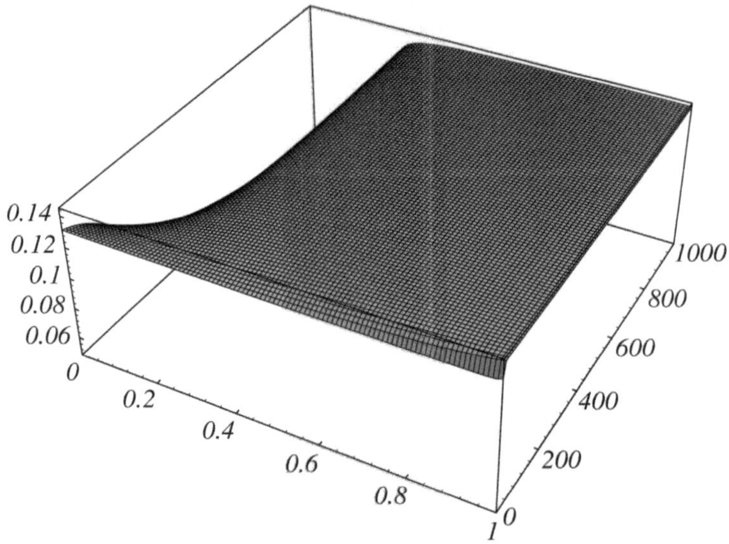

Fig. 1. mode $k = 0, (m_N \times \lambda_0) \in [0, 1000] \times [0, 1]$

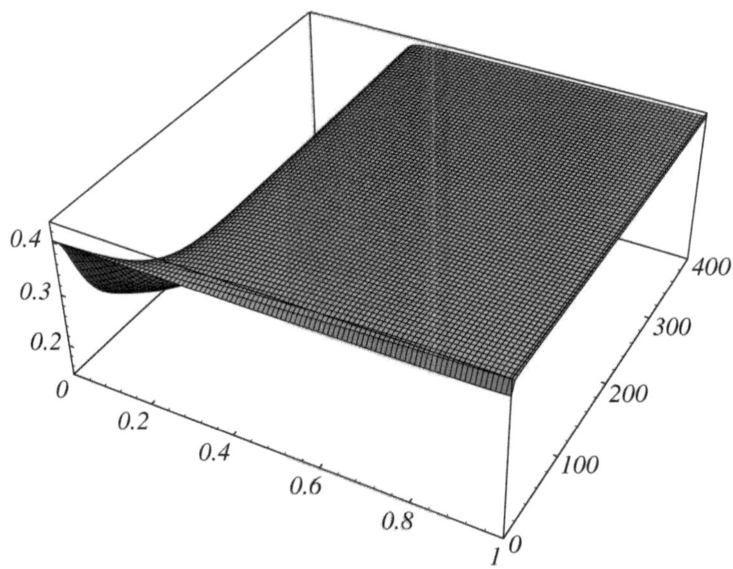

Fig. 2. mode $k = 1, (m_N \times \lambda_0) \in [0, 400] \times [0, 1]$

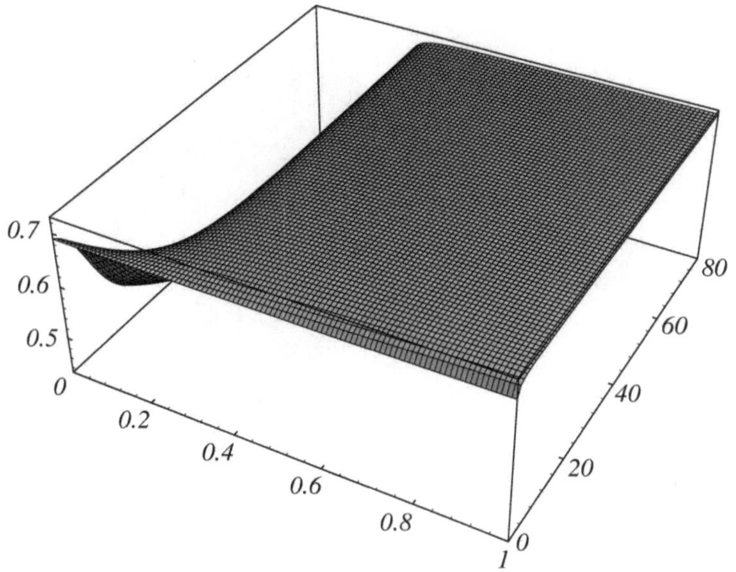

Fig. 3. mode $k = 2, (m_N \times \lambda_0) \in [0, 80] \times [0, 1]$

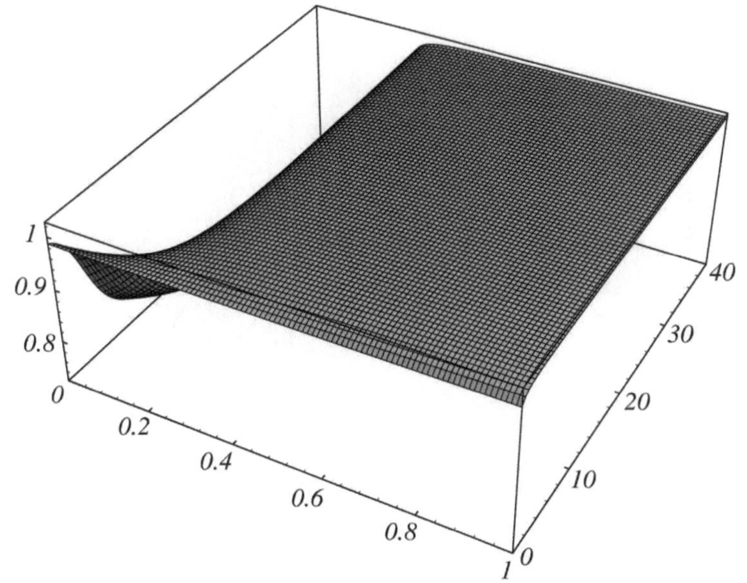

Fig. 4. mode $k = 3, (m_N \times \lambda_0) \in [0, 40] \times [0, 1]$

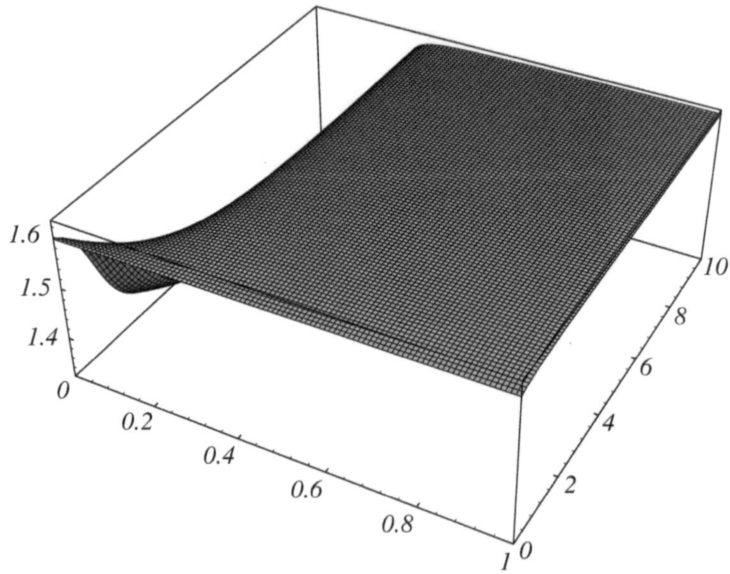

Fig. 5. mode $k = 5, (m_N \times \lambda_0) \in [0, 10] \times [0, 1]$

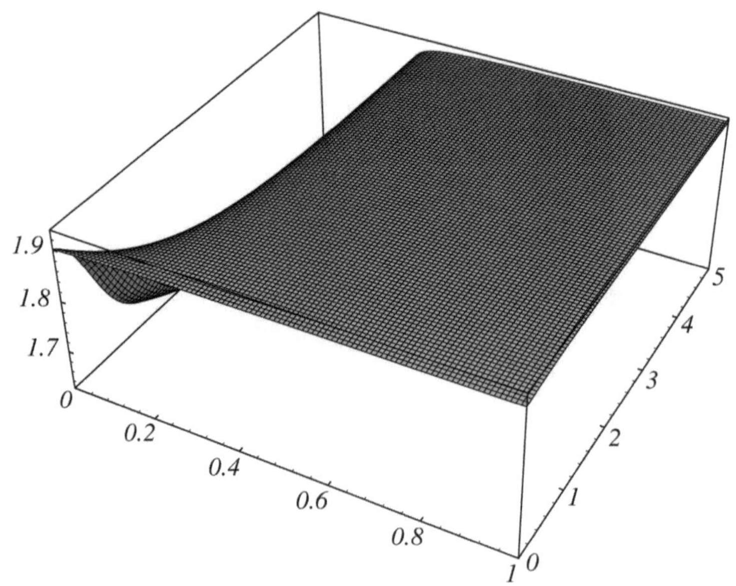

Fig. 6. mode $k = 6, (m_N \times \lambda_0) \in [0, 5] \times [0, 1]$

4 Conclusion

Analysis of compliance effect on deformable system shows high sensitivity of the system to this parameter because of modification of power distribution over resonant modes. It is shown that there exists a threshold compliance value above which resonance frequencies are invariant versus tip load and are independent of mode number. This threshold value is easily technically realizable as it corresponds to a current value of axial spring stiffness representing the compliance. A further improvement has been observed when adding compliance at link tip. This observation would seem to oppose to common idea that higher stiffness is required when the system becomes more deformable. In fact existence of compliance power sink allows better internal regulation of power distribution over the resonant modes, which will not have to be taken care off by an outer controller.

References

1. W.J. Book (1984) "New concepts in leightweight arms", *Proceedings of the 2nd International Symposium on Robotics research*, Kyoto, Japan, 203-205
2. R.W. Clough and J. Penzien (1975) "Dynamics of structures", Mc Graw Hill, New York

Index

Understanding Complex Systems

Edited by J.A. Scott Kelso

Jirsa, V.K.; Kelso, J.A.S. (Eds.)
Coordination Dynamics: Issues and Trends
XIV, 272 p. 2004 [3-540-20323-0]

Kerner, B.S.
The Physics of Traffic:
Empirical Freeway Pattern Features,
Engineering Applications, and Theory
XXIII, 682 p. 2004 [3-540-20716-3]

Kleidon, A.; Lorenz, R.D. (Eds.)
Non-equilibrium Thermodynamics
and the Production of Entropy
XIX, 260 p. 2005 [3-540-22495-5]

Kocarev, L.; Vattay, G. (Eds.)
Complex Dynamics in Communication Networks
X, 361 p. 2005 [3-540-24305-4]

McDaniel, R.R.Jr.; Driebe, D.J. (Eds.)
Uncertainty and Surprise in Complex Systems:
Questions on Working with the Unexpected
X, 200 p. 2005 [3-540-23773-9]

Ausloos, M.; Dirickx, M. (Eds.)
The Logistic Map and the Route to Chaos:
From The Beginnings to Modern Applications
XVI, 411 p. 2006 [3-540-28366-8]

Kaneko, K.
Life: An Introduction to Complex Systems Biology
X, 366 p. 2006 [3-540-32666-9]

Braha, D.; Minai, A.A.; Bar-Yam Y. (Eds.)
Complex Engineered Systems: Science Meets Technology
VIII, 394 p. 2006 [3-540-32831-9]

Baglio, S.; Bulsara, A. (Eds.)
Device Applications of Nonlinear Dynamics
XI, 260 p. 2006 [3-540-33877-2]

Aziz-Alaoui, M.A.; Bertelle, C. (Eds.)
Emergent Properties in Natural and Artificial Dynamical Systems
X, 280 p. 2006 [3-540-34822-0]

FSC
www.fsc.org
MIX
Papier aus verantwortungsvollen Quellen
Paper from responsible sources
FSC® C105338

If you have any concerns about our products,
you can contact us on
ProductSafety@springernature.com

In case Publisher is established outside the EU,
the EU authorized representative is:
Springer Nature Customer Service Center GmbH
Europaplatz 3, 69115 Heidelberg, Germany

Printed by Libri Plureos GmbH
in Hamburg, Germany